CW01654472

Flower Shower

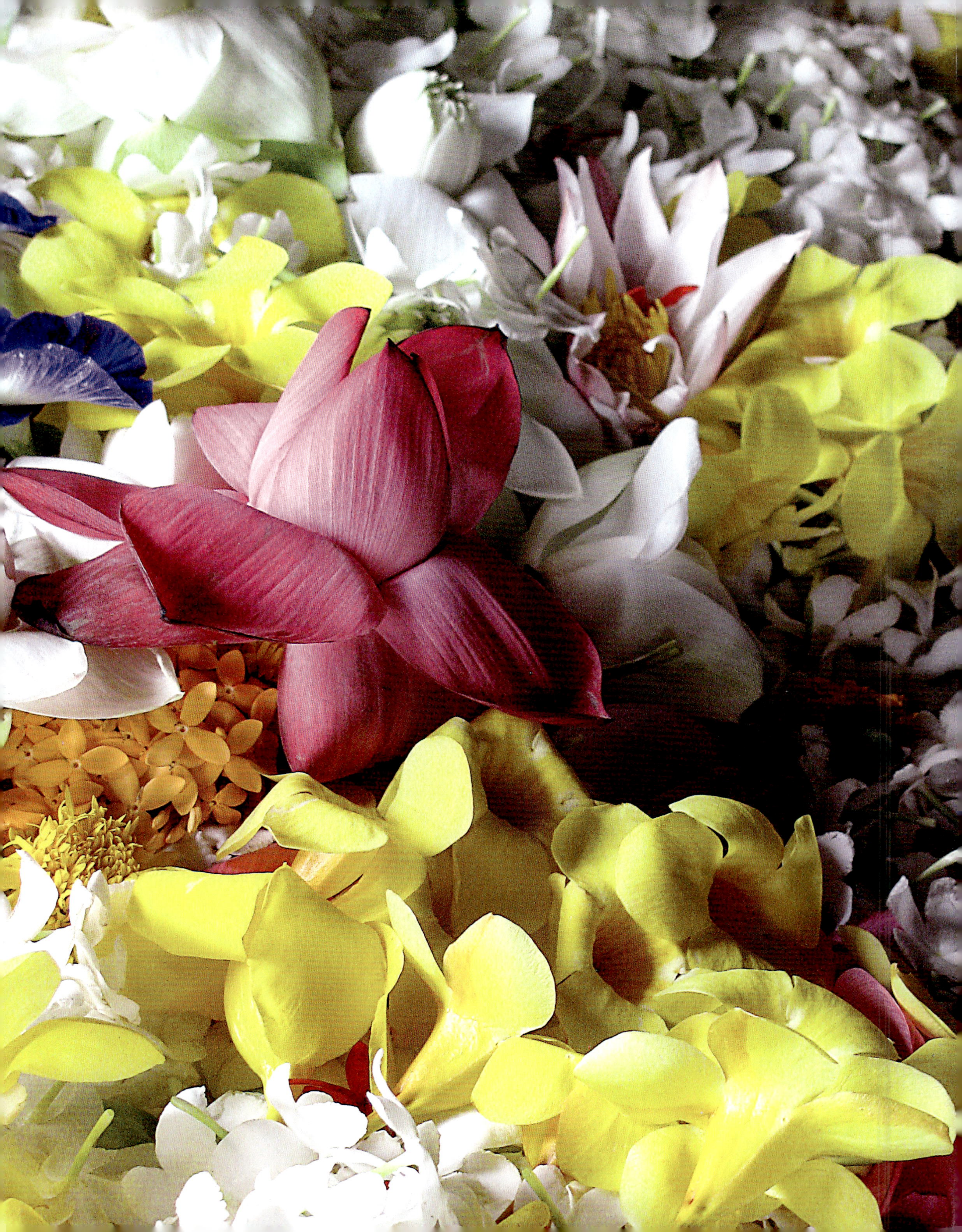

Flower Shower

The Culture of Flowers in India

ALKA PANDE

NIYOGI BOOKS

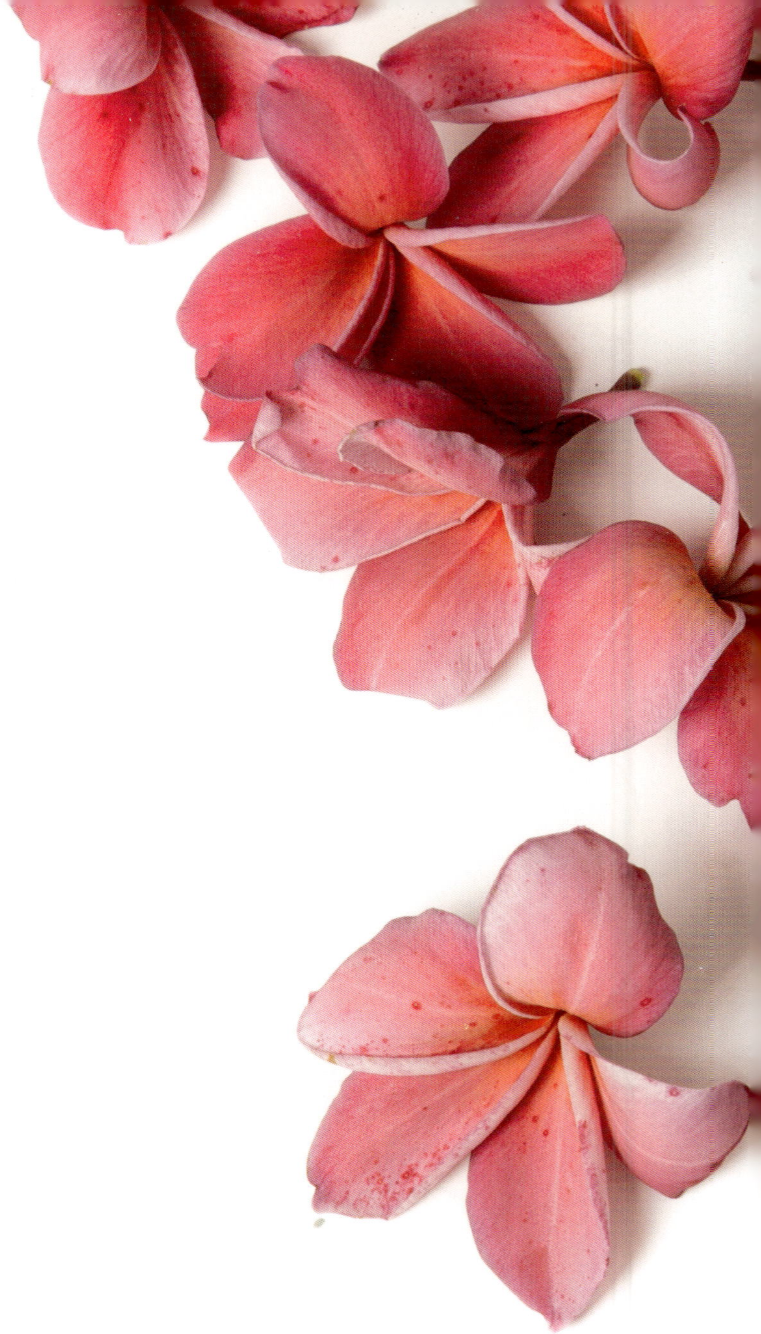

Published by
NIYOGI BOOKS
Block D, Building No. 77,
Okhla Industrial Area, Phase-I,
New Delhi-110 020, INDIA
Tel: 91-11-26816301, 26818960
Email: niyogibooks@gmail.com
Website: www.niyogibooksindia.com

Text © Alka Pande

Editor: K.E. Priyamvada
Design: Misha Oberoi

ISBN: 978-93-85285-95-0
Publication: 2019

All rights are reserved. No part of this publication may be reproduced or transmitted in any form or by any means, electronic or mechanical, including photocopying, recording or by any information storage and retrieval system without prior written permission and consent of the Publisher.

Printed at: Niyogi Offset Pvt. Ltd., New Delhi, India

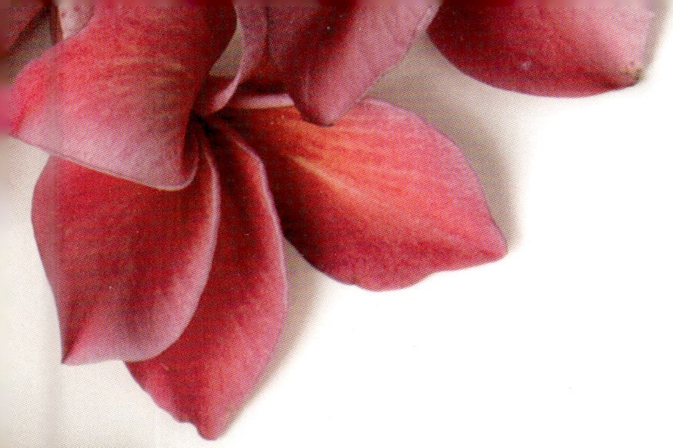

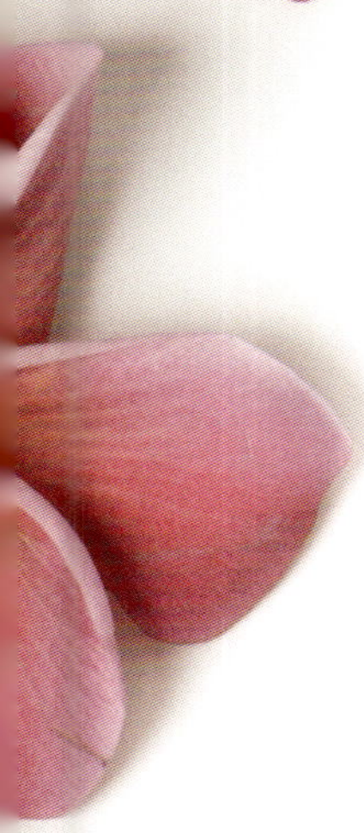

Contents

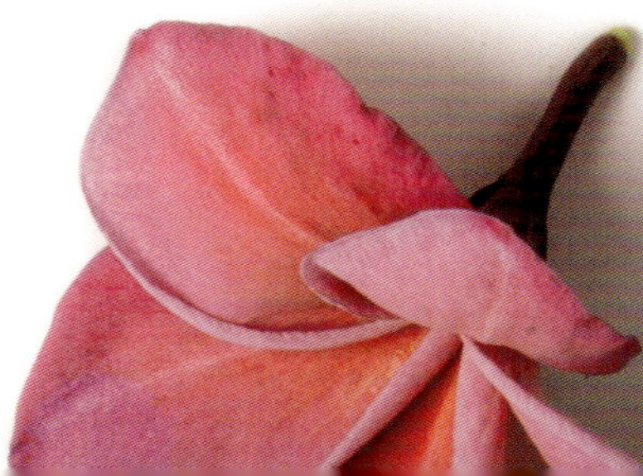

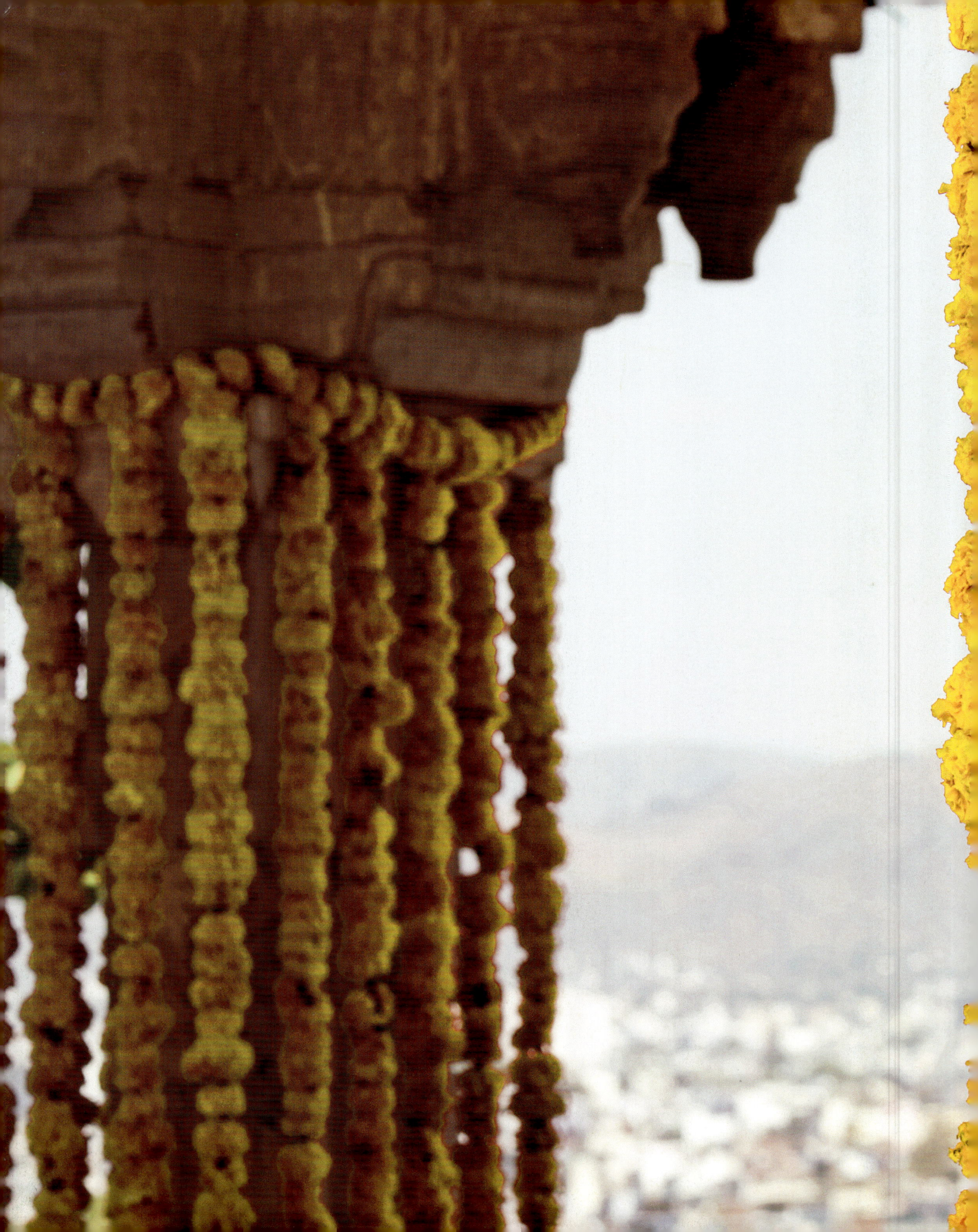

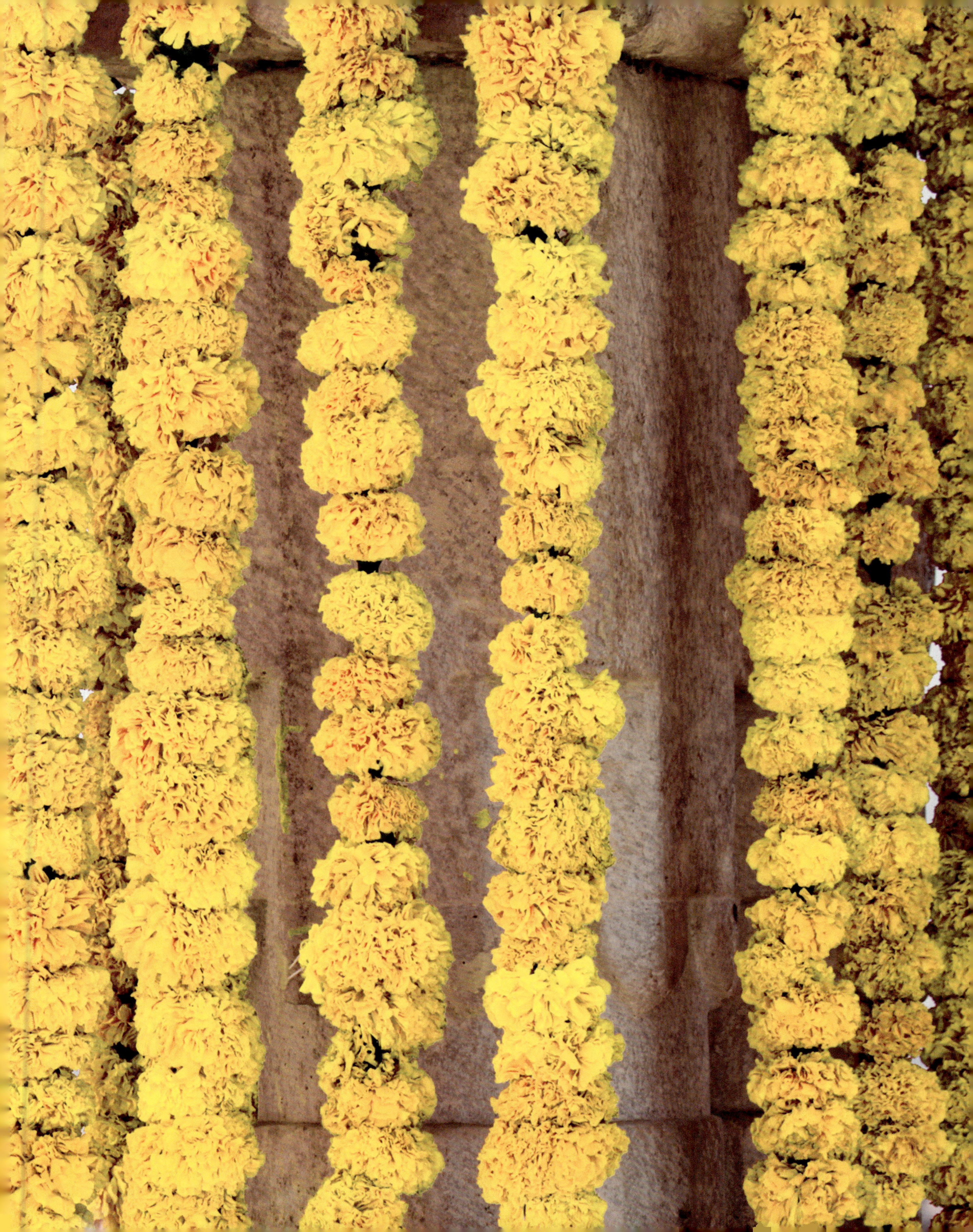

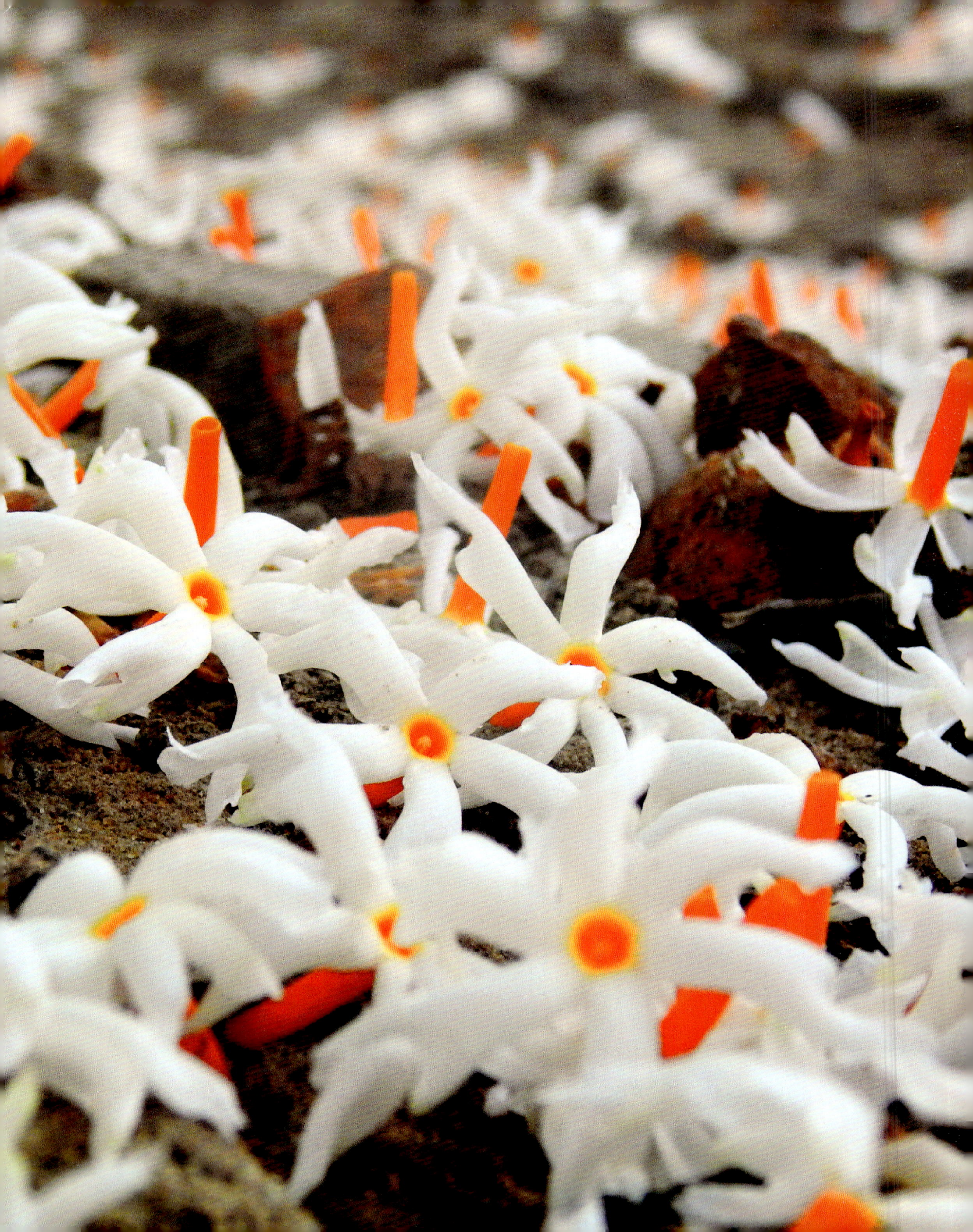

Introduction

'True love is night jasmine,
a diamond in darkness, the heartbeat
no cardiologist has ever heard.
It is the most common of miracles, fashioned
of fleecy clouds—a handful of stars tossed into
the night sky.'
—Jim Bishop (American writer, 1907–1987)

The night-flowering jasmine is my favourite flower, till date, and every home of mine has always had the parijat tree. From August to December, I never fail to run into my garden and gather the delicate pearl-like blossoms with an orange stem and gently float them in a silver cup of water in front of the gods on my personal altar. As a Brahmanical Hindu growing up in a traditional ritualistic household, spirituality was associated with the ritual of prayer. As a young child I remember we could not eat food without saying our prayers and even today, I do not sip my favourite drink of strong milky Indian tea from a big ceramic cup—which is the high point of my day—without first performing my daily prayers.

Aah...the parijat tree was my earliest memory of the happiest time of my childhood, in my maternal grandmother's home in Meerut. It was my job to pick up the fragile flowers and hand them to my grandmother who would then tell me to place them in the pooja room, the most intimate and precious room in the sprawling haveli, where I spent many a glorious autumn vacation of my young life.

Facing page: Parijat flowers, also known as shiuli

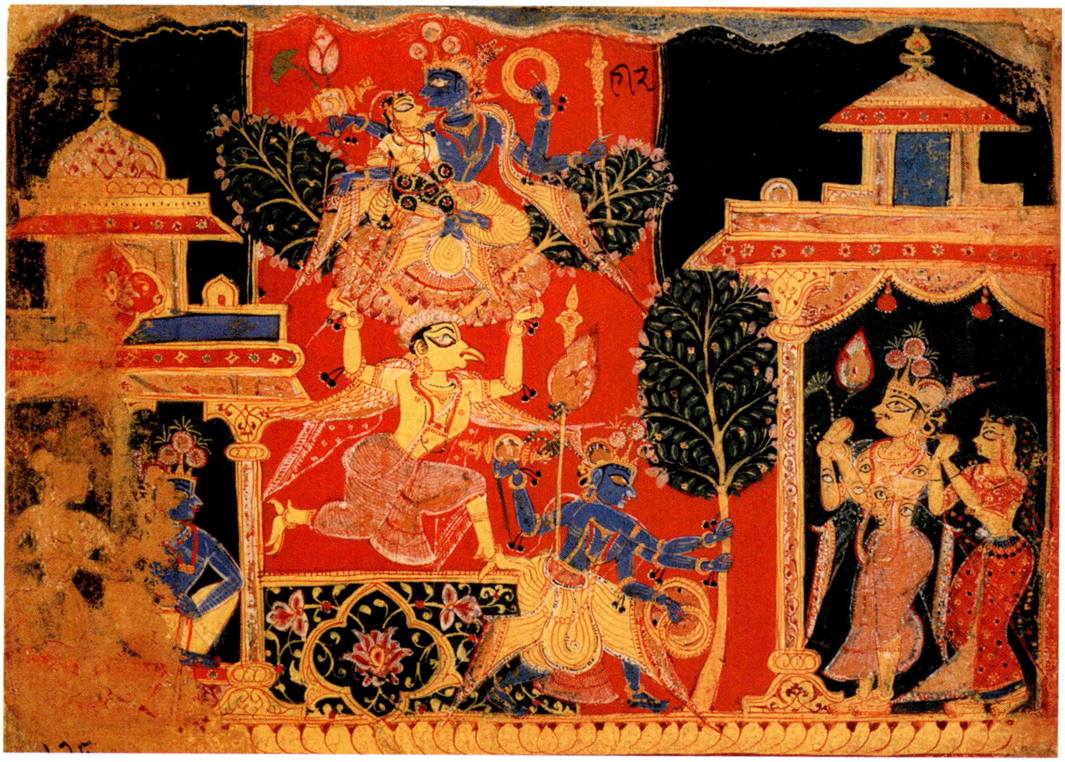

'Krishna uproots the parijata tree', Folio from *a Bhagavata Purana*, opaque watercolour and ink on paper, 1525-1550

My love and fascination for these flowers is as ardent today as it was more than 50 years ago. *Nyctanthes arbor-tristis* is really a small tree with flaky grey bark. The flowers are simple with either a five- to eight-lobed white corolla with a flame orange-red centre. They are grouped in clusters of two to seven, with each flower opening at dusk and dropping off the tree at dawn. Later I learned why the tree's scientific name is *arbor-tristis*—sad tree or the tree of sorrow—because it loses its brightness during the day. But what really stays in my heart is the more romantic story of how the Princess Parijataka's heart was broken by the Sun God Surya as he flamed across the sky from the east to the west. Since the Sun God did not reciprocate her feelings, she killed herself in despair and from her cremated ashes, the parijat tree arose. Since the Princess Parijataka is unable to bear the sight of her love during the day, she blooms only at night. The flowers are shed at the touch of the first rays of the sun, and their fragrance then floods the entire region as a sign of the undying love of the princess for her lover, the Sun God.

Known also as *shiuli* in West Bengal, the parijat is the official flower of the state. There are many more myths surrounding my favourite flower and I am happy to share them, since they are not well known. It is also related to the *Kalpavriksha*, and in the *Mahabharata, Bhagvatha Purana* and *Vishnu Purana*, it is said that the parijat tree appeared as a result of the *Samudra Manthan* or churning of the milky ocean.

The parijat has yet another interesting story. It was brought by Narada from Indralok and given to Krishna, who then gave it to his favourite wife Rukmini. The trouble maker Narada went and squealed to Satyabhama, Krishna's more possessive wife, who then coveted the tree. A series of incidents followed, and ultimately there was a battle between Indra and Krishna, where the former lost. Being a bad loser, Indra cursed the tree so that it would not bear any fruit and thus the parijat tree, till date, remains fruitless. The tree was thus brought from Indralok to Dwarka by Krishna and an amicable solution found. The tree was planted in Satyabhama's garden but the flowers fell

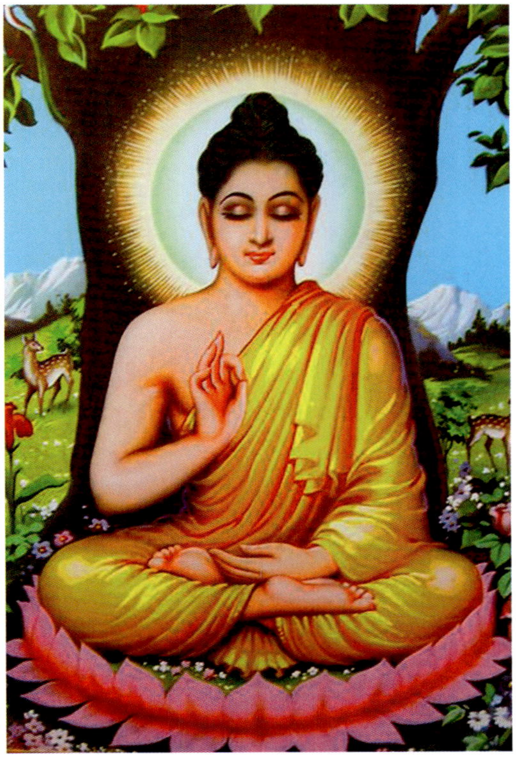

Above right: Image of Buddha sitting on a lotus flower

Below right: A lotus flower, a symbol of wonder

into Rukmini's home, which was adjacent to Satyabhama's home.

Within India's mythic terrain, indigenous knowledge abounds, and till date, I enjoy reading these myths and legends which abound with traditional wisdom and offer much practical advice and solutions. It depends on each one of us how to interpret them and apply them in our lives. I simply adore the allegories and metaphors of the cultural context of flowers within Indian culture. Each flower has its own stories, its own gods, its own benefits and its very own natural and phantasmagorical landscapes.

The lotus is very special to India and seems to be present in art, thought, and literature, literally permeating the cultural consciousness of the land and the people. It is the underlying thread which connects the sacred, the divine, the pleasurable and the joyous, literally the *navarasas* of life within the Indian subcontinent.This is the reason I thought it would be appropriate to devote a separate chapter to this symbol of the *adbhutam* (wonder). Different attributes ascribed to the flower are subjectively classified to reflect their manifestations in philosophy, art, architecture, mythology, the historical narrative, the divine and the sacred and its many cultural meanings.

The lotus petal has been used as a metaphor for describing the beauty of the eyes, *mukula* or lotus bud for a woman's private parts, and the lotus pond as a symbol of sensuality and the lovers' meeting place. Indian Rajput and Pahari miniature paintings are embellished with *champa* (plumeria) trees and lotus ponds and that is the reason why I have separate chapters for the lotus, rose, *champa* and *genda* (the Indian marigold) in this work.

Indian myths are still present in contemporary India, and so I go back to the cultural historical context of flowers. In the two great epics, the *Ramayana* and the *Mahabharata*—which are still being constantly interrogated or delved into as muses for inspiration—flowers and trees abound in the physical and sensory spaces of the humans depicted.

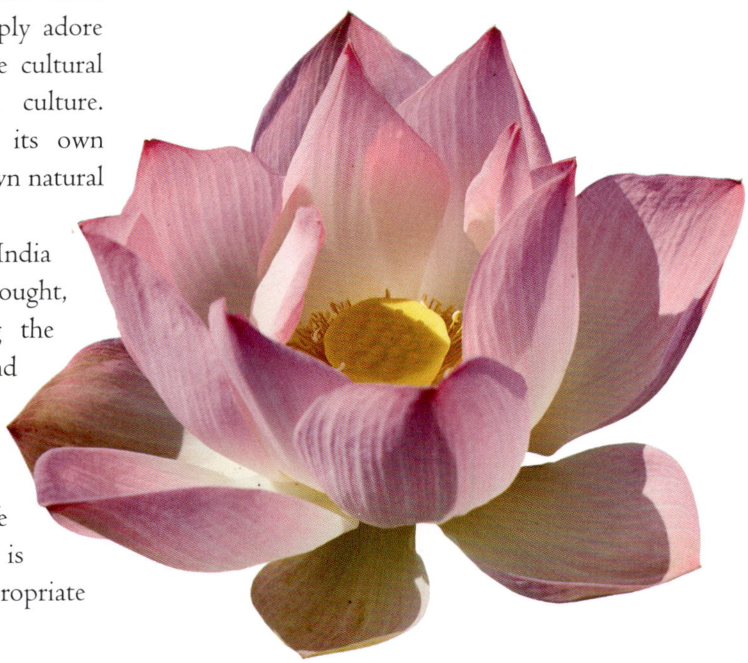

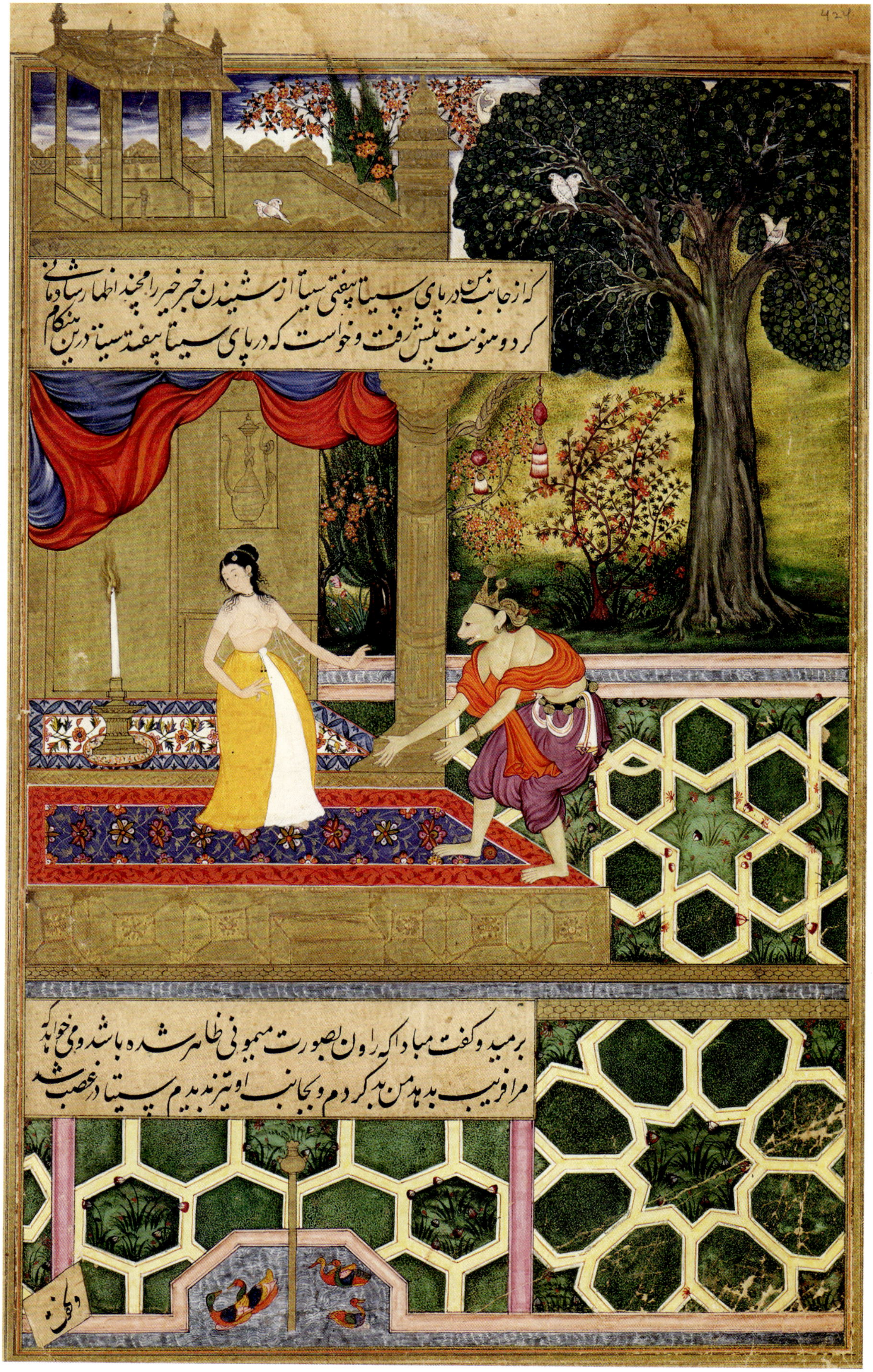
424
که از جانب من در پای سیتا بیفتی سیتا از شنیدن خبر خیر رامچند اظهار شادمانی
کرد و ممنونت پیش رفت و خواست که در پای سیتا بیفتد سیتا درین هنگام
برمید و گفت مبادا که راون بصورت میمونی ظاهر شده باشد و می خواهد که
مرا فریب بدهد من بدگردم و بجانب او تیز ندیدم سیتا در غضب شد

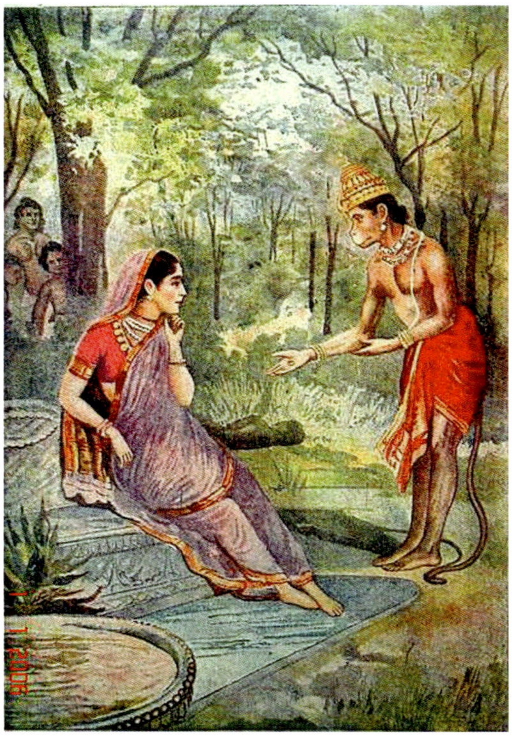

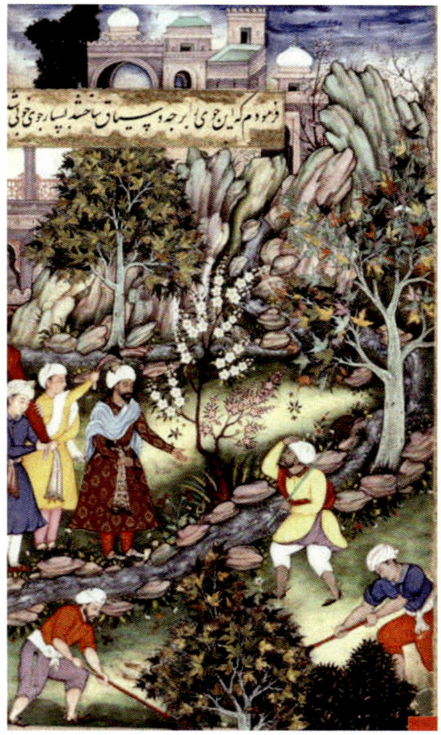

Above right: 'Sita in the Ashoka Vatika visited by Hanuman', bazaar art, early 1900s

Above far right: 'Babur watching men altering the course of a stream', *Baburnama*, c. 1590

Facing page: 'Sita shies away from Hanuman, believing he is Ravana in disguise', Mughal miniature, 1594

Below: Garland of rose, marigold and jasmine

The concept of planting trees as an enclosed space is seen in the *Ramayana*, where Sita is kept in confinement in the *Ashoka Vatika* (Ashoka grove) by Ravana in Lanka.

In the 'Sabha Parva' of the *Mahabharata*, there are many descriptions of beautiful gardens, parks and lakes in Indraprastha, the celestial capital city of the Pandavas. Flowering trees, like *arjuna*, banyan, *peepul* and cassia are also mentioned.

Pre-modern India saw flowers represented in the temples. The Buddha was born under the shade of the *ashoka* tree in a garden at Lumbini, in present-day Nepal; he gained enlightenment under the *Bodhi* or *peepal* tree; and in Buddhism, the mantra '*Om mani padme hum*' (the jewel in the lotus) makes the lotus a living icon of the Buddhist faith. In all the Buddhist monasteries at Nalanda and Taxila, aromatic and beautiful gardens were laid out.

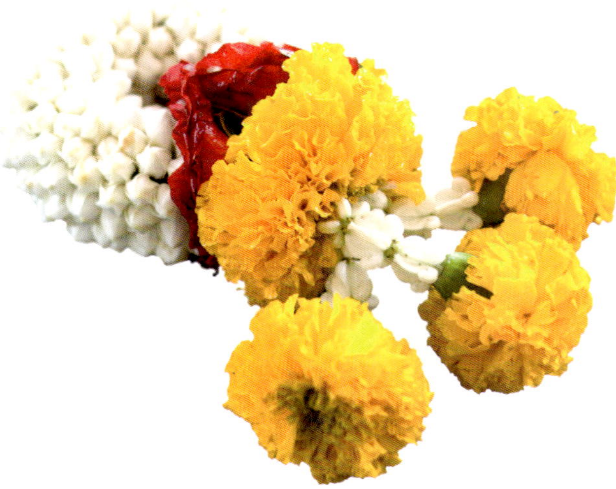

Babur, the first Mughal emperor, was the one who also brought in a variety of exotic and rare flowers and flowering trees from Persia and Central Asia to India, the most well-known being the *chinar*, which has become synonymous with the landscape of Kashmir. From Babur to Shah Jahan, all the Mughal Emperors had a passion for creating exquisite pleasure gardens known as Charbaghs. They made it part of their language of painting and architecture, while the Mughal Empire translated it into tamed and untamed gardens.

In the modernity of India, the British, too, encouraged the exchange of rare seeds and herbs between their colonies and their homeland, with Kew Gardens becoming the site of the Imperial collection of flora from across the continents.

From the modern to the post-colonial, flowers became signifiers of class, caste, identity and, of course, the politics of poetics. *Van* (forest), *upvan* (mini forest), *vatika* (groves or small personal gardens), to the much larger 'company gardens' with their herbaceous borders, and now the entire global cosmopolitan culture of cut flowers, as bouquets, have turned the cultural metaphor of flowers into a signifier of an increasing cosmopolitanism of which India is so much a part.

Flowers have always been part of the pleasure principle. Their beauty and fragrance makes them the symbol of hedonists and pleasure seekers. Used for decoration, both personal and environmental, flowers bring beauty and fragrance to a garden and when cut, they bring these attributes to the interiors.

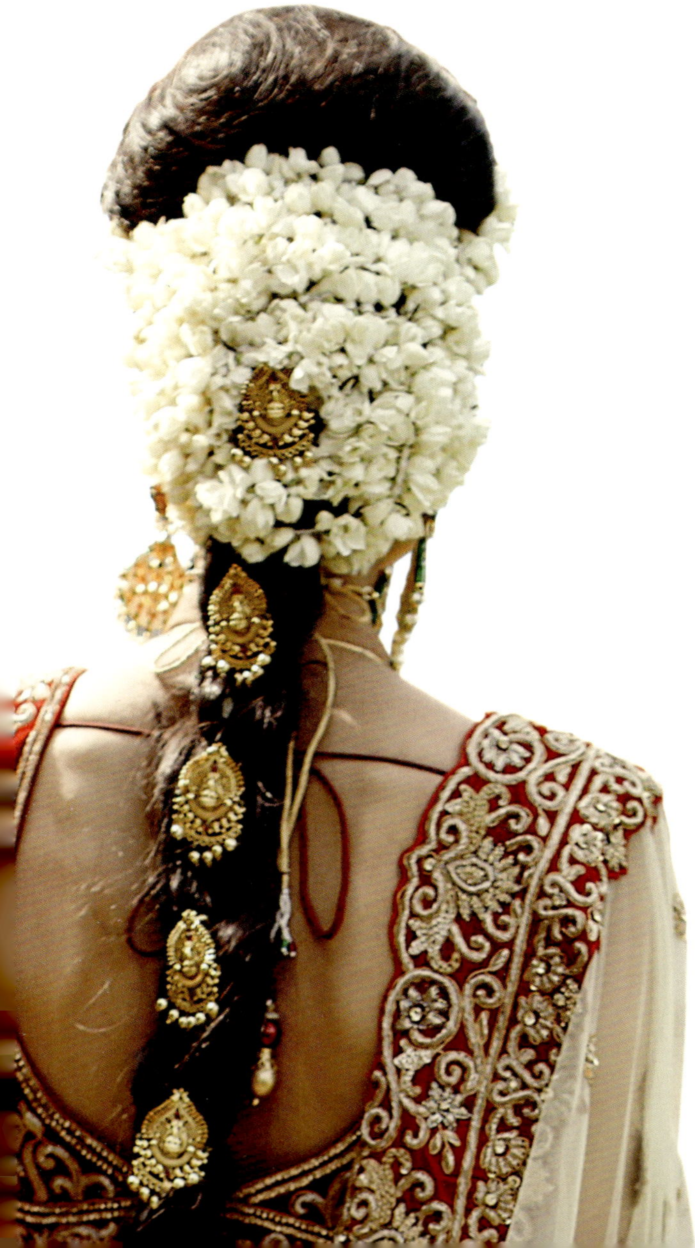

Flowers are also used as ornaments—consider the famous Kashmiri *phoolon ka gehna*, as hair adornments or *mallipus*, (jasmine flowers worn as hair adornments), in many parts of South India. Mughal India popularised the jasmine and rose as fragrances and as garlands for kings returning victorious from battle. These flowers have also long been used as sacred symbols of exchanging vows of marriage—threaded to form the *jaimala* or wedding garland, which is an essential part of Indian marriage ceremonies.

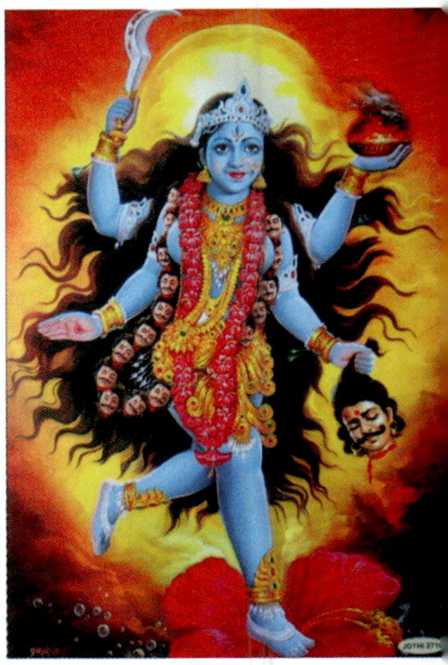

Flowers are an integral part of the ritual of worship. In India, every god and goddess has a favourite flower or a leaf. For example, the Devi loves the red hibiscus, while Shiva is happy with the delicate parijat which was offered to him by Kunti, the mother of the Pandavas, the heroes of the *Mahabharata*. And, of course, the lotus, is seen in painting and sculpture as the leitmotif of the Buddha.

Within the Indian context, the idealised beauty also comes through nature. Whether it is the description of the *nayika*, where her lips are as red as pomegranate seeds, or her eyes like lotus petals.

India, with its huge biodiversity, has an entire landscape of flowers which, by itself, is overwhelming. Flowers are part and parcel of our lives, from birth to death, as offerings, as blessings, as adornment, as medicine, as food and also as a sign of love and sharing. Over the years, with the critical readings of fragrances, we see how aromas can evoke memory, how they are responded to by different genders and how floral perfumes also become mediators of desire and pleasure. And it is through an exploration of the floral essence that new perceptions and interpretations of flowers can be read in the book.

Above: Devi wearing a garland of red hibiscus

Left: Jasmine flowers adorn the hair of a bride

Below: Flowers for worship

Bottom: Jasmine flower garlands for adorning women's hair, for sale in South India

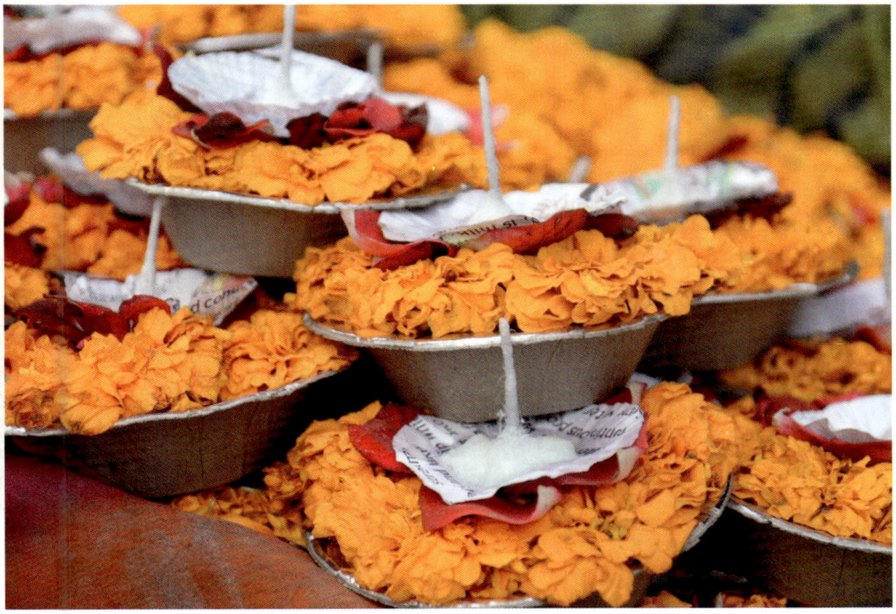

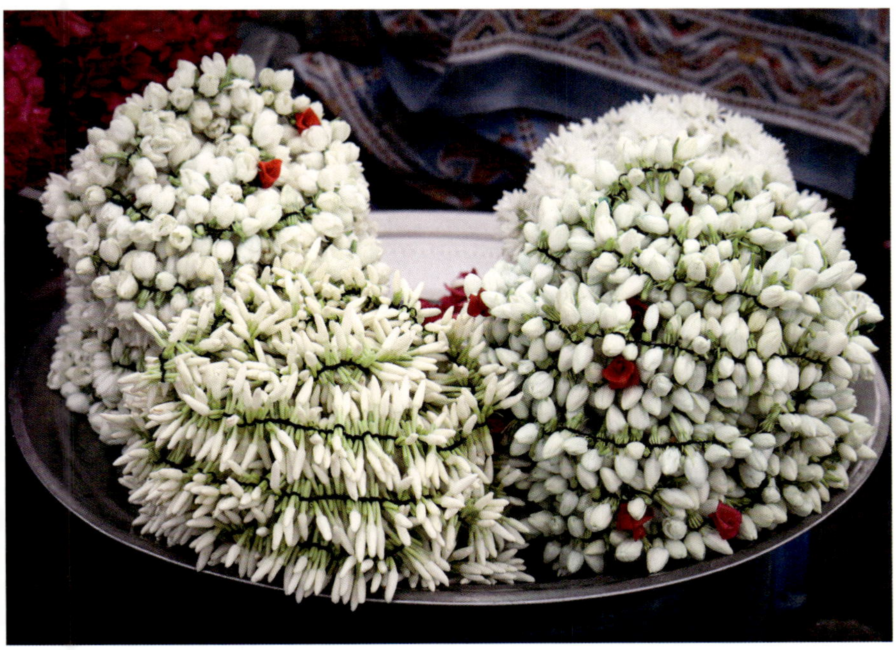

The revival of the 13th century festival of flower vendors, the fabled *Phool Waalon Ki Sair*—the more formal name being *Sair-e-Gul Faroshan*—by our first Prime Minister, Jawaharlal Nehru in the 1960s, is a pointer to the plural culture of the nation which is seen through the lens of flowers. This week-long festival is held annually in October or November during the crisp and fresh season soon after the end of the monsoon and the onset of the rigorous cold Delhi winter. The secular procession starts from Chandni Chowk and ends at the shrine of the 13th century Sufi saint Khwaja Qutubuddin Bhaktiar Kaki visiting the temple of the powerful goddess Devi Yogmaya in Mehrauli along the way. It's an incredible sight—the city's florists march through the historic site of Mehrauli, with enormous fans decorated with flowers of every colour, after seeking blessings from both a Sufi shrine and a Hindu temple. The *sair* concludes with a *qawwali* in the impressive 16th century monument. Flowers once again become the unifying glue of two diverse religions of India, bringing in the shared heritage of the unique plural culture of the land. Flowers are ephemeral, yet their

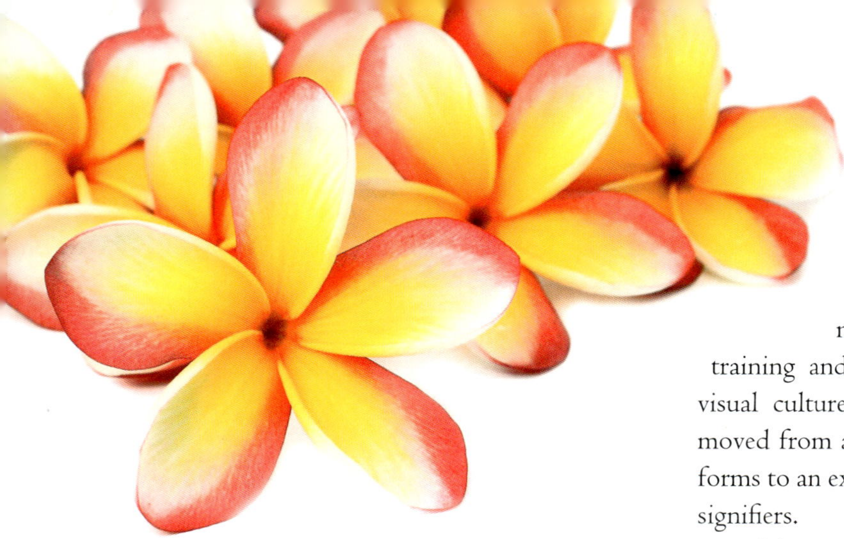

beauty lives in the representation of artists and poets. Across centuries and continents, the flower remains a recurring leitmotif. From eulogistic delights, romantic symbols to tragic metaphors, flowers are present everywhere. At the same time, it is interesting to note that there is a huge absence of flowers in the genre of the witty and the humorous.

As I started my fragrant journey with flowers, a plenitude of doors opened for me. Being an art historian by training and then moving into the area of visual culture, my understanding of flowers moved from an appreciation of their idealised forms to an exploration of their role as cultural signifiers.

I began by a general reading of literary writers, to translations of Indian *'kavya'* literature, which is heavily embellished, idealised Sanskrit poetry. I then started entering the mental landscapes of postcolonial writers like Antiguan-born Jamaica Kincaid, to the *Garden Plots* of Shelly Saugauro[1] to *Seven Flowers* by Jennifer Potter,[2] I looked at the critical narratives of flower power; to the study of the flower in Indian art, the *Flora Indica* of the famed William Roxburgh,[3] to the magnum

Above left: Frangipani blossoms

Below: Flowers with fragrance - Jasmine sambac (*mogra*) with gerbera flowers

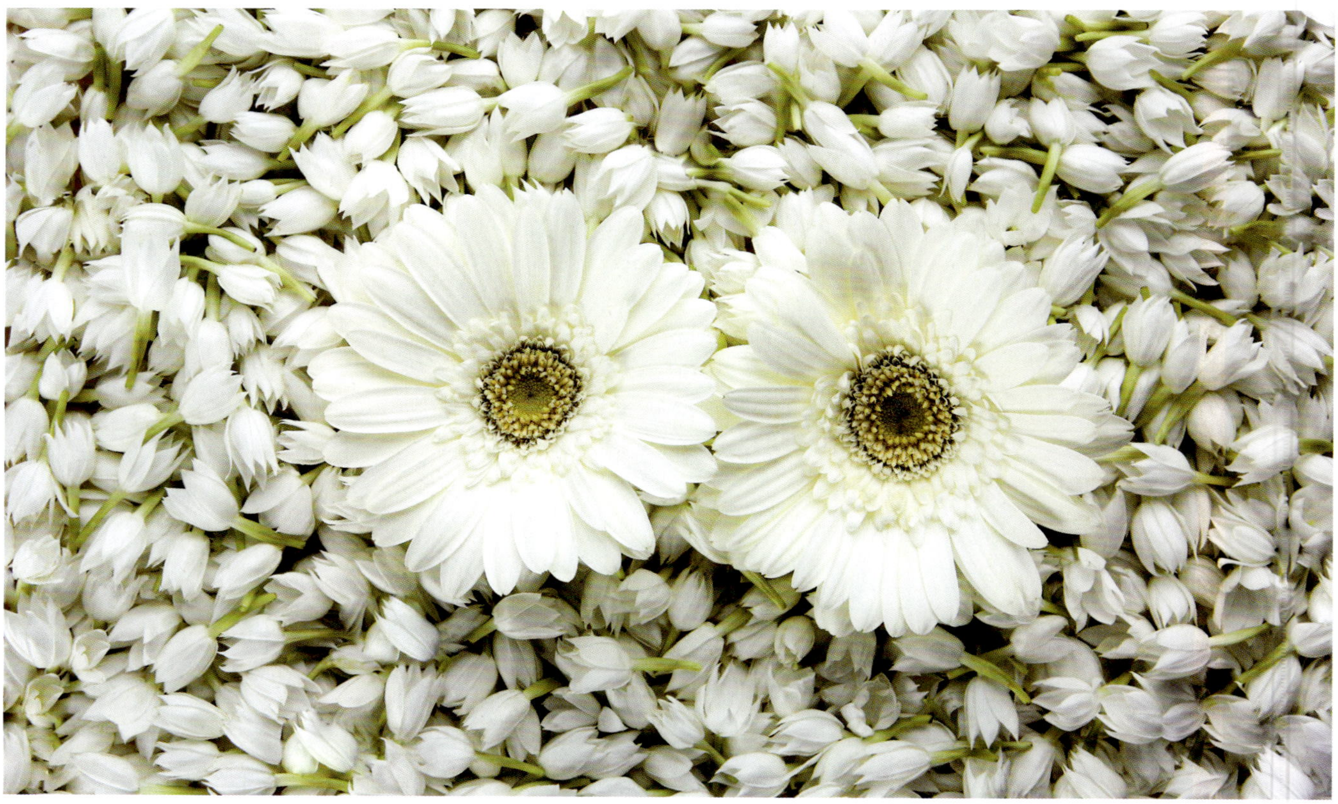

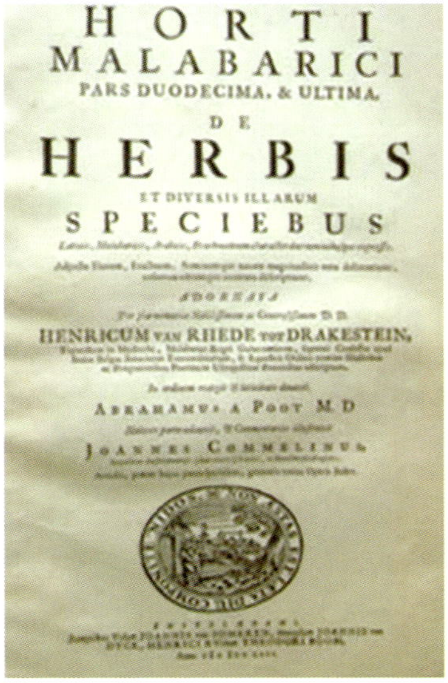

HORTI
MALABARICI
PARS DUODECIMA, & ULTIMA.
DE
HERBIS
ET DIVERSIS ILLARUM
SPECIEBUS

ADORNATA
HENRICUM VAN RHEEDE TOT DRAKESTEIN,
ABRAHAMUS A POOT M.D
JOANNES COMMELINUS.

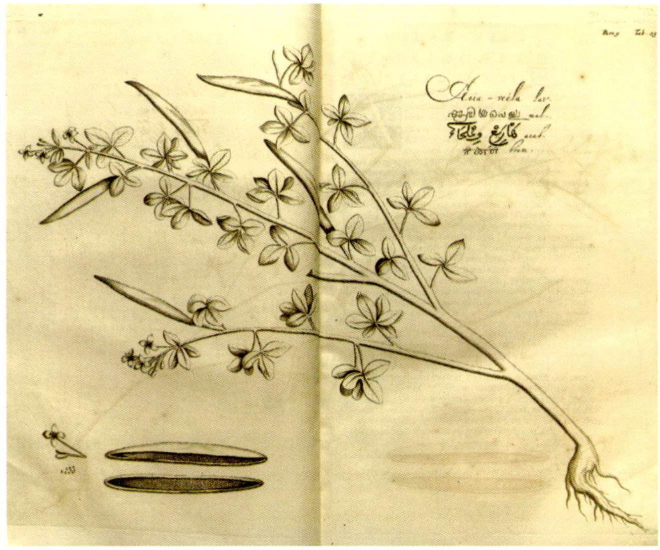

Above right: Title page of the *Hortus Malabaricus* (Garden of Malabar) a compendium of medicinal plants of Kerala

Above far right: Detail from the *Hortus Malabaricus*

opus of medicinal plants in Kerala, *Hortus Malabaricus* (Garden of Malabar);[4] to Gustav Krumbigel's journey, which my friend Suresh Jayaram in Bangalore has so appropriately documented in his book *Whatever he Touched, he Adorned*.[5] This also led me to revisit Kew Gardens time and again to view the 'stealth of the empire,' which in itself proved to be an extremely illuminating insight. These studies once again led me to salute the English for their tremendous archiving farsightedness in preserving cultures. It is a magical landscape which is often the muse of writers, artists and, naturally, the horticulturists.

Flowers can be viewed through multiple meanings, from poetics to botany and from the physicality of gardening to the abstraction in thought. My approach to this floral bouquet is from a cascade of ideas which engage me. The book is a personalised story of flowers, more in the diaristic genre, where it is part memory, part research, part aesthetics and part lived experience.

ENDNOTES

1. Shelley Saguaro, *Garden Plots: The Politics and Poetics of Gardens* (Burlington, VT: Ashgate, 2006).

2. Jennifer Potter, *Seven Flowers: And How They Shaped Our World* (London: Atlantic Books, 2013).

3 William Roxburgh, *Flora Indica, or, Descriptions of Indian Plants*, ed. William Carey (Serampore: Printed for W. Thacker. 1812).

4. Hendrik van Rheede, Ranga Bhat, Vinayaka Pandit, Appu Bhat, et al, *Hortus Malabaricus* (Amsterdam: 1678-1693).

5. Suresh Jayaram, *Whatever he Touched, he Adorned* (Bangalore: Goethe Institute, 2011).

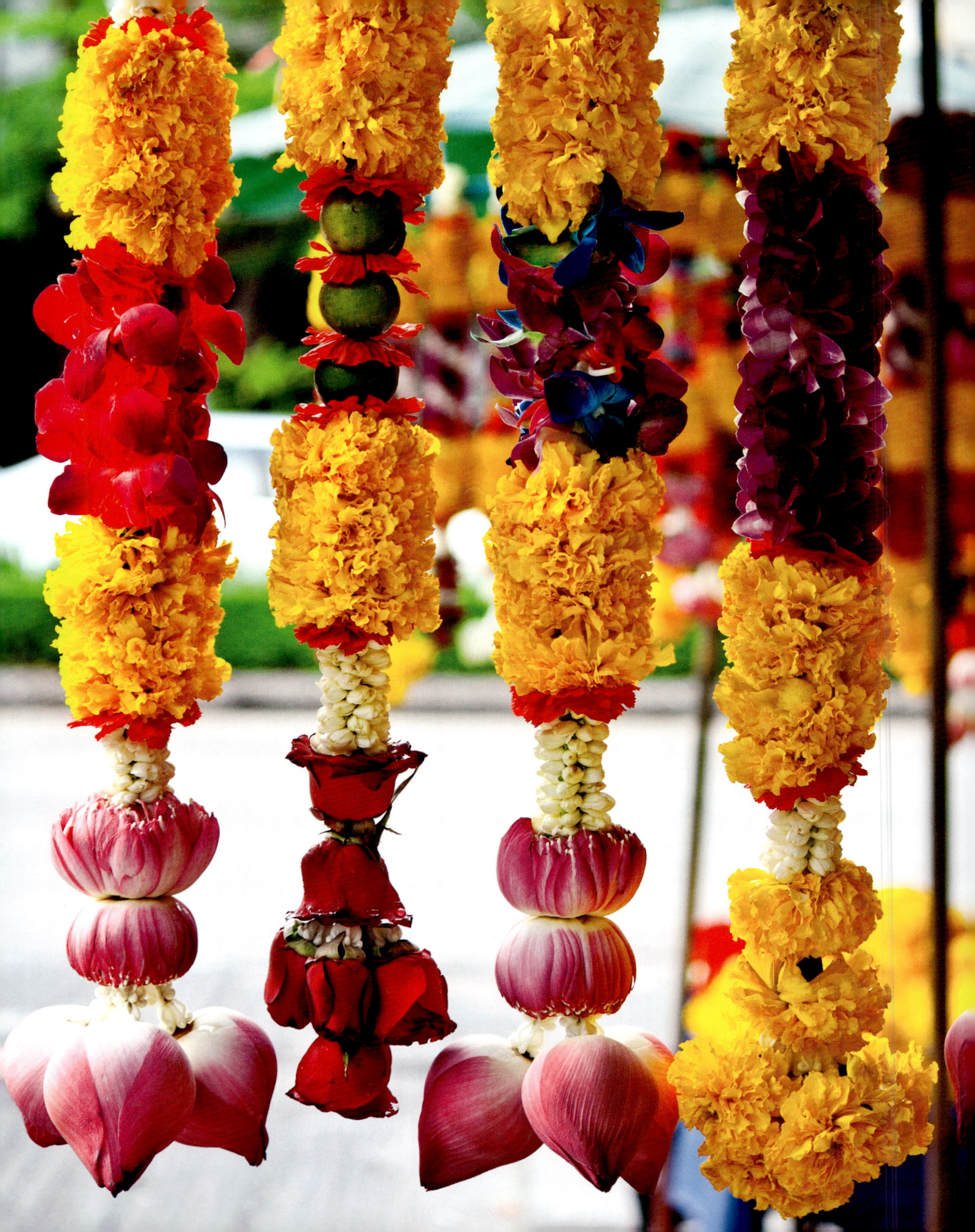

Flowering of Consciousness

Who are you, reader, reading my poems a hundred years hence?
I cannot send you one single flower from this wealth of the spring,
one single streak of gold from yonder clouds. Open your doors
and look abroad. From your blossoming garden gather
fragrant memories of the vanished flowers of a hundred years before.
In the joy of your heart may you feel the living joy that sang
one spring morning, sending its glad voice across a hundred years.
—Rabindranath Tagore

A flower is a representation of nature or *prakriti*. The prefix *pra* means superior and the suffix *kriti* denotes creation. In the *Devi Bhagavata Purana*, the Vedic sage Narada describes *prakriti* as made up of three syllables—*pra* signifies the *sattva guna*, the most exalted quality (goodness, balance, harmony); *kri* denotes the *rajo guna* (passion or activity) and *ti* denotes the *tamo guna* (disorder, imbalance). Thus when we offer flowers to the deities, it signifies a 'Flowering of Consciousness'.

Flower, fleur, flora, is to blossom, bloom or reach the heights of development, be it in beauty, intellect, purity or spirituality. Flowers have been one of the prime symbols of nature to inspire philosophers, artists, poets, writers, designers and couturiers. Flowers enthral the senses in every way—sight, smell, taste or feel. The sheer variety in shape, size, colour and form throws up an entire universe of inspiration. A flower represents the purest gift, be it to the divine or to the beloved. It represents the most exquisite symbol of beauty to seduce every muse. It pervades life from birth to death, and makes its presence felt at every occasion, from the joyous to the sacred and the sorrowful.

Facing page: Flower garlands for a Hindu religious ceremony

CELEBRATION OF THE DIVINE

At their peak, flowers are colourful, and frequently sweet-scented. From these qualities emerge the symbolic meanings of divinity associated with floral goddesses in mythology. In India, certain flowers are often associated with specific gods. Flowers have medicinal qualities used in Ayurveda and are deeply embedded in Indian myths with allegorical meanings.

The great philosopher and art historian Ananda K. Coomaraswamy wrote,

> The extant remains of Indian art cover a period of more than two thousand years. During this time many schools of thought have flourished and decayed, invaders of many races have poured into India and contributed to the infinite variety of her intellectual resources; countless dynasties have ruled and passed away. But just as through all Indian schools of thought there runs like a golden thread the fundamental idealism of the Upanishads, the Vedanta, so in all Indian art there is a unity that underlies all its bewildering variety.[1]

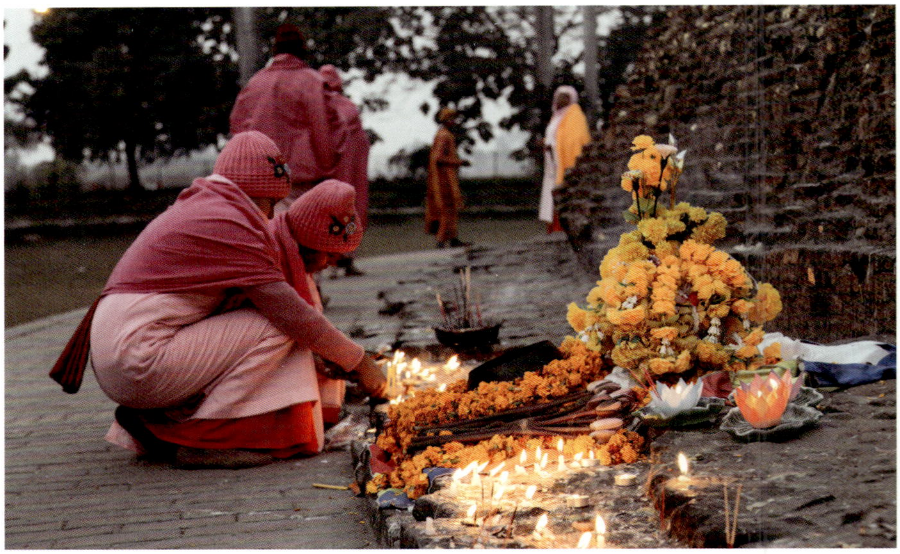

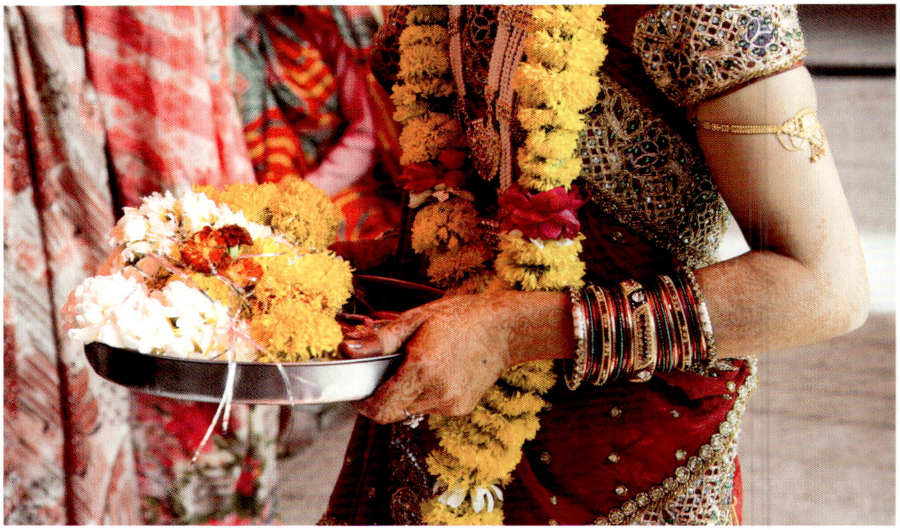

Unity in plurality is a defining characteristic of the Indian ethos, and amidst the vast array of styles and influences in classical Indian art, the one unifying strand is the celebration of the Divine—and what better symbol for this than the flower? *Pushpa, phool, suman, kali* are words redolent with the association of worship, offering, blessing, ecstatic union, and more. Flowers have always been associated with the Creator, religion and worship, and with innumerable myths and legends across various cultures.

In the subcontinent, *pooja*, the word for the worship ritual can be directly translated as the phrase 'floral act.' Flowers can be said to be made up of the *pancha bhoota* which are the five elements of Nature according to Hindu philosophy—air, water, fire, earth and ether. As Sri Aurobindo aptly put it,

> Earth's flowers spring up and laugh at time and death. In its simplicity, it carries the vibrations of the *Akasha*—the ethereal element itself—that is all that is most abstract, pure and perfect. It is, above everything else, the form, behind which is the sound, the all-powerful creative Mantra.'[2]

Flowers have lent inspiration to architects and thinkers through the ages. A few fields

Top: Lotus flower offerings at Ramabhar Stupa, built at the site of Lord Buddha's cremation, Kushinagar, Uttar Pradesh

Above: Flowers play a role as auspicious items in a Hindu wedding ceremony

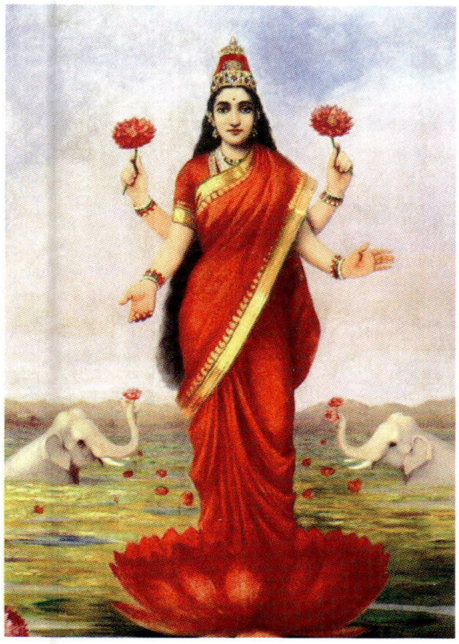 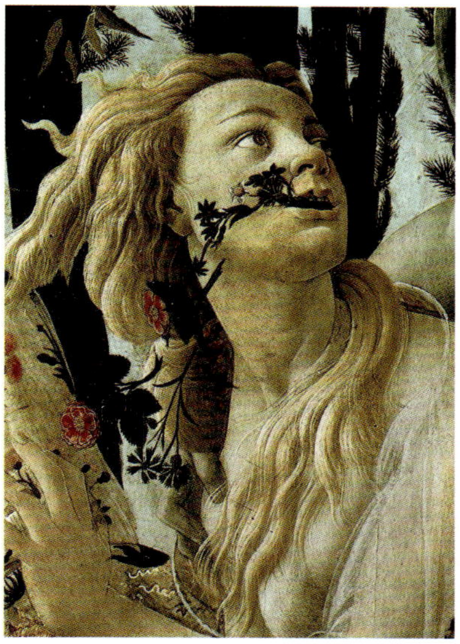 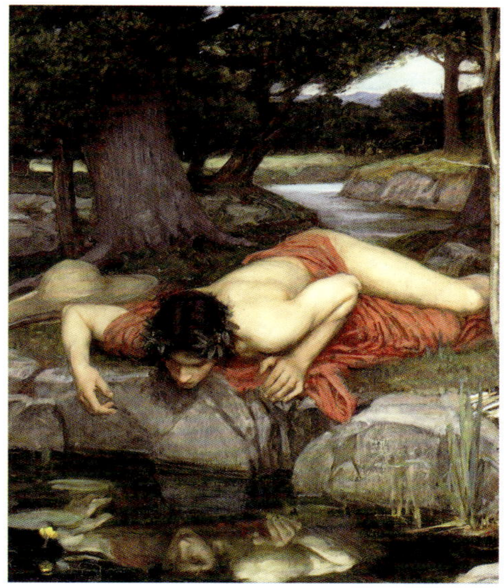

Above: 'Goddess Lakshmi', Raja Ravi Varma, oil on canvas, 1896, coll. Maharaja Fateh Singh Museum, Lakshmi Vilas Palace, Vadodara, Gujarat

Above middle: The nymph Chloris from Sandro Botticelli's 'Primavera' (c. 1482), Uffizi Gallery, Florence, Italy

Above right: Narcissus, detail from 'Echo and Narcissus', by John William Waterhouse (1849–1917), oil on canvas, Walker Art Gallery, Liverpool, UK

Below: Stained glass window depicting Immaculate Heart of Mary with white lily

that took their cue from flowers are religion, the yogic arts, the field of adornment and ornamentation, beautification, costuming, the visual arts, cuisine and the healing arts.

FLOWERS IN MYTHOLOGY

Flowers from around the world appear in mythology, and are associated with stories or myths from various cultures. Flowers are a significant part of the Christian faith. The red anemone is said to be a very early symbol of the Trinity, the colour red representing the blood shed by Christ when he was crucified. The lily signifies Mary, and is also considered a symbol of the empire of Upper Egypt.

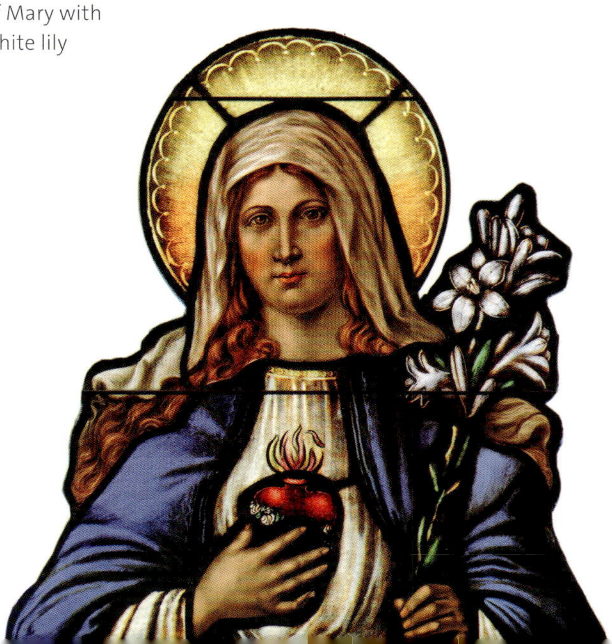

Greek and Roman mythologies are rich in symbols of nature. Chloris, is a Greek nymph associated with flowers, nature and springtime. The beautiful youth Narcissus is known both for his vanity and for lending his name to the eponymous flower. The Roman goddess Flora is the deity of flowers, while the Roman god, Zeus' wife, Hera, the goddess of women and marriage, is associated with the lily. Just as the lotus is linked with the birth of Lakshmi, it is also connected to the birth of the Egyptian goddess Isis. In the Indian subcontinent, two major religions—Hinduism and Buddhism—revere the lotus, which carries within itself multiple meanings and metaphors.

FLOWERS AS AN AESTHETIC MOTIF

The multicultural nature of the Indian subcontinent, throws up a rich tapestry of the presence of flowers as an aesthetic motif. Various flowers, many of which are now unknown, are strung through the verses of the great epic poems (the *Ramayana* and the *Mahabharata*), Jain and Buddhist literature, Sangam literature of the South, and later the works of Kalidasa, Bilhana, Jayadeva and other great Indian poets. Further cultural encounters

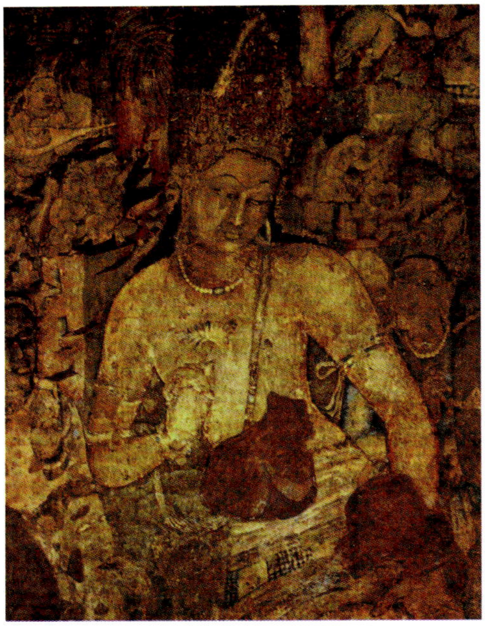

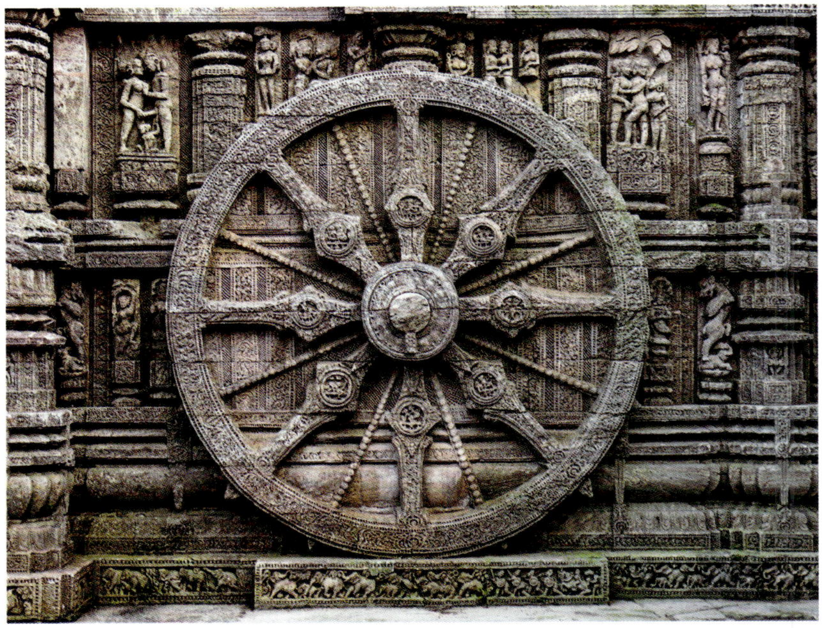

with Mughal and British rule have introduced once-foreign imagery, but soon adapted into the Indian mainstream, to enrich the pages of literature and poetry.

One of the best-known representations of a flower in ancient Indian art is the painting of the Padmapani in Ajanta. This luminous work created around 470 CE depicts the Avalokiteshwara holding a lotus in his right hand and looking down with an infinitely compassionate expression. The lotus flower is also used as a recurrent motif in the murals, elaborately sculpted pillars and ceilings of the *viharas* and *chaityas* of the later Ajanta caves as well as at Ellora. The Sanchi stupa is another treasury of motifs taken from nature, including details of flowers, plants and vines. The lotus has great relevance in the Buddhist as well as Hindu faith. It carries within itself multiple meanings and metaphors and was named the national flower of India after Independence.

As cave temples gave way to the elaborate construction of temples in the medieval age, painting and sculpture flourished alongside. We see various interpretations of the floral motif in the grand temples of Khajuraho in Madhya Pradesh, Bhubaneswar and Konark in Odisha, Kanchipuram and Mahabalipuram in Tamil Nadu, to name a few. With the advent of the Mughal School of Art, floral interpretation was found in built heritage as well as the famous miniature paintings. Modern Indian artists have celebrated flowers in bright colours and abstract forms, such as the works of A. Ramachandran and his wife Chameli Ramachandran, Mrinalini Mukherjee, Jayashree Burman, Suhasini Kejriwal, Badri Narayan, Karl Antao, and Jiten Thukral and Sumit Tagra.

Above left: Fresco of the Bodhisattva Padmapani holding a lotus, Ajanta, Cave 1, Maharashtra

Above: Stone wheel engraved with floral motifs at the Konark Sun Temple, Odisha, 13th century

Below: Engraved flowers and plants on the gate of the Sanchi Stupa, Madhya Pradesh

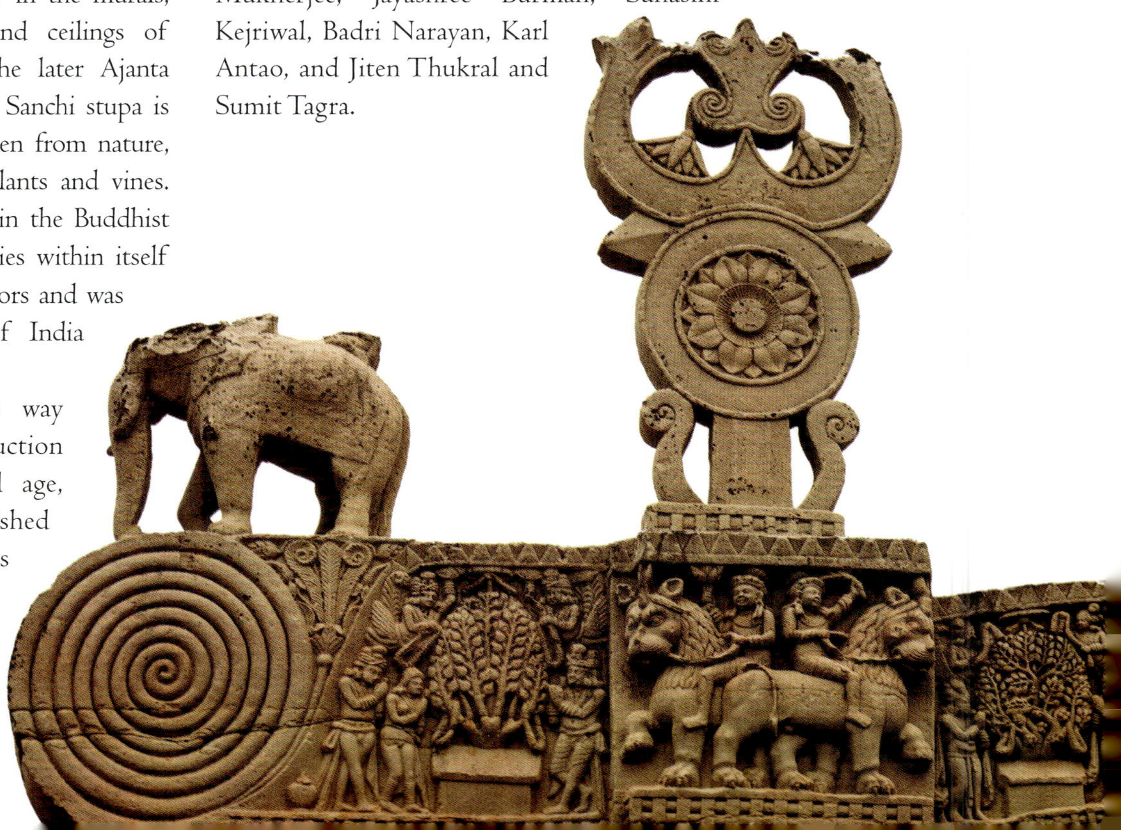

EVERY STAGE OF LIFE

Flowers have myriad meanings in stories spanning all the stages of human existence that feature recurring floral leitmotifs. From symbolising the purity of the virgin to the ardour of sexual congress, from the celebration of birth to sympathy for a loss at death, its interpretations are plentiful.

Flowers are an almost universal symbol for new life as well as for death. Springtime stirs in our hearts the joys of blossoming life as flowers bloom anew. However, flowers also have delicate lifespans, withering and perishing, which is why they are connected to death, especially white flowers that are symbols of purity as well as passing away. Hence they are used in wreaths and funeral tributes at both personal and public occasions.

There are some plants that bloom for only a few weeks, often in the spring or early summer. The *nargis* or narcissus, *parijat* or night jasmine, *champa* or frangipani, and *chameli* or jasmine, are some ethereal reminders of the passing of the seasons, and day into night. The shapes and colours of the blossoms also signify much. Yellow flowers are signs of celebration in Indian culture for their golden hue, but in Mexico they are considered funeral flowers. Chinese Taoists were depicted with gold flowers sprouting from the crowns of their heads to signal that they had achieved enlightenment. The shapes of flowers are significant as well, for example, the lotus shape is considered very auspicious throughout the history of India and a man or woman with eyes shaped like lotus petals was considered the epitome of beauty in Indian culture.

Within the Indian context, flowers are associated with offerings to gods and goddesses, and are vital ingredients for *attars* or natural perfumes extracted from them. They are also used for decoration such as *rangoli* and *torans*, in Ayurvedic medicine, for heightening the flavour of food, and most importantly, for worship and adornment.

Several Indian flowers find common resonance with other flowers found globally, such as carnation, hyacinth, lily, narcissus, poppy, rose, sunflower and violet. The intangible beauty and *je ne sais quoi* of flowers, a combination of their allure, shape, fragrance and vibrant hues affords them pride of place in many cultures across the globe.

Below: White flowers are symbols of purity as well as condolence. Hence they are used in wreaths and funeral tributes at both personal and public occasions

Following pages: '*Somnium Genero - Anima Turbo 4*', Thukral & Tagra, acrylic on canvas, 2012

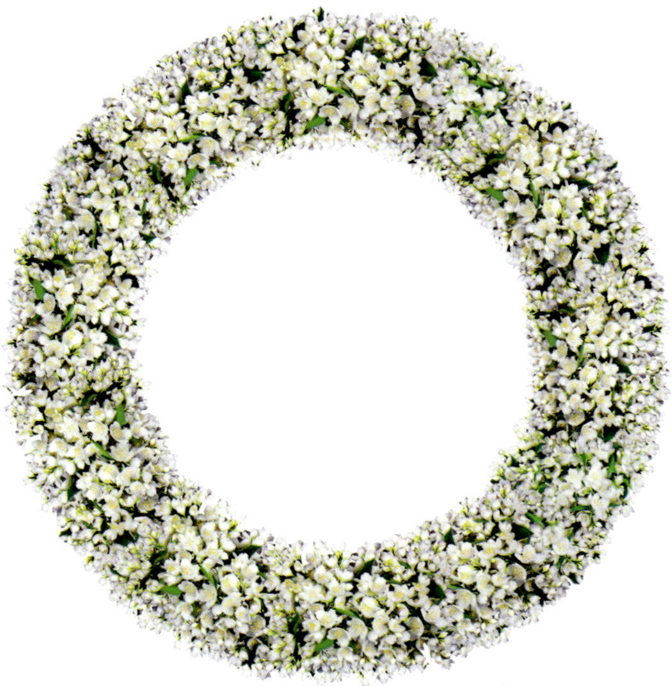

ENDNOTES

1. Ananda K. Coomaraswamy, *Essays on National Idealism* (New Delhi: Munshiram Manoharlal Publishers, 1981).
2. https://www.lntecc.com/homepage/documentspanchangam/2011_12.pdf (Accessed 18 June 2016).

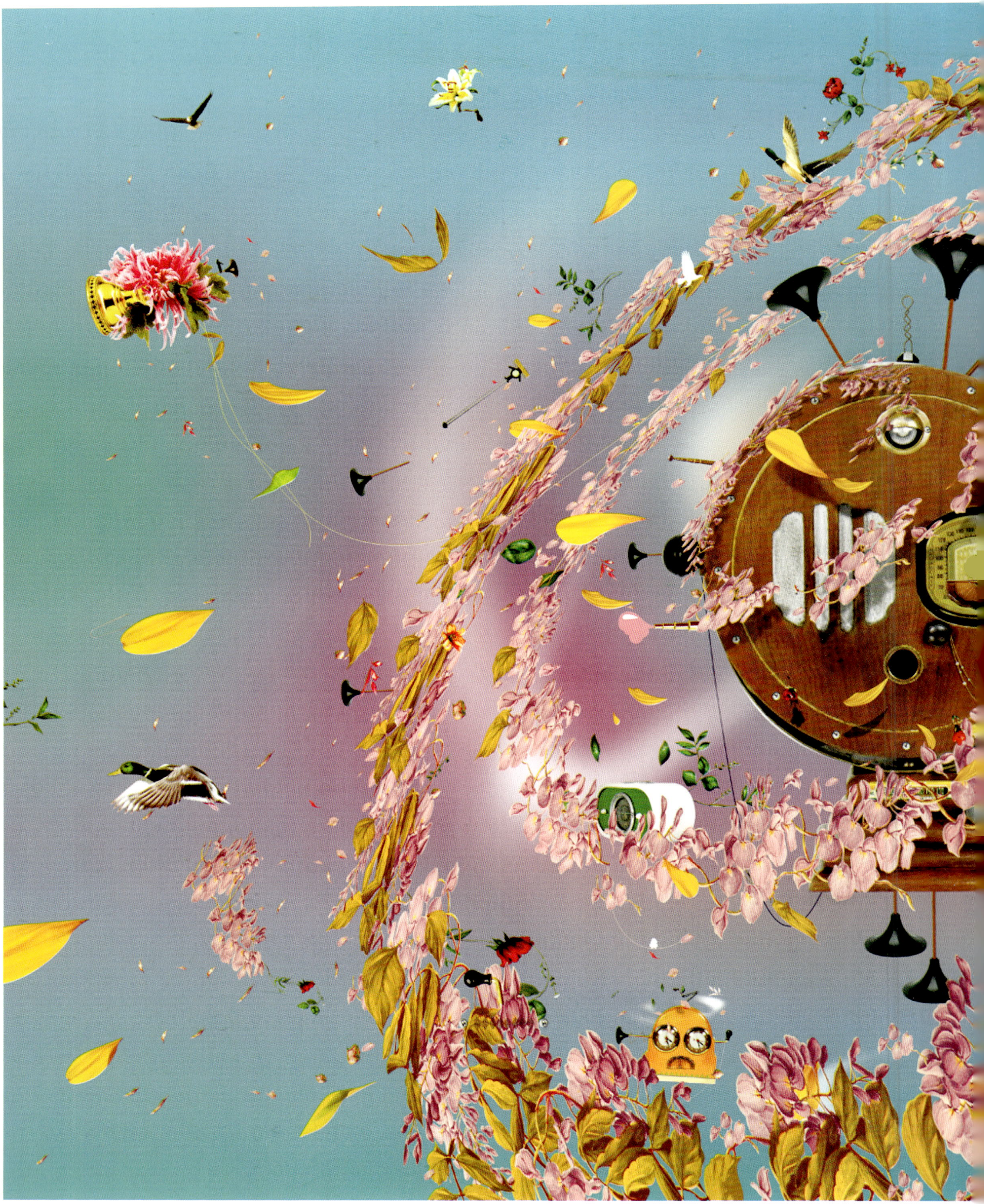

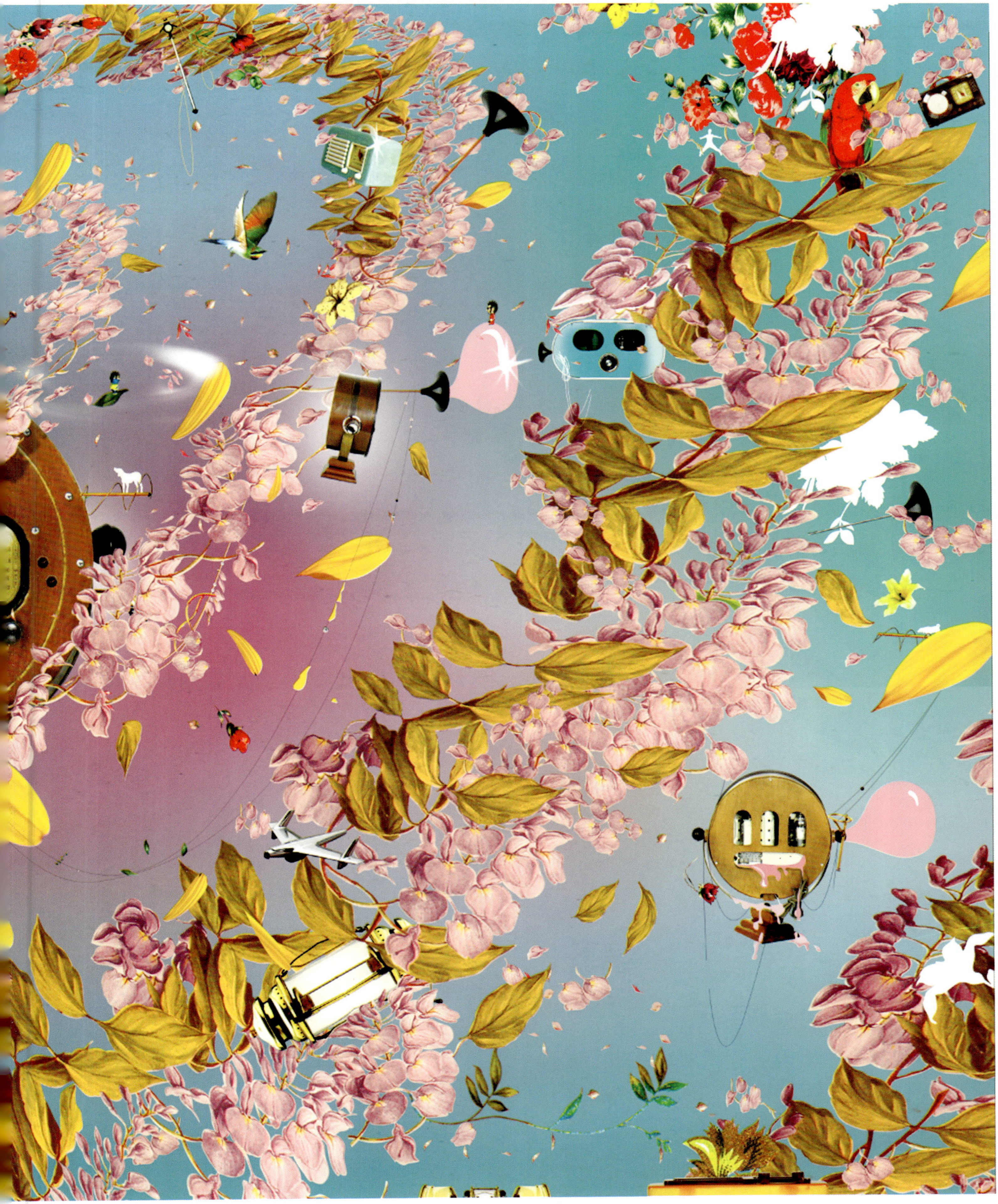

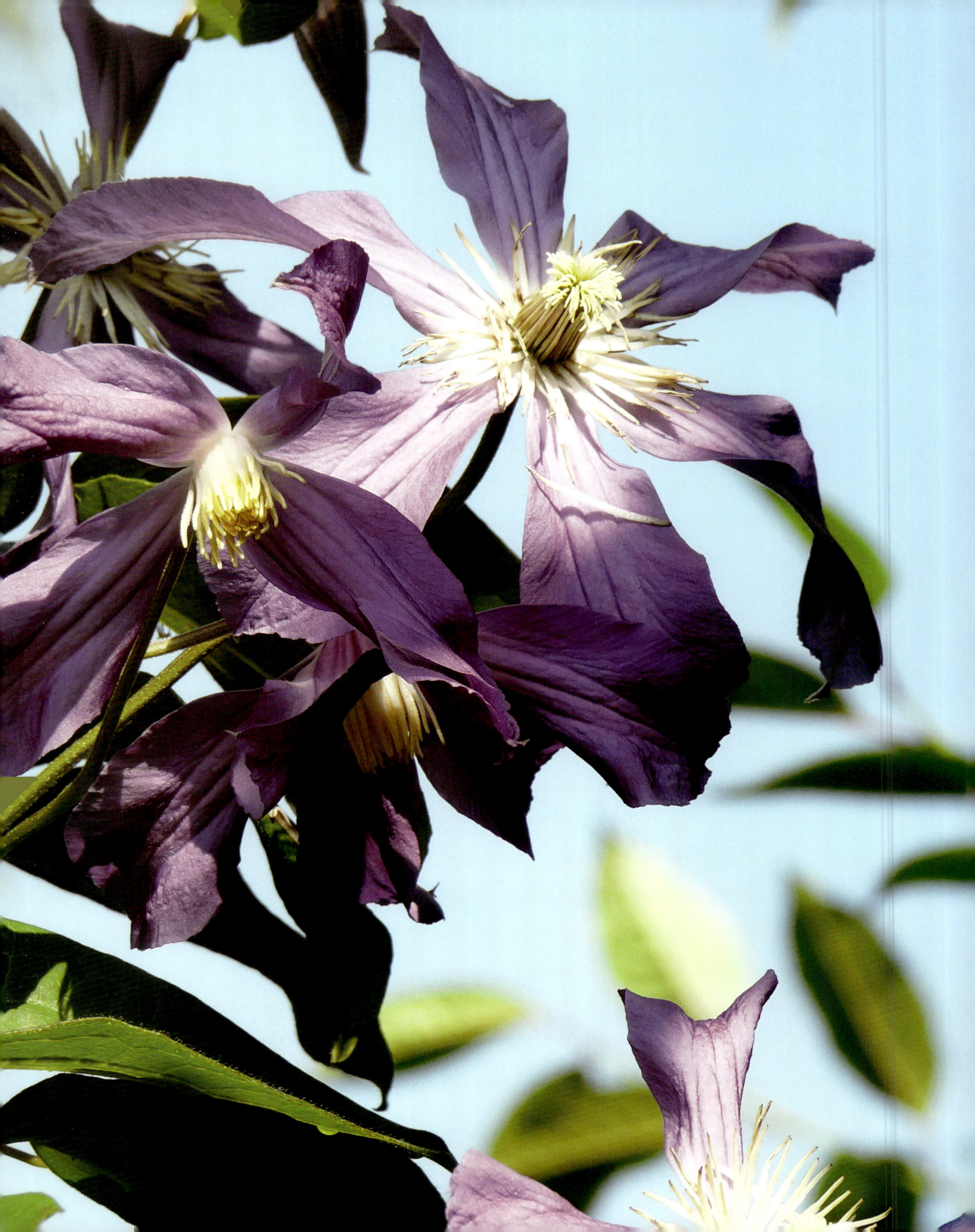

Artistry of Flowers

In the middle of the sky temple
Blooms a flower
Its petals are down
Its roots are up.
—Kabir[1]

From pottery featuring a snake and floral design in the earliest archaeological remains of the Indus Valley Civilisation, to the depiction of floral motifs in Hindu temples and Buddhist stupas and the exquisitely realistic floral renditions in Mughal miniatures, flowers in Indian art are depicted with loving tenderness. Be it painting, sculpture, architecture or folk art—flowers are one of the most popular motifs. They are present on every surface: frescoes, murals, walls, wood carvings, pottery, palm leaf manuscripts, paper, canvas and textile. Wherever you look, the flower is not far to find.

FROM TIMES IMMEMORIAL

Traditional Indian art has always been an expression of the sublime. But along with spirituality, nature has also played a vital role in its inspiration. Going as far back as the Indus Valley Civilisation, one observes that simple patterns from nature were utilised for ornamentation. Examples from Mohenjo-Daro, Harappa and Lothal (2500–1500 BCE) range from the

Facing page:
Lilac flowers of
climbing clematis

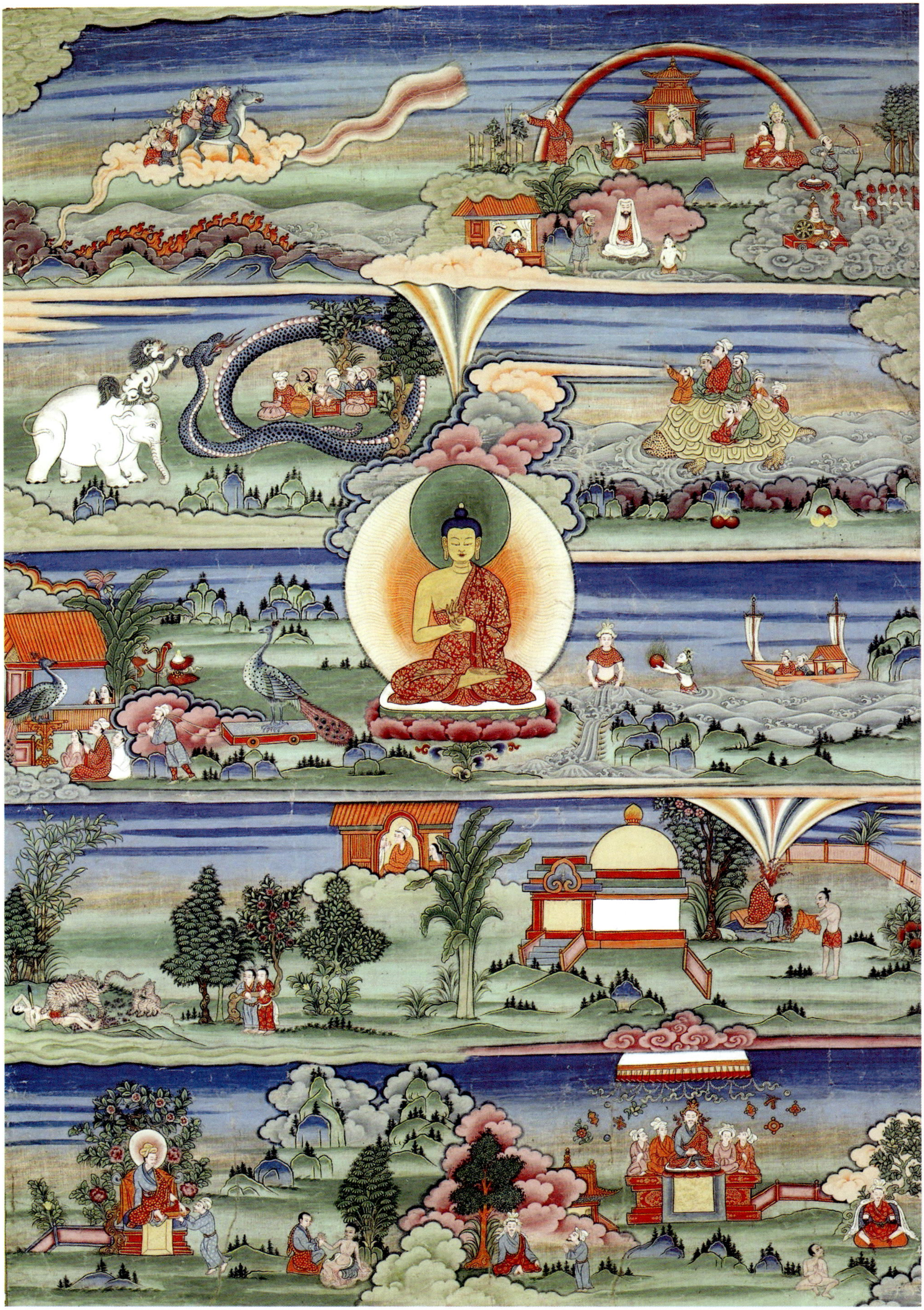

Right: Terracotta pot decorated with a floral motif, from the Indus Valley Civilisation

Far right: Sculpture of a priest-king from Mohenjodaro wearing a robe with a trefoil clover-leaf design

Below: Stone sculpture of Khasarpana Lokesvara standing on a lotus pedestal, Pala dynasty, c. ninth century CE

Facing page: Painted thangka of the Jataka Tales, Phajoding Gonpa, Thimphu, Bhutan, 18th-19th century

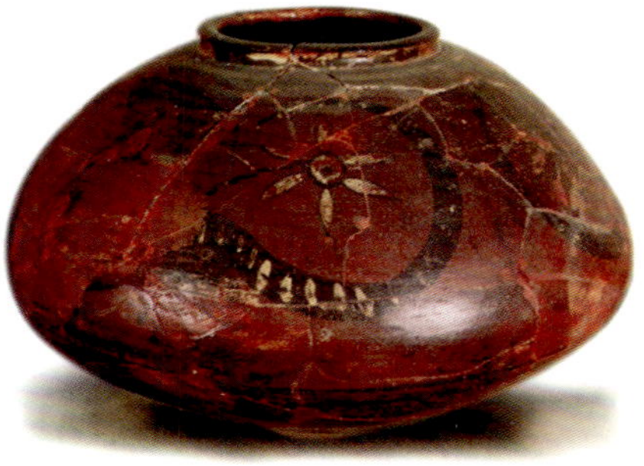

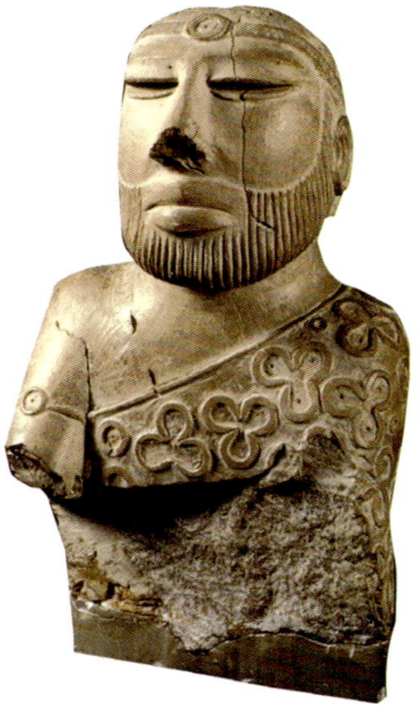

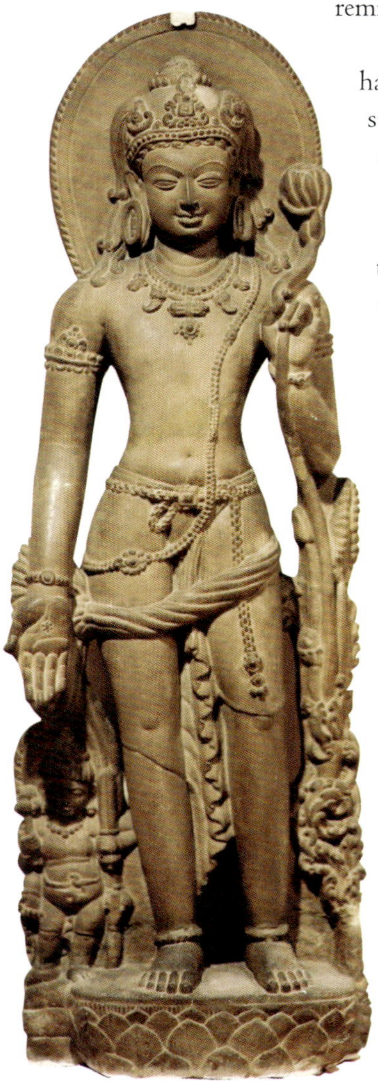

distinctive three-petalled trefoil of clover found on the clothing of a limestone figurine of a priest-king to the animal and plant motifs on the seals from the time. Even the pottery was judged to be covered in creepers, the patterns reminiscent of flowers and leaves.

Tree worship, an age-old practice which had close ties with nature has offered ample subject matter for art motifs throughout the early history of India. Indian craftsmen appear to have possessed an affinity for natural wonders such as flowers and trees, incorporating them as they did in decorative motifs, giving them a rare and noble beauty. The fine descriptive nature hymns of the *Rig Veda*, the oldest sacred book of the Hindus, and later the Jataka tales that feature stories on the life of the Buddha, use symbols and imagery based on flowers, leaves and fruits in great profusion.

The depiction of flowers and buds can be seen in Buddhist art as well as Hindu temple architecture. Flora and fauna form an important part of the decoration in temple architecture, as well as stupas, *viharas* and *chaityas*, celebrating nature with its blessings and rhythms. These motifs are interwoven in the structure for auspicious reasons, as well as for the beautification of the temple. Different kinds of floral motifs are amalgamated with geometrical forms, as are interlacing patterns of creepers (an integral motif in Gupta art). A significant Indian flower motif is the lotus (*padma* or *kamal*), which is considered the most sacred flower for many religions and signifies the epitome of pure beauty. Besides, the lotus is used as a pedestal for most Buddhist and Hindu deities in temple sculptures and paintings.

BUDDHIST INSPIRATIONS

The sculptures in the stupas of the early Buddhist period (250–100 BCE), use specific trees to symbolically represent the Buddha, such as the *pipal* (*bodhi*) tree under which he gained enlightenment, the sal and the *ashoka* tree. Other trees appearing in these sculptures are the scarlet flowering silk cotton, the banyan and the mango. Each tree is shown beautifully portrayed with its distinctive foliage, flowers or fruit, as the case may be.

One of the best-known ancient monuments of Indian art is the Sarnath pillar, one of 19 pillars erected by the Mauryan emperor Ashoka (268–232 BCE) to spread the Buddha's teachings over north India. The pillar was broken during the Turkish invasion and while the base stands at the original site, the intricately carved upper portion is displayed at the Sarnath Archaeological Museum. The base of the capital is an inverted lotus, above which stand four animals (horse, bull, elephant and lion) to indicate the four cardinal directions. The crowning glory is the capital with four lions standing back to back, and this has now been adopted as the official emblem of India. At first glance, the lotus may be outshined by the bold animals, but this is a floral symbol that recurs over and over again in Buddhist art, be it in pillars, ceilings of *chaityas* (prayer halls), *toranas* (gates) or stupas (a place of meditation where sacred relics are consecrated). The lotus is used to represent the purity of the body and mind, rising above the muddy waters of material attachment.

Another landmark of Buddhist art and architecture is the Sanchi Stupa, the foundations of which were laid by Emperor Ashoka. The monolithic pillar here too has a lion capital standing on a base ornamented with small lotus petals. Underneath it, on either side, there springs a lotus bud, the stalks of the lotus being hung with garlands. The large bell shaped member is also made up of lotus petals turned downwards. Between the abacus and the member is a small convex moulding with regular hollows depicting the seed compartments of the lotus radiating from the centre like the spokes of a wheel.

In the varied sculptures at the Bharhut Stupa of the Sunga period (184–74 BCE), many types of flowering trees are depicted. The railings and pillars of the stupa have carved medallions that use flower compositions, chiefly the lotus in different forms. Nothing of this ruined stupa remains at the original site, except for the foundations. The main gates, pillars and railings can be found on display at the Indian Museum, Kolkata. The decorative medallions are ornamented with the lotus buds emerging from a bowl; the fully opened lotus and lotus leaves in various forms and elephants offering lotus buds and flowers to the Buddha.

The Vidisha pillar, also known at the Heliodorus Pillar, named after the Buddhist wife of the Emperor Ashoka shows an abundant use of flowers. At the same site, another carving depicts a bunch of flowers in the hands of a female figure. Such female figures known as *Yakshis* are nature spirits associated with fertility that can be seen in all

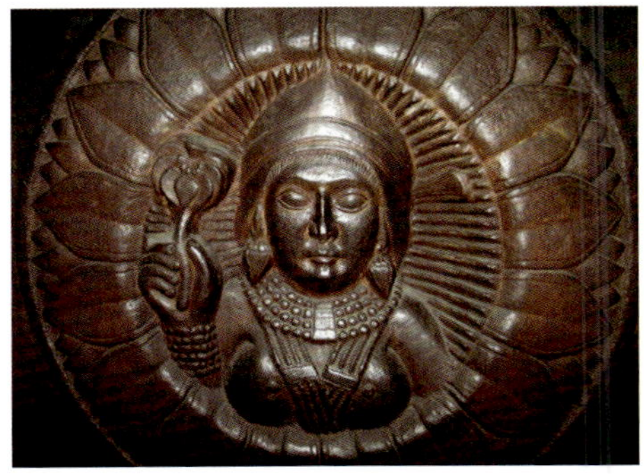

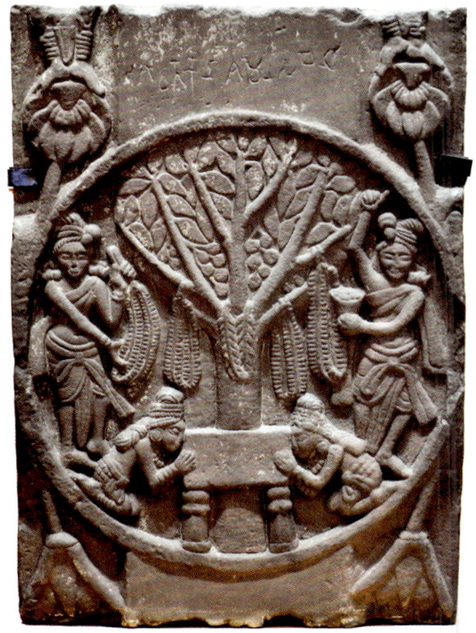

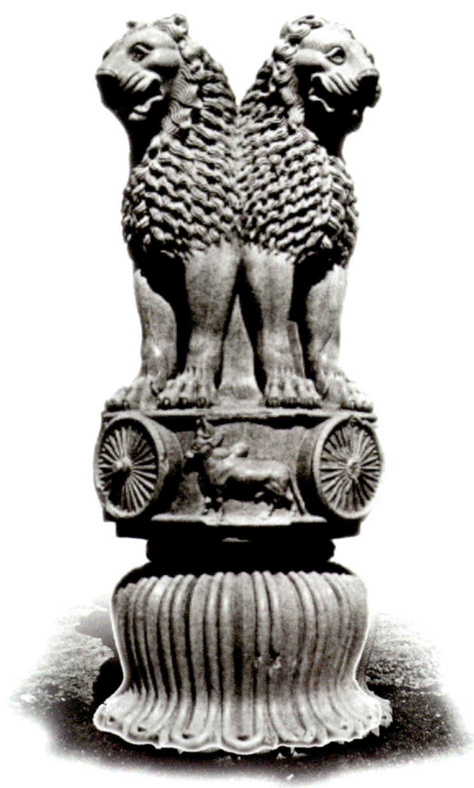

Top: Female figure holding a lotus from the Bharhut Stupa, at Indian Museum, Kolkata

Above: Worship of the Bodhi tree, with Yakshini, Bharhut Stupa of the Sunga period.

Left: The Lion Capital at Sarnath stands proudly on a lotus-shaped base

the Buddhist stupas of this period. They are entwined with flowery trees, symbolising the idea that budding plants long for the touch of a beautiful woman.

Over the centuries the design of the lotus began evolving and taking on different meanings. In the rock-cut caves of Karla in Maharashtra (2nd to 5th century AD) the seed vessel of the lotus is made much more conspicuous and symbolises the holy city at Lake Mansarovar. It is enclosed in a quadrangle shrine, the openings on each side representing the four 'gates' of the lake. Above this shrine, prominent upturned petals of the lotus symbolise the snow-capped Himalayan peaks. On the top are four Buddhist divinities mounted on elephants, symbolising inhabitants of the heavenly spheres that rise from the summit of the Mount Kailasha. It is believed that Mount Kailasha is at the heart of the six mountain ranges and is also symbolic of the lotus.

Below: Four Buddhist divinities mounted on elephants, rest on a lotus pedestal, Karla Caves, Maharashtra

Right: The Heliodorus Pillar at Vidisha features carved petals and floral motifs, 2nd century BCE, Vidisha, Madhya Pradesh

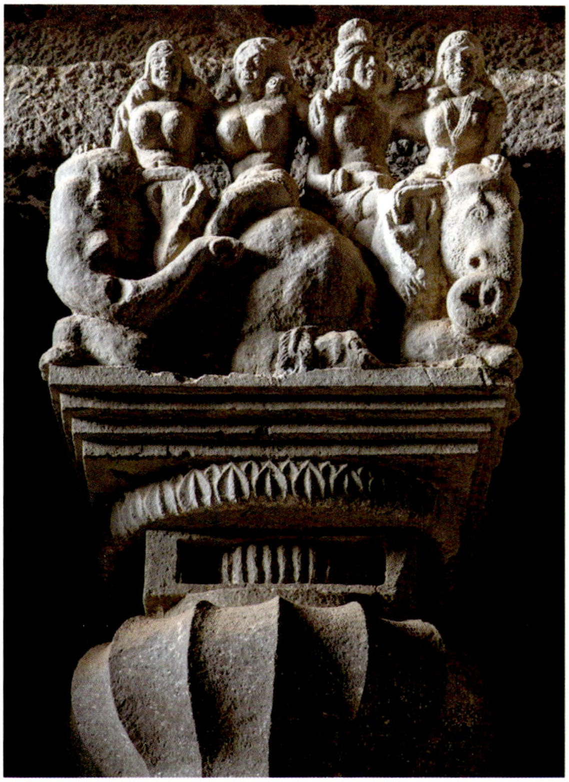

Closely connected to the lotus is the water pot, *kalash* or *kumbh*, which holds the nectar of immortality, churned by the gods and demons from the cosmic ocean. The design of the pillar of the World Lotus became modified in later Indian art. At Ajanta, some six or seven centuries after Ashoka's time, the original symbolism of the World Lotus is barely recognisable. The conspicuous turned down petals of the lotus that formed the 'bell' in the old capitals have disappeared and the seed vessel with the upturned petals resembles a sacrificial water vessel, the lota.

OUTSIDE THE BUDDHIST FOLD

One of the greatest collections of Gandhara art is the Kronos Collection at the Metropolitan Museum in New York. The first item here is a rare and extraordinary silver rondel depicting the goddess Hariti, a fertility goddess often said to be connected to children. Hariti reclines in a chair, the back and side panels of which are covered in an intricate pattern comprising waves and dots while two birds perch along the back. Two lotuses, one in full bloom and the bud of another, are visible alongside her chair. The lotus motif is also evident in the chair leg which resembles an upturned lotus. A distinctive motif called the egg

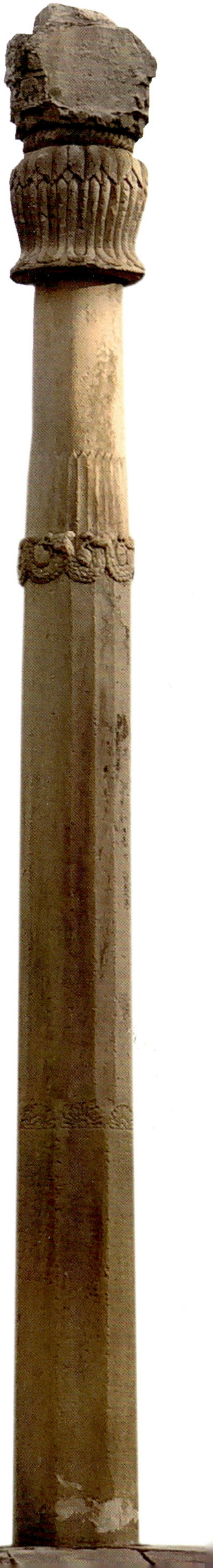

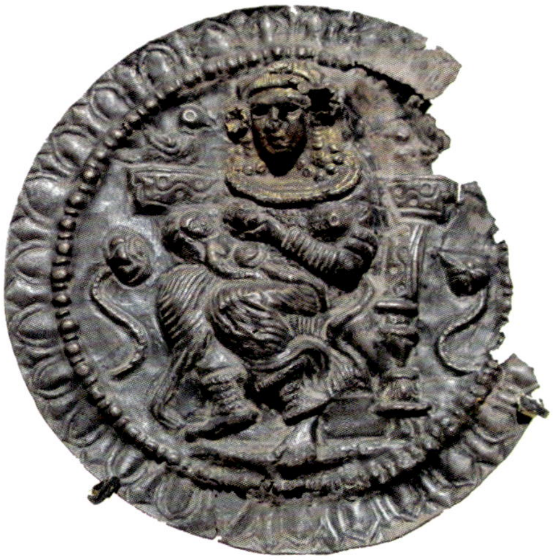

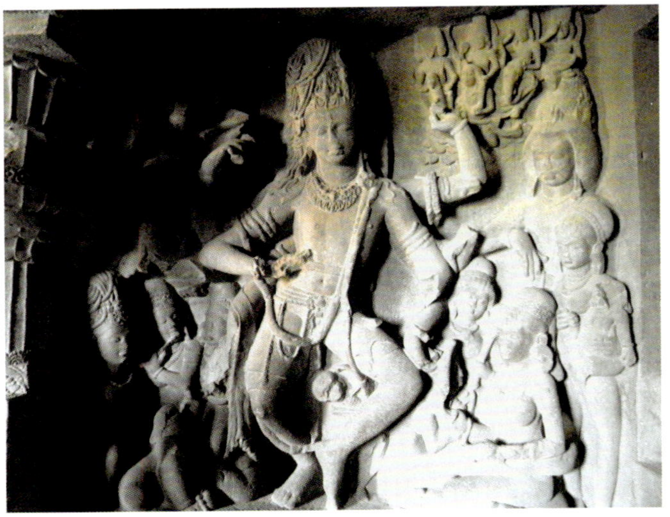

and dart runs along the inner perimeter of the whole artwork. Here, again, the lotus pattern re-emerges, edging the outside of the rondel in a stylised lotus petal trim.

Some very interesting examples of the use of the flower motif can be seen at the Mansar site, dating around the 5th century, in the Ramtek Tehsil near Nagpur. A large diadem of tiny flowers and pointed leaves adorns the bun of a female figure: lying obliquely on one side, it slips towards the top of the head and is richly decorated. The Samalaji Sculptures of Western India dating from the 6th century and now mainly found in the Baroda state museum show interesting formations of leaves and flowers ornamenting headdresses. In the group of sculptures in cave 21 at Ellora that depicts Ganga and the dancing Shiva, there are also several sculptures of women adorned with flowers and leaves.

The Khajuraho group of temples (circa 950–1050) are a rich storehouse of Indian decorative sculpture using flowers, leaves and creepers as ornamentation. Among the various classical forms of female figures displaying flowers as visual imagery are *darpana* (looking into a mirror), *alasya* (relaxed and indolent), *ketaki-bandha* (adorning with ketaki flower), *nartaki* (dancer) and *vinyasa* (pensive).

The sculptures of the 10th century Mukteswara temple in Odisha are stunning, incorporating numerous floral motifs. Many believe that Odiya temples are offerings to the holiest of the deities and not simply places to house their likenesses. They are adorned with magnificently carved likeness of gods and goddesses, monarchs and motifs that incorporate flora and fauna. The depictions of the floral motif in the Kamakhya temple in Assam can be considered an important example from the north-eastern part of India.

Above left: Rondel with lotus trim depicting the Goddess Hariti, Gandhara, c. 1st century

Above: Dancing Shiva and women adorned with flowers, Cave 21, Ellora, Maharashtra

Below: Detail of floral motifs in Mukteswara temple, Odisha

Facing page: 'Animals of Hindustan, small deer and cows', illustration from the *Baburnama*, late 16th century

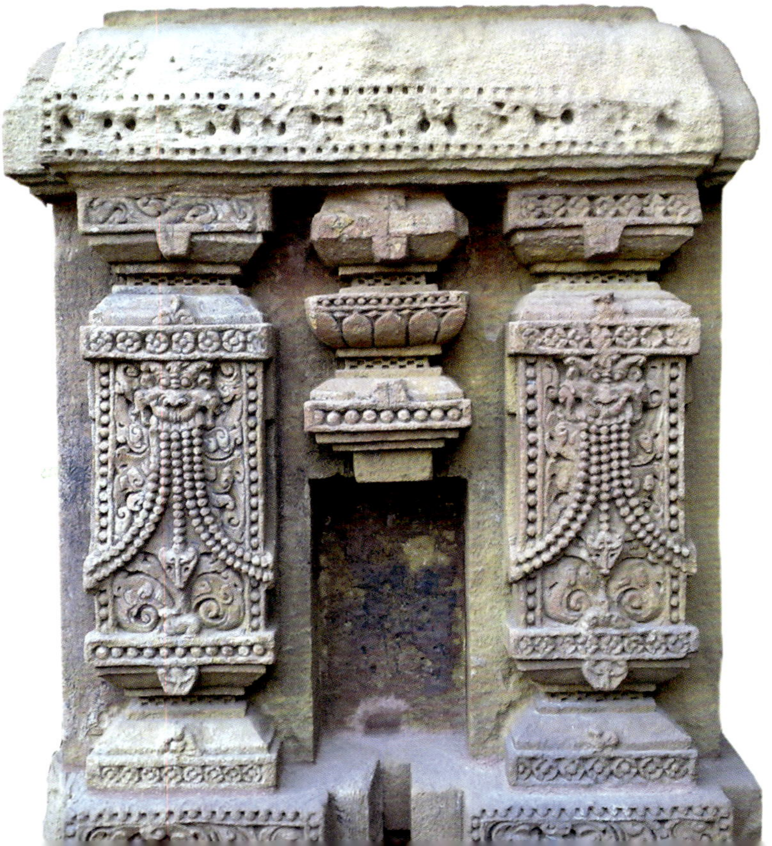

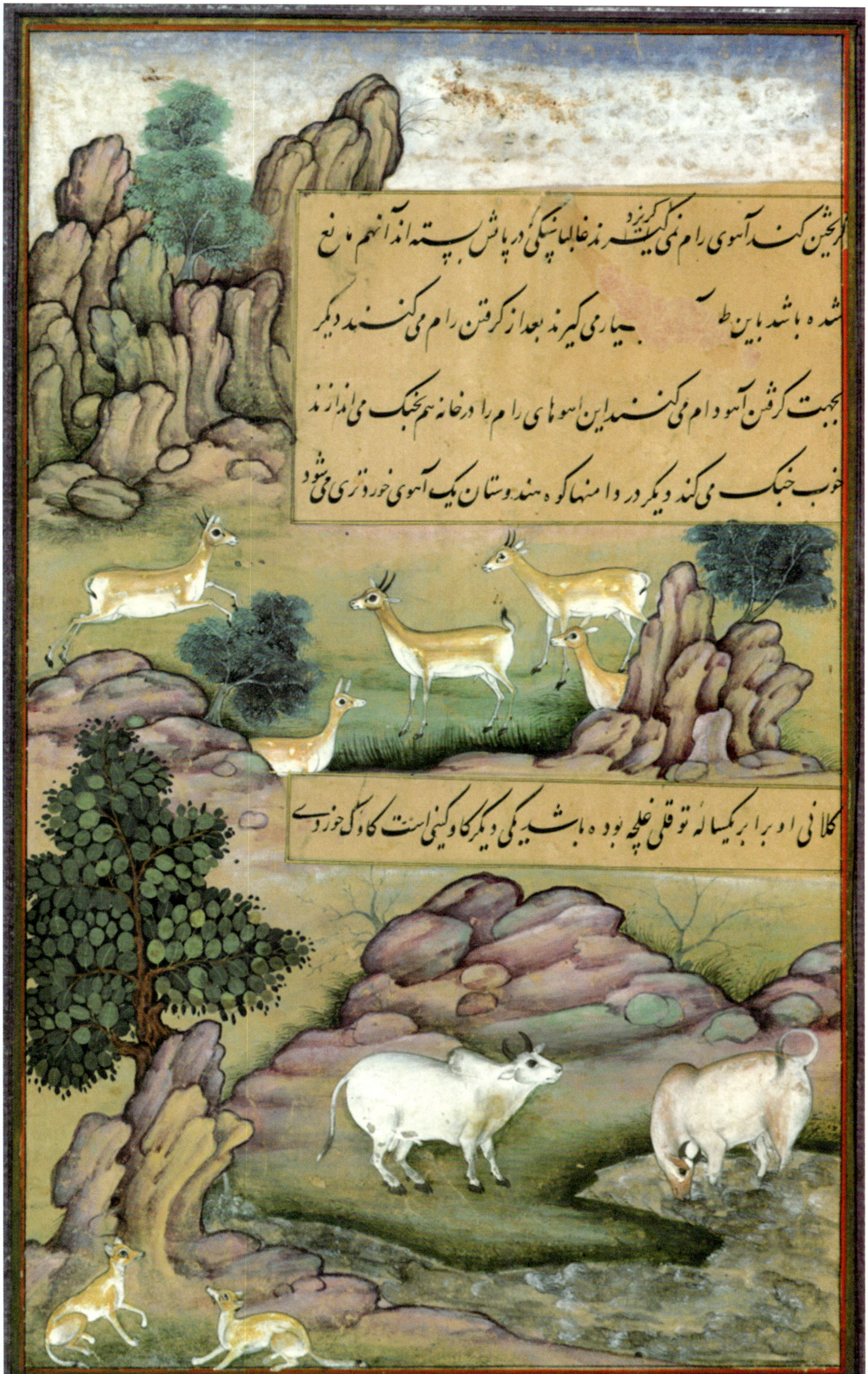
اگر چنین کند آهوی رام نمی گریزد غالبا سنگی در پاش بسته اند آنهم مانع
شده باشد باین ط بیاری می گیرند بعد از گرفتن رام می کنند دیگر
بجهت گرفتن آهو دام می کنند این آهوهای رام را در خانه هم بجنگ می اندازند
خوب جنگ می کند دیگر در دامنها کوه هندوستان یک آهوی خوردتری می شود
کلانی او برابر یکساله توقلی غلچه بوده باشد یکی دیگر گاوگینی است گاوک خوردی

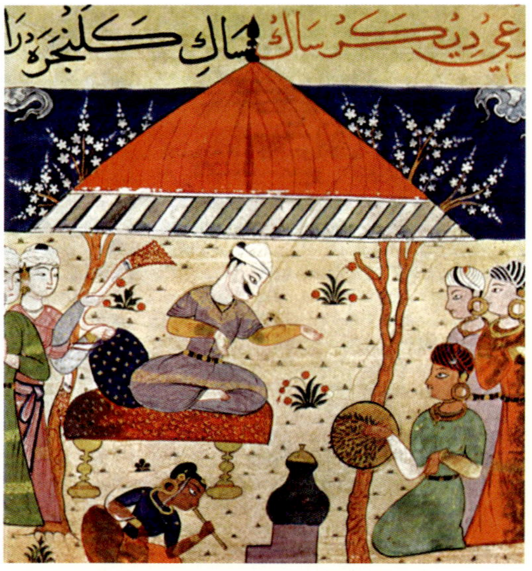

Scene from the *Nimat Nama* manuscript

NIMAT NAMA

Preceding the Mughal period, an illustrated cookery book called *Nimat Nama* was commissioned in the time of Ghiyasuddin Khilji also known as Ghiyath Shah (1469–1500) of the Malwa Sultanate in Central India. The manuscript shows Ghiyasuddin Khilji supervising food being prepared by maids. In the *Nimat Nama*, the clouds, the tufts of grass and the trees with flowers in bloom make apparent the Persian inflection. Indian elements are noticeable in the use of colours, and in some female figures through their costumes and ornaments. In this manuscript, one can notice the first attempts towards the evolution of a new style of painting by the fusion of the Persian style of Shiraz with the indigenous Indian style.

FLOWERING OF THE MUGHAL SCHOOL

In 1526, Babur, a military adventurer from Central Asia, established the Mughal dynasty in India. During their 330 year reign, the Mughals gained an international reputation for their wealth and intellectual and artistic pursuits. The oftentimes nomadic nature of his life was not unfamiliar to Babur, a practice that his forbears had adopted, setting up camp in the recesses of nature from where they could reign.

One of Babur's first acts in India was to establish landscaped flower gardens in Kashmir, a practice his successors continued across much of India with increasing grandeur. Each of Babur's successors patronised the arts in a significant way and the blend of Persian techniques with Indian themes and traditions gave rise to the world-famous Mughal School of Art, deeply impacting the culture of South Asia and remembered as the apex of refinement and luxury.

As mentioned earlier, the Mughal emperors patronised the depiction of Nature

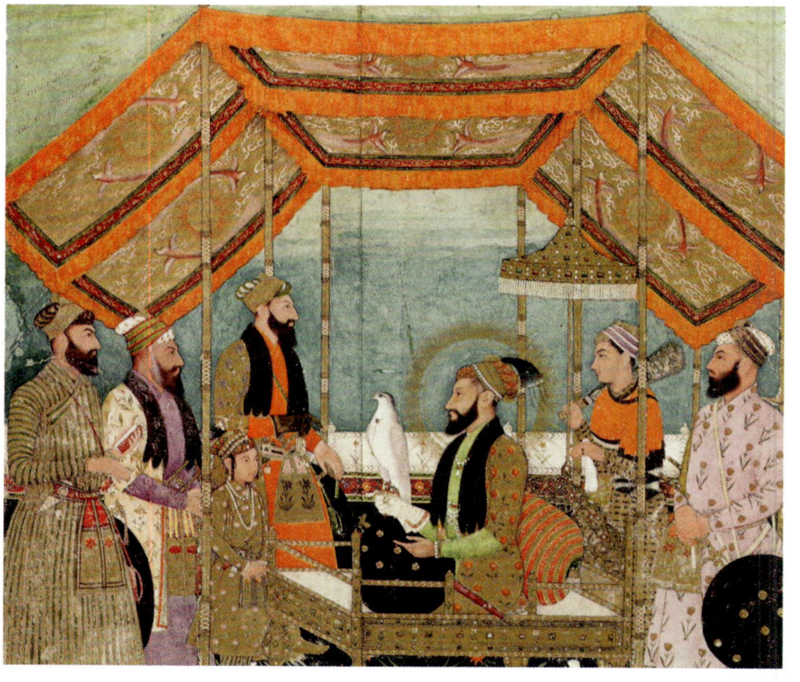

Above: 'Aurangzeb holds court', painting attributed to Bichitr, c. 1660; note the floral motifs on the robes of Aurangzeb and his courtiers

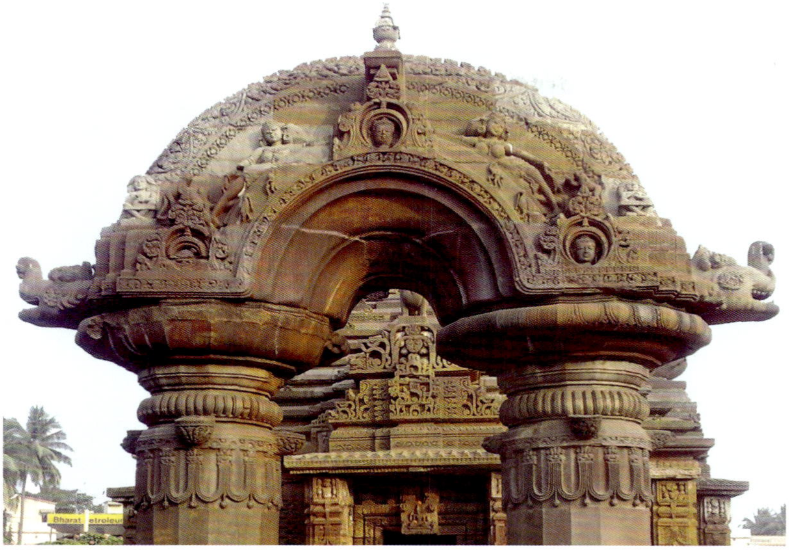

Above: Carved flowers and vines on the Torana, Mukteshwara Temple, Odisha

through painting in a big way. Babur's memoirs or the *Baburnama,* is a rich repository of illustrations of flora and fauna. The interest for art was kept up by Jahangir whose deep abiding passion for nature inspired and gave rise to portraits of the avian variety as well as floral and animal studies. His queen Nur Jahan also shared his love for nature. Artists were encouraged to paint anything beautiful, whether flowers, plants or animal studies, bringing about a new awareness in nature study. Jahangir encouraged his leading court painter Ustad Mansur to do individual studies of flowers he had seen in Kashmir, and the master painted over a hundred floral varieties.

The art of painting reached a new peak of refinement under Jahangir's rule.An array of portraits and court scenes known as *durbar* art as well as depictions of the fauna were popular subjects, as the emperor gave great impetus to plant and animal studies.

It is no wonder that one of the most noteworthy manuscripts illustrated then was a book of fables featuring animals. It was known as *Ayar-i-Danish.* Some pages of the *Ayar-i-Danish* are preserved in the Cowasji Jahangir collection, Bombay, and others in the Chester Beatty Library, Dublin. The *Anwar-i-sunavli,* is another, similar tome of animal fables that contains a treasure trove of illustrations and can now be found in the British Museum, London. These works date back to the early 17th century.

By the mid-17th century, the taste for the naturalistic portrayal of flowers reached an apogee of artistic expression Akbar, who was an ardent lover of flowers, fruits and plants, introduced a new realism in art. Even though the Persian conventions were more predominant, in due time, artists from various parts of India assembled at the Imperial court, where out of the 17 celebrated painters of Akbar's Court, 13 of them were of Hindu origin. Thus Indian flora blossomed in the artistic panorama and the depiction of flowers—buds, half-blooming and full-blown—opened new vistas in the art world.

ETCHED IN STONE

The Mughals are mainly remembered around the world for the architectural marvels they created all over India, the Taj Mahal being the pinnacle of decorative design in marble. Architectural decoration was inspired by traditional Persian and Indian flora, as well as by European botanical artists. An extremely elaborate art form, floral motifs and arabesques were found sometimes to cover whole surfaces. The pattern formed a kind of infinite rhythm, an effect that came about from the constant reuse of the wavy lines. The effect was soothing, the pattern one that could be varied by adjusting an element such as colour or line. Tablets and panels featuring encrusted jewels and floral and plant patterns in low relief often enhanced the architecture.

The flower motif goes back in time to the early representations based on philosophical and metaphysical concepts. *Toranas* are the gateways to Indian and Buddhist shrines, and are the harbingers of good luck. Often made of flowers and leaves, they are erected as temporary gates in households. Examples of long-lasting stone *toranas* can be seen in Sanchi and Barhut in Central India; the Rajarani Temple and the Jagannath Temple in Orissa, or the *gopurams* of the Meenakshi Temple in Madurai, and other elaborate temples in Tamil Nadu. In the ancient Kamakhya Temple in Assam, there is a strong continuity with Classical Gupta art, which is seen in the singular style of the carved creeper motif. A number of floral motifs are amalgamated with the geometrical designs at Kamakhya, where religious and secular motifs, flora and fauna become an integral part of surface ornamentation.

While flora and fauna had religious allegories associated with them in Hindu art, in Islamic art floral designs became even more integral to the artistic representations, since depicting the human form was completely banned in Islamic ideology.

Calligraphy played another major role in the decoration of mosques, gateways and palaces. This was the way the Islamic rulers immortalised the verses of the Quran in praise of the divine Maker. The interiors as well as exteriors of the Delhi Jama Masjid have been decorated extensively by calligraphy, while earlier examples can be seen in the Purana Quila. One of the prettiest examples of the art of calligraphy meeting floral motifs is Itimad-ud-Daulah's Tomb in Agra, built between 1622 and 1628.

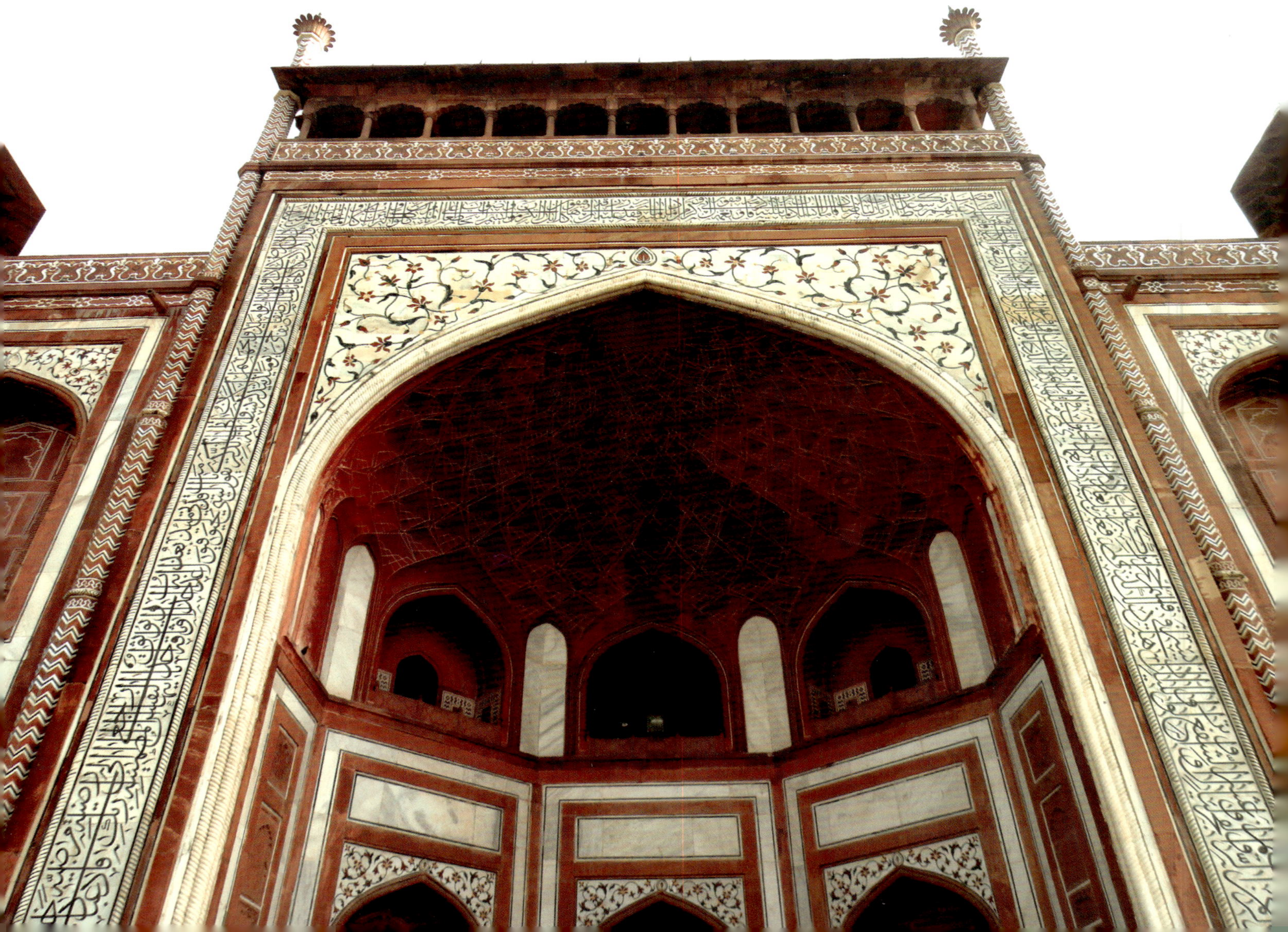

Floral inlay work on the Great Gate (*Darwaza-i rauza*)—entrance to the Taj Mahal

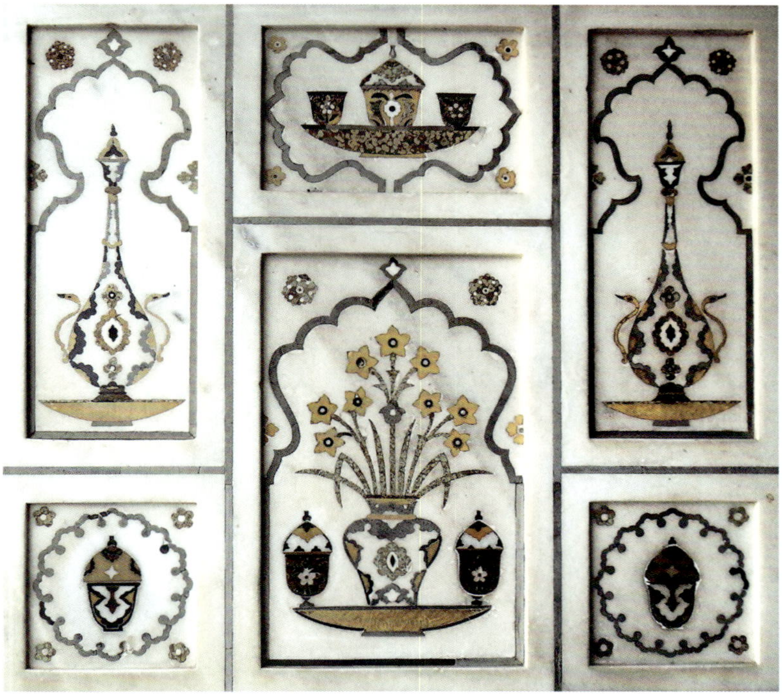

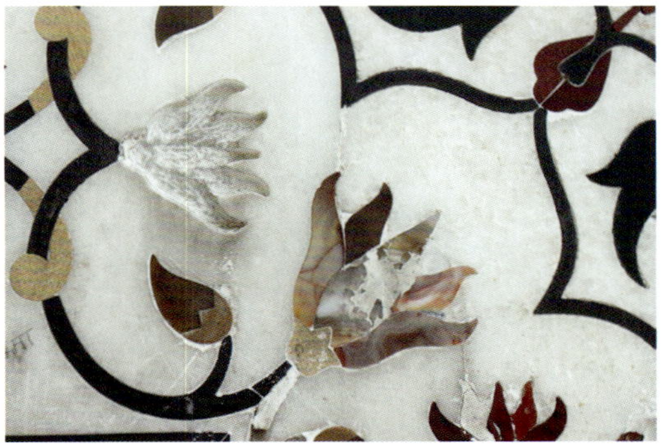

Above: Floral motifs in the tomb of Itimad-ud-Daulah, Agra

Right and Below right: Inlay with semiprecious stones called pietra dura or *parchinkari*, at the Taj Mahal

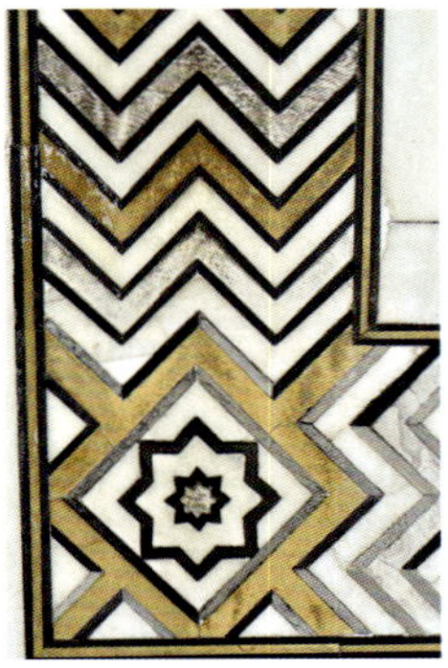

This 'jewel box' of decorative architecture marks the transition between the earlier grand monuments hewed out of red sandstone (Humayun's Tomb in Delhi and Akbar's tomb in Sikandra) to the later masterpieces built of white marble and pietra dura inlay (the hallmark being the Taj Mahal).

According to the French traveller Tavernier, who claims to have been a witness to the construction of the Taj Mahal from the beginning to the end, some of the most exquisite patterns that formed the floral designs was seen in the meticulously crafted pietra dura in the surface decoration of the Taj Mahal. The beauty of the pietra dura of the Taj, and the forts at Agra and Delhi, still serve as a source of inspiration to many artists and calligraphers alike. The extraordinary delicacy of these floral motifs which embellish the marble surface of the Taj is astounding. The flowers are sculpted in marble in *Munabbat Kari,* which is an inlay with semi-precious stones called *parchinkari,* which produces luminescent reflections. A variety of flowers, like dancing arabesques over the imperial cenotaphs, point to the brilliant virtuosity of the Mughal lapidaries. In Islamic culture, roses in particular, along with other flowers, are also metaphors for the kingdom of God.

The paradisiacal imagery of heaven is evident in the motif of flowers, which are inlaid on the funerary chambers of the mausoleum. Accompanied by fruit, the flower vases express the enormous abundance which awaits the faithful.

> They set stone flowers in the marble,
> That by their colour, if not their perfume,
> Surpass real flowers.

So wrote the court poet Abu Talib Kalim, in a tribute to the grandeur of the imperial lapidaries.

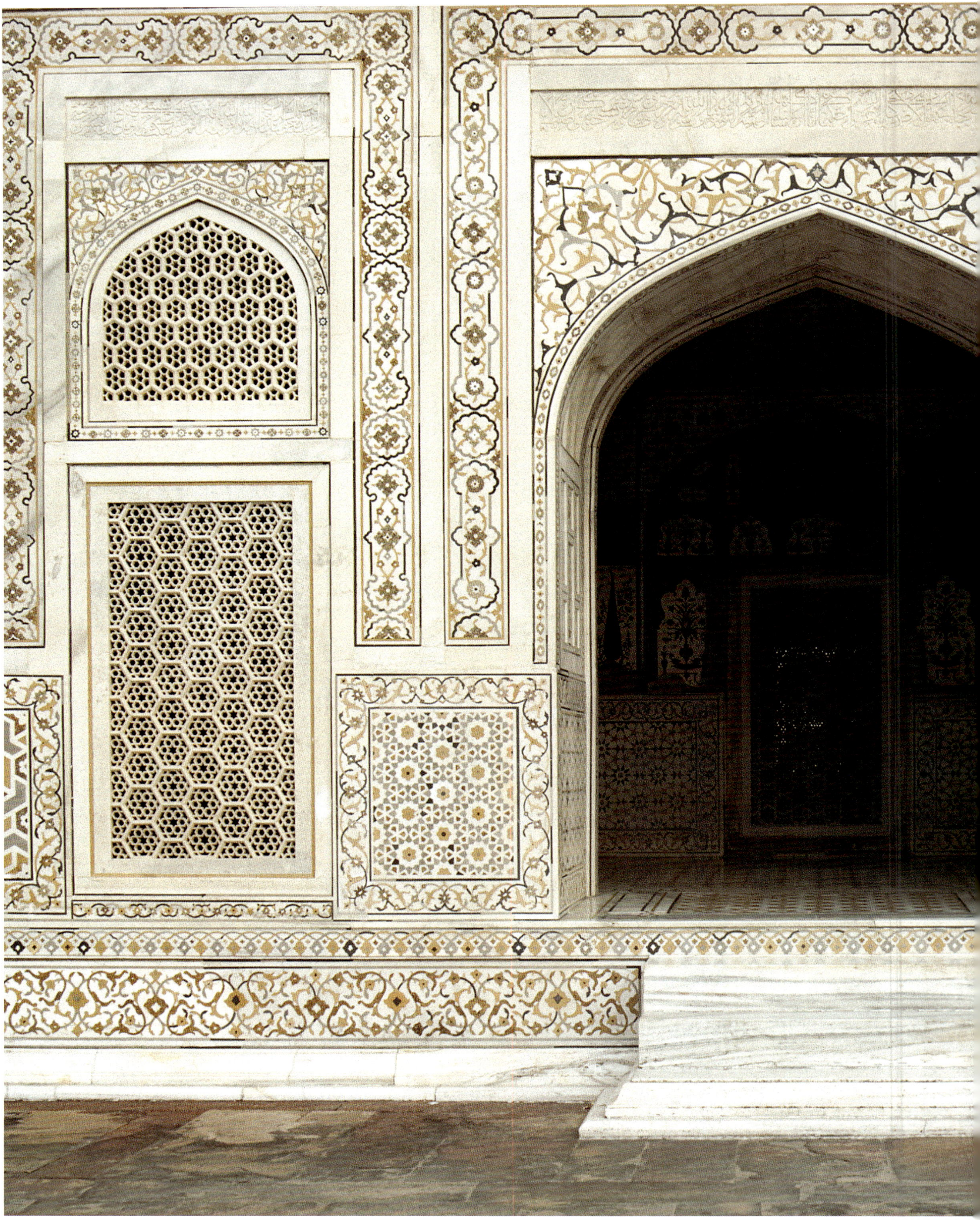

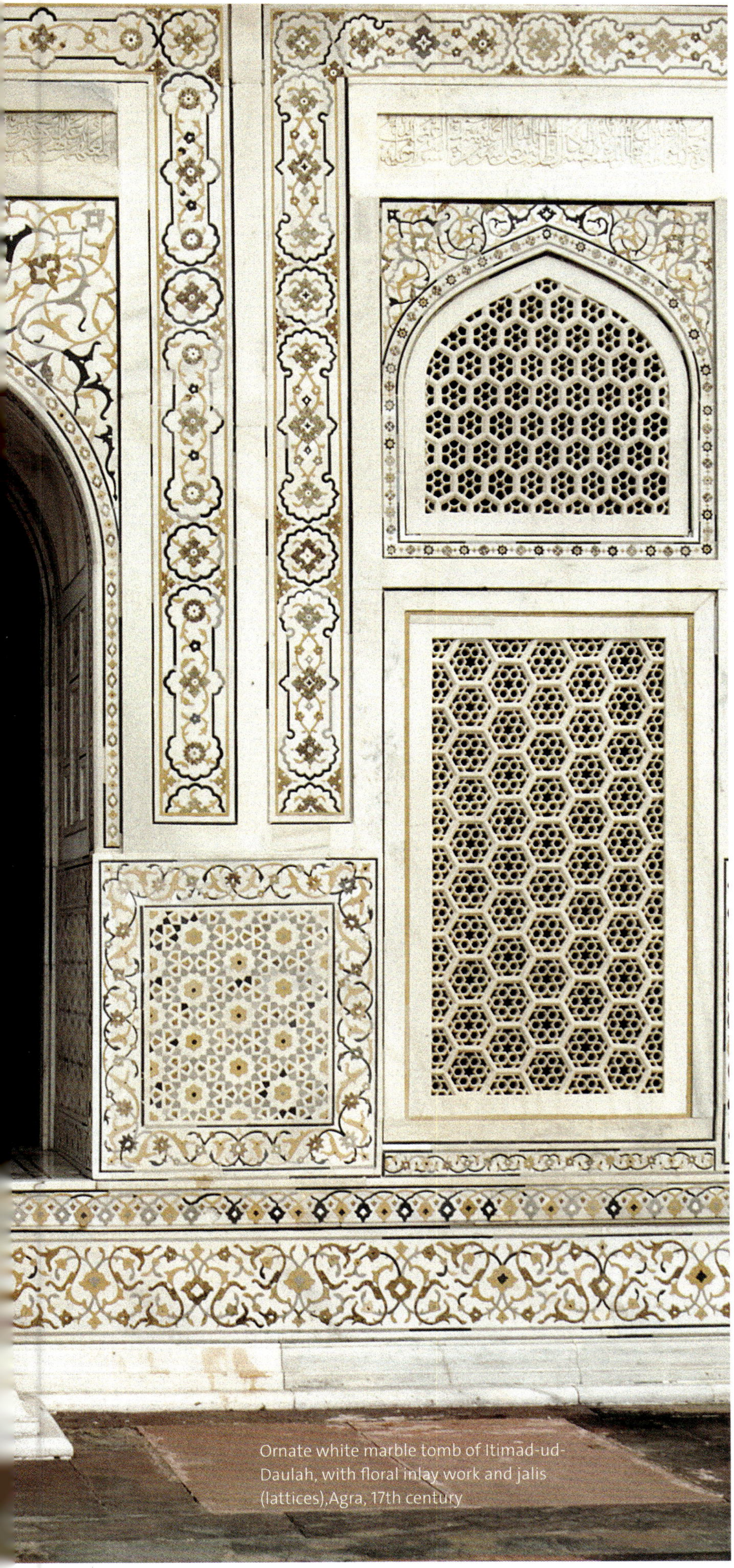

Ornate white marble tomb of Itimad-ud-Daulah, with floral inlay work and jalis (lattices), Agra, 17th century

Calligraphy, geometry, and floral/vegetal designs are often seen as unified visual components in Islamic art. There is often a strong element of calligraphy against geometric patterns, floral and vegetal designs. The arabesque motif was a beloved embellishment, recalling intricately interlocked tendrils of foliage. The place and time of the embellishment offered up its own unique characteristics; for example floral designs in medieval India, Turkey and Iran would differ greatly from one another as each would be inflected by the nuances of the time and place that gave rise to it. Naturally artists from certain places had a greater affinity for a certain design over others leading to a proliferation of a certain pattern over others.

The slow evolution in the principles that had been introduced by the Mughals was demonstrated in the architectural approaches to the tombs, mosques and gateways. However, there was a rich give and take between the indigenous design ethos of India and the Islamic settlers who made the land their home, the product of which gave rise to a new aesthetic that wove itself into the foundations of the monuments that went on to be built.

MINIATURE TREASURES

The lotus, the *champa* (plumeria) and the *genda* (marigold), to name a few, have been a recurring leitmotif in the art of India over the centuries. This seeps into the miniature paintings that display the richest variety of flower paintings. The earliest examples of miniature painting in India are found in the religious texts on Buddhism executed under the Pala Dynasty of eastern India, and the Jain texts created in western India during the 11th to 12th centuries AD. Numerous illustrated manuscripts on palm leaf contain images of Bodhisattvas or Jain saints, with decorative borders and panels using the lotus and other flower patterns. The Muslim invasion in the first half of the 13th century led to mass-scale destruction of Buddhist monasteries, and with it the decline of this school of art. However, the Western India School of Art continued to flourish in Gujarat, Rajasthan and Malwa during the 12th to 16th centuries, due to patronage of wealthy Jain merchants. This school was also the repository of many miniature paintings found in illustrated texts. Among the most famous texts are *Kalakacharya-Katha* (circa 1400) which is now housed in the Prince of Wales Museum, Mumbai, and the *Kalpasutra* (circa 1439) from Mandu, housed in the National Museum, New Delhi.

The pietra dura of the Mughal school especially feature flower motifs, as depictions of the body were prohibited. A sensitive example of this is Mansur's painting of scarlet Kashmiri tulips roughly dating back to the 17th century. In the 'Portrait of a Ruler of a Rajput Royal Family' by the Kishangarh artist, Amar Chand, the subject is seated on a terrace, outlined against the blue sky, holding a flower, possibly a rose, in front of him. Another instance is the Jammu work 'Portrait of Maharaja Gulab Singh,' the subject of which is not only depicted reverently inhaling the fragrance of the flower in his hand but is also seated on a *jama* that bears a rose pattern. His turban is also adorned with a flower-shaped ornament.

In the south the Deccani School flourished in Bijapur, under the patronage of Ali Adil Shah I (1558–80) and his successor Ibrahim II (1580–1627). The *Najum-al-ulum* (Stars of Sciences) was an encyclopaedia illustrated in 1570 during the reign of Ali Adil Shah I. Containing nearly 900 miniatures, it has been preserved in the Chester Beatty Library, Dublin. The statuesque and lean women featuring in the works are clad in South Indian dress. The 'Throne of Prosperity' is one

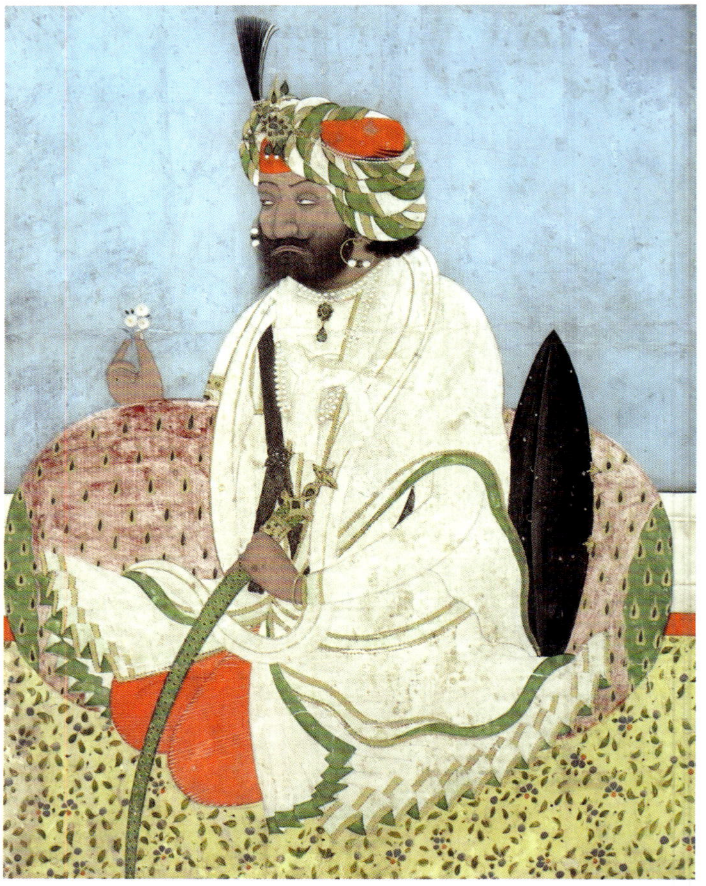

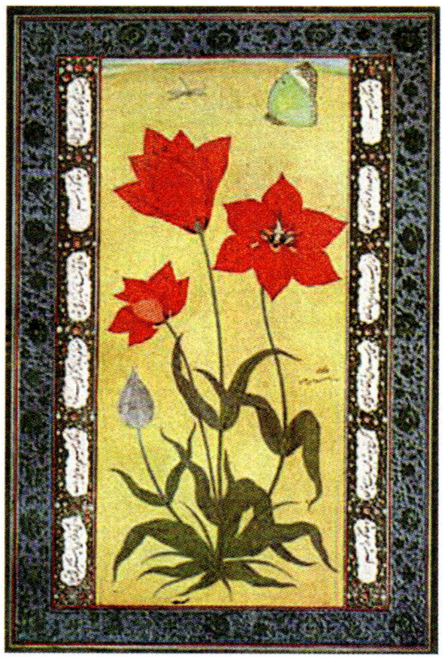

Above: Portrait of Maharaja Gulab Singh of Jammu and Kashmir, holding a flower, c. 1846

Left: Painting of a tulip by Ustad Mansur, Mughal court painter, c. 1600s

Facing page: A page from the *Najum-al-ulum* (Stars of Sciences) depicting flora and fauna, Deccani School, 1570

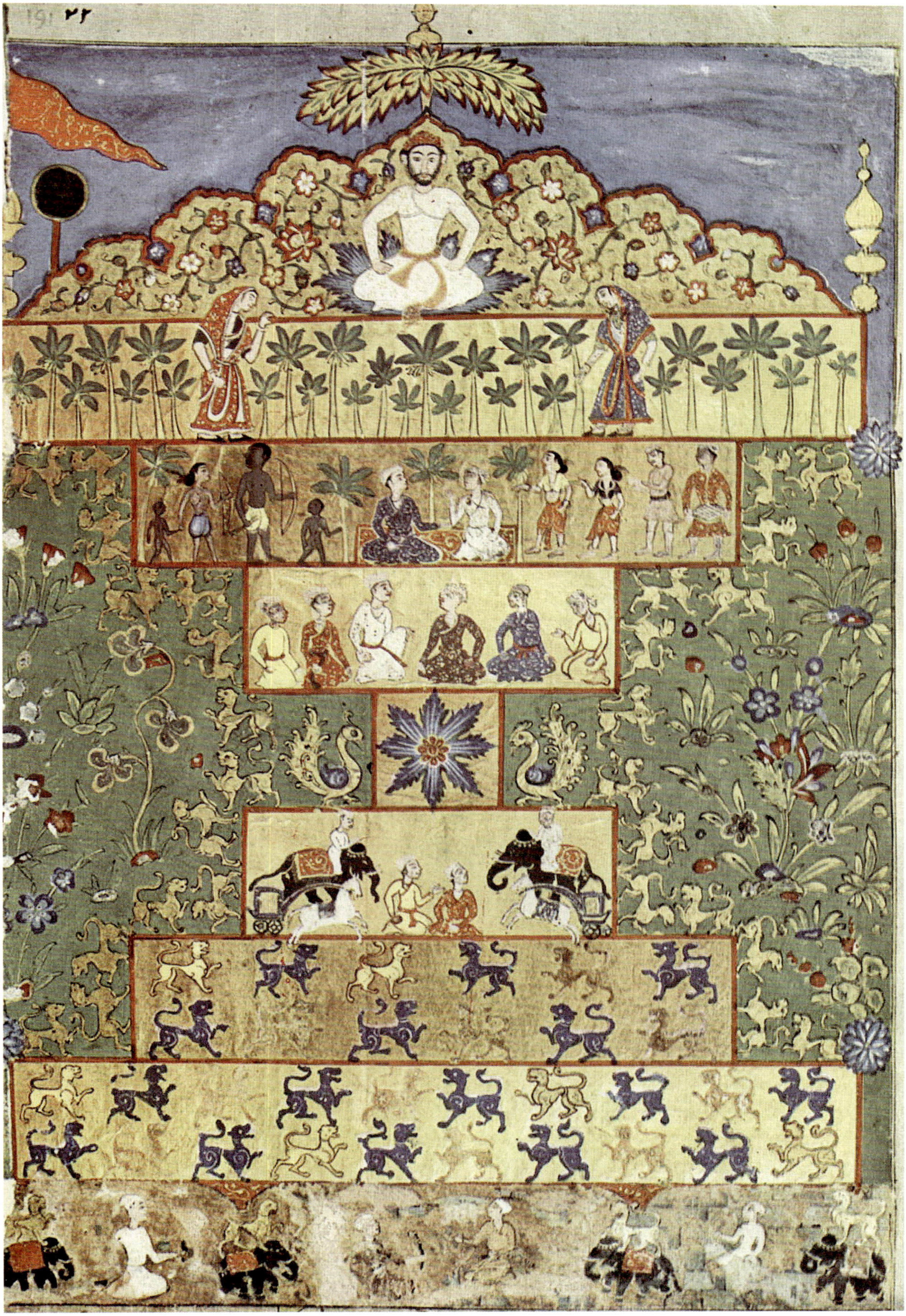

of the subjects that has been depicted. Persian influences are apparent in the use of gold, the arabesque patterning and the flora used atop the throne. Many salient characteristics of the Deccani tradition are also manifest as can be demonstrated by the distinctive colour palette, the palm trees, the wildlife and men and women portrayed.

Miniature painting reached its peak with the Mughal School of Art, which was formally established in Akbar's court. Under the patronage of various successors of Akbar, this school was an elegant and refined synthesis of indigenous Indian styles of painting and the Safavid school of Persian painting. It is defined by its love of naturalism derived from close observation of nature. Famous manuscripts of Akbar's era containing beautiful miniatures are the *Akbarnama* (circa 1600) now in the Victoria and Albert Museum, London, and the *Hamzanama*, now part of a private collection in Switzerland.

A famous portrait of Jahangir (1615–20) is part of the miniature collection at the National Museum, New Delhi. Unusually, it shows Jahangir with a picture of the Virgin Mary clasped in his hand. Along the golden borders are profusions of flowers. Another miniature (circa 1650) at the National Museum is from Shah Jahan's court and shows a gathering of *sufi* saints in an open setting. The background shows green foliage against a golden sky, while the borders have floral designs in gold.

Miniature art from the 17th to the 19th centuries, stands apart from the secular Mughal tradition, and is rooted in Hindu mythology and traditions. These are the Bundi, Mewar, Malwa, Pahari and other schools of art. The themes are from the *Ramayana*, *Mahabharata*, *Gita Govinda* of Jayadeva, *Rasamanjari* of Bhanudatta, *Rasikapriya* of Kesavadasa, and more.

A *Rasikapriya* illustration of the late 17th century Bundi School shows a scene of Krishna being tricked by a *gopi* with an empty butter vessel. In the background are trees and in the foreground is a river with flowers and a pair of aquatic birds. The painting has a border in brilliant red. The definitive characteristics of the Bundi School evident in this miniature, are the rich and glowing colours, the rising sun in golden colour, and the crimson-red horizon. The Mughal influence is visible in the refined drawing of the faces and an element of naturalism in the treatment of the flora.

Facing page: Jahangir holding the picture of the Madonna, 1620, coll. National Museum, New Delhi; note the profusion of flowers along the borders

THE FOLK IMAGINATION

The canvas of Indian folk painting is spread far and wide across the country and is rich with the use of symbols. Some of the major folk forms are Madhubani from Bihar, Warli from Maharashtra and Gujarat, Kalamkari from Andhra Pradesh, Patachitra from Orissa, Phad from Rajasthan, and so on. Folk art can be found on walls of homes, floors and courtyards, palm leaf, fabric, earthenware, and recently paper and canvas. In spite of the different styles, the subjects of folk art find a common inspiration, namely popular gods and goddesses, and events from daily life, such as harvest, weddings and festivals. Intertwining and embellishing these themes are symbols from nature, such as the sun and moon, plants and flowers, animals and birds.

Probably most famous now are the Madhubani paintings created by the women of the Mithila region. They are typically characterised by exaggerated figures and patterns inspired from nature, etched out in bold colours and outlined in black. Another famous form is the Warli paintings dating back to the Stone Age, but thriving even

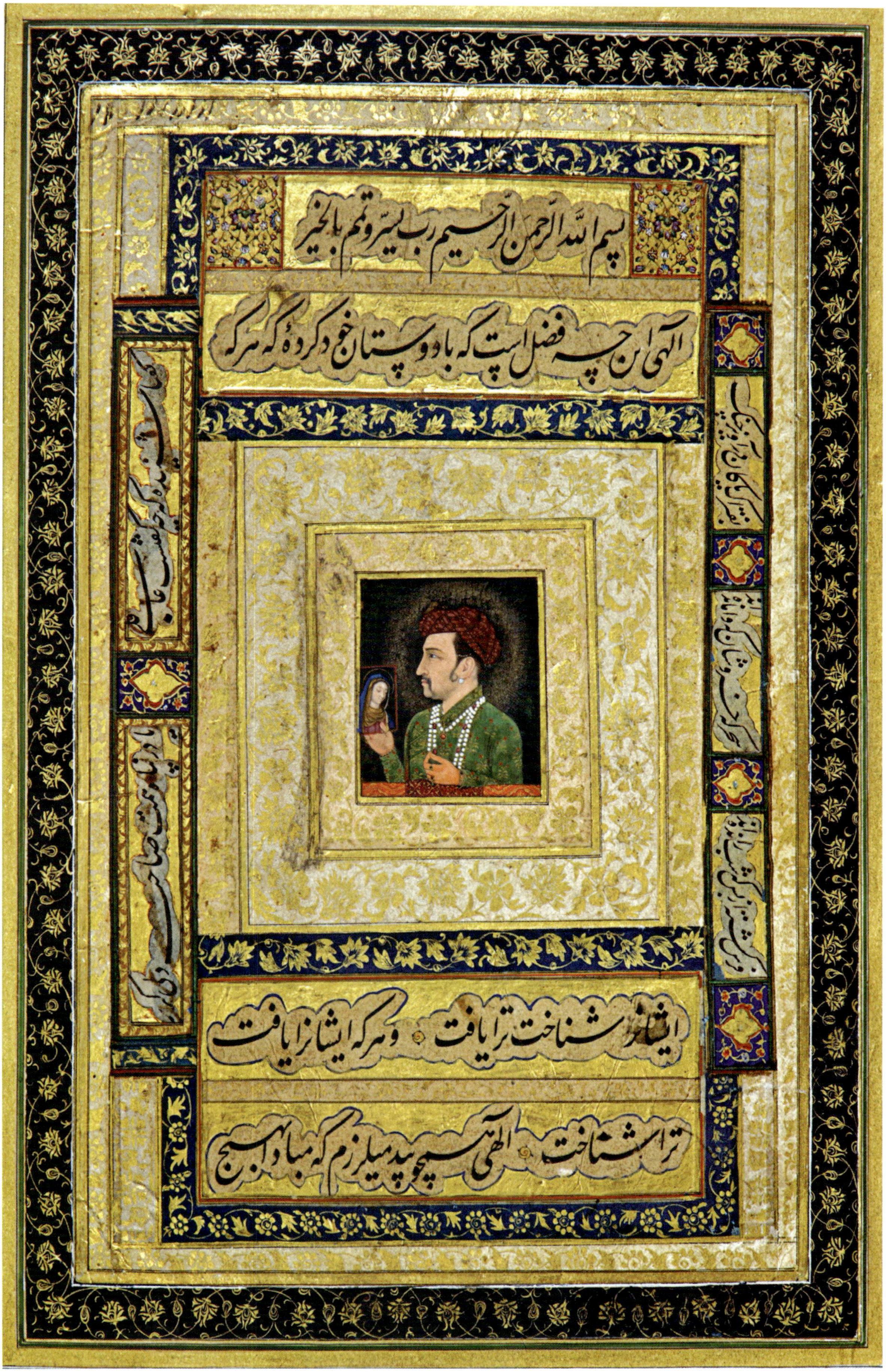
بسم الله الرحمن الرحیم رب یسر و تمم بالخیر
الهی این چه فضل است که با دوستان خود کرده که هر که
ایشانرا شناخت ترا یافت • و هر که ایشانرا یافت

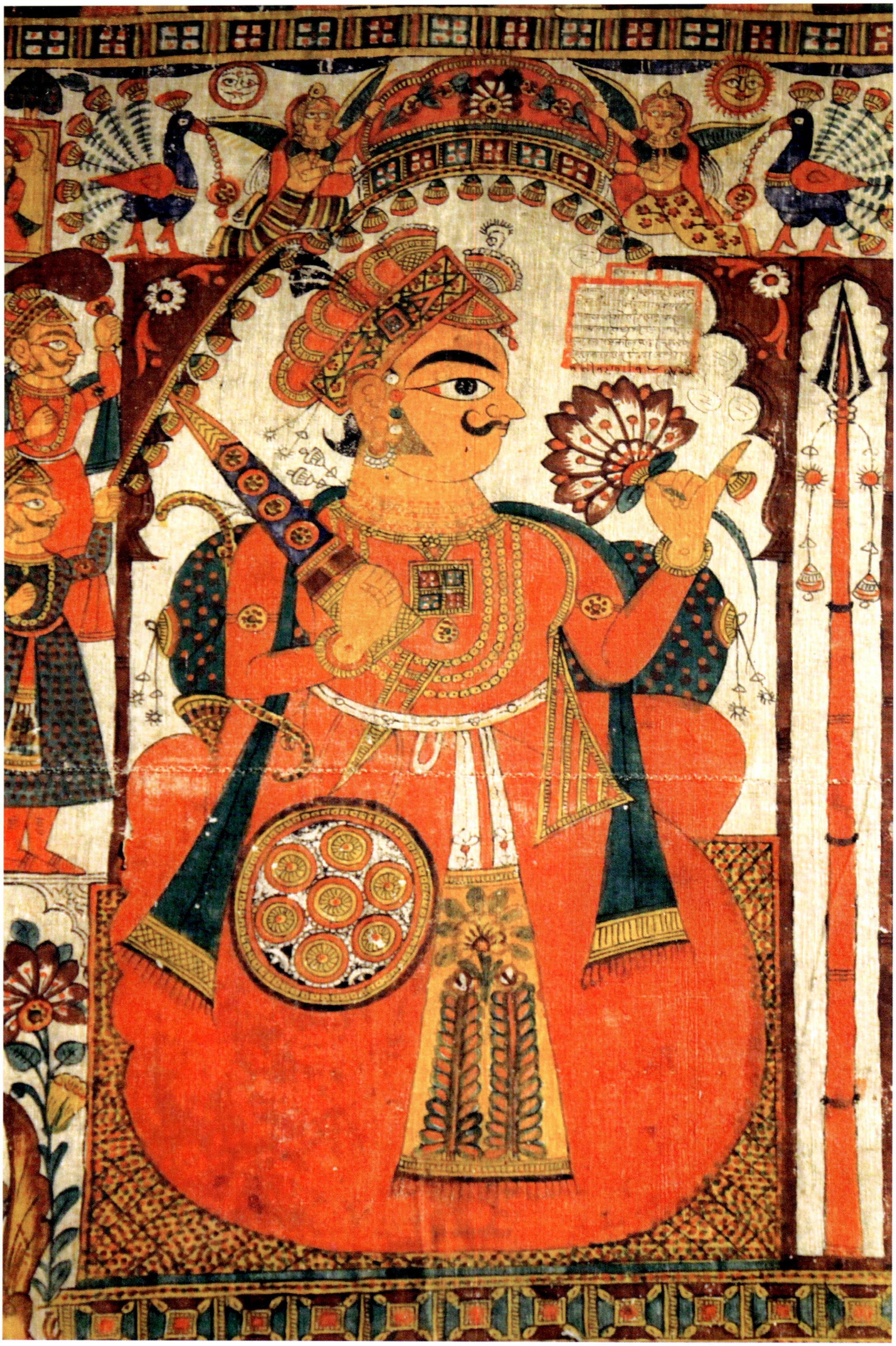

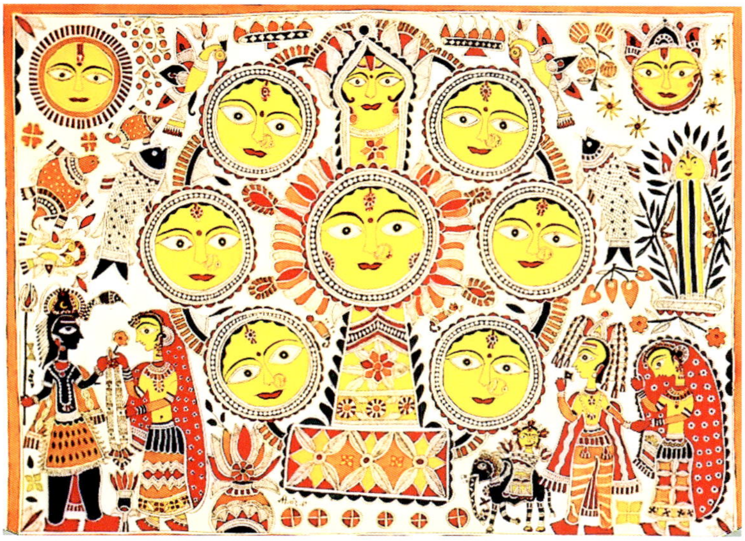

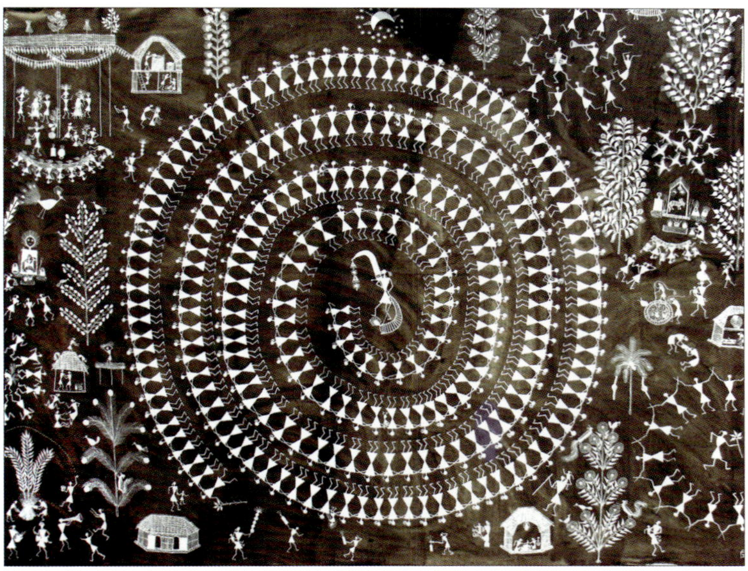

Top: 'Kohbar of Mithila' includes scenes inspired by nature, by Padma Shri Sita Devi, 2016

Above: 'Dancers and musicians', Warli painting depicting scenes from rural life by Jivya Soma Mashe, Thane district, Maharashtra, 2010

Facing page: A scene from '*Pabuji ka phad*', coll. National Museum, New Delhi; note the details of birds and plants in the background

today in regions of Thane, Nasik, Dhule, Valsad, Daman and Diu. The fertility of nature and the Mother Goddess is the main theme of Warli art. The three basic shapes of triangle, circle and square are used to etch out designs in earthy colours such as ochre, black and brick red. Kalamkari paintings from Masulipatnam display the intricate use of flowers, foliage and creepers to embellish the main subject of gods and goddesses. The influence of the Persians, and later Dutch colonialists, led to intricate detailing and realism in this style. The Phad paintings of Rajasthan are created on cloth scrolls and depict everyday rural life, flora and fauna, animals and birds in vibrant vegetable dyes.

There has been an unbroken tradition in the representation of flowers in the Indian creative ethos. Rarely is a flower absent in Indian art forms, be it built heritage, sculpture, painting, folk art or textiles. In fact it is part of the inner and outer spaces of every Indian home. From the spiritual to the decorative, from the sculptural to the sensual, flowers are an integral part of the Indian imagination.

ENDNOTE

1. Neeti M. Sadarangani, *Bhakti Poetry in Medieval India: Its Inception, Cultural Encounter and Impact.* (Sarup and Sons. 2004), p. 101.

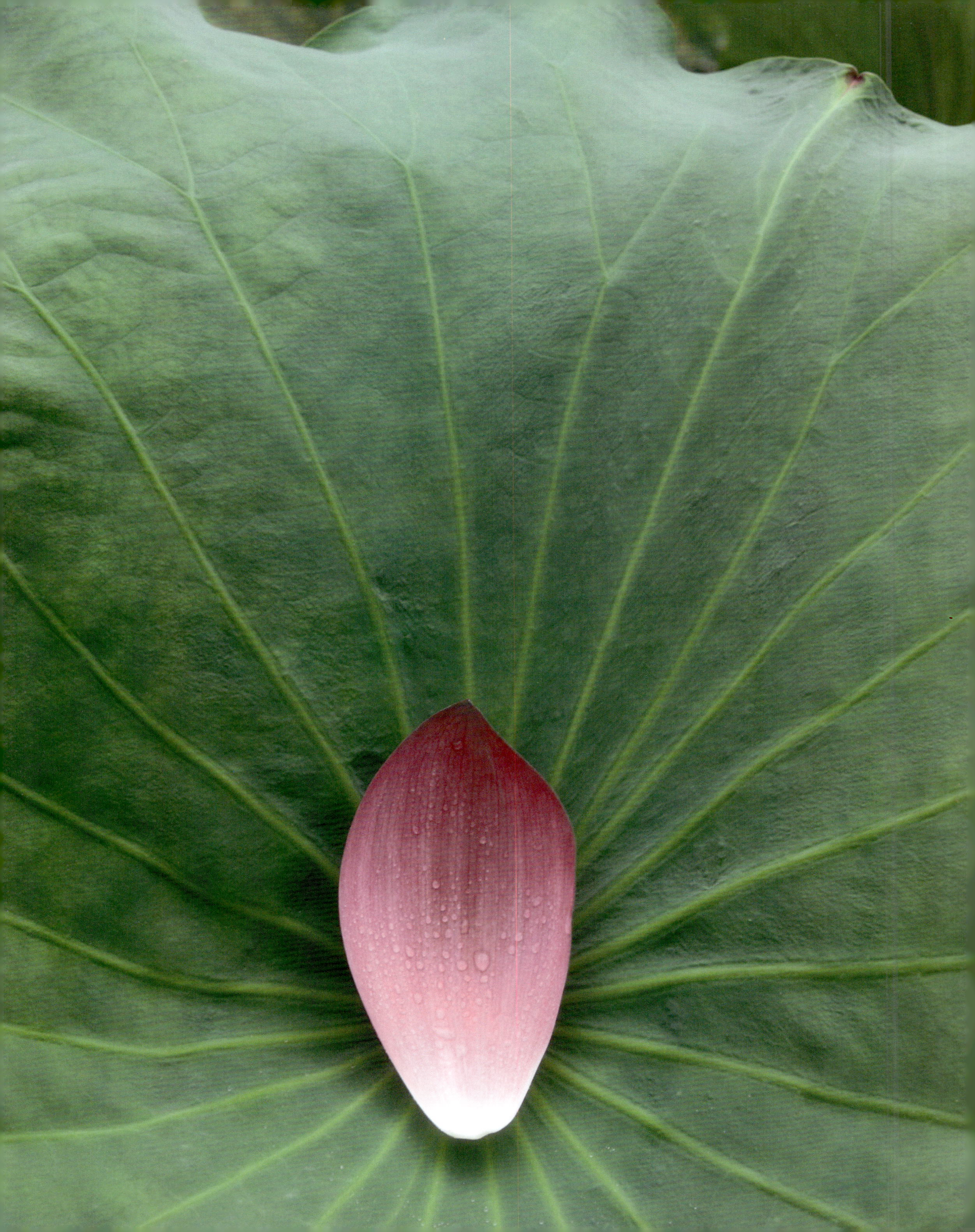

Floral Inscription

Paint, O son of Yadu, with fingers cooler than sandal-wood,
place a circlet of musk on this breast,
which resembles a vase of consecrated water, crowned with fresh leaves,
and fixed near a vernal bower to propitiate the God of Love.
—Jayadeva, *Gita Govinda*[1]

What could be more intoxicating than this love play of Radha and Krishna? The *Gita Govinda*, by the 12th century Odiya poet Jayadeva, with its voluptuous imagery and passion, has been called the 'Indian Song of Songs' by Sir Edwin Arnold, the 19th century scholar and Indologist. Flowers are very much a part of the allegories and metaphors which translate the wisdom of traditional India into the cultural consciousness of her people. It is amazing to witness the lavish use of floral imagery seen in the writing of Indian poets and authors, be it in classical Sanskrit poetry, particularly in medieval and *Ritikaleen* poetry and modern and vernacular writings. This use of the metaphor of flowers with beauty and love is probably unparalleled in any other genre of world literature. Throughout the centuries, flowers have remained central to the Indian literary tradition. The pre-modern, early medieval and late medieval literary texts are replete with flower motifs and metaphors. The attempt to immortalise the flowers within these texts gives subtle shape to the literary techniques. Tracing the lineage of the flower from Buddhist and Hindu mythology to its multiple usages in the form of motifs, metaphors and even devotional decoration, brings us closer to its heterogeneous interpretations.

Facing page: A lotus petal on its leaf—this implies Zen

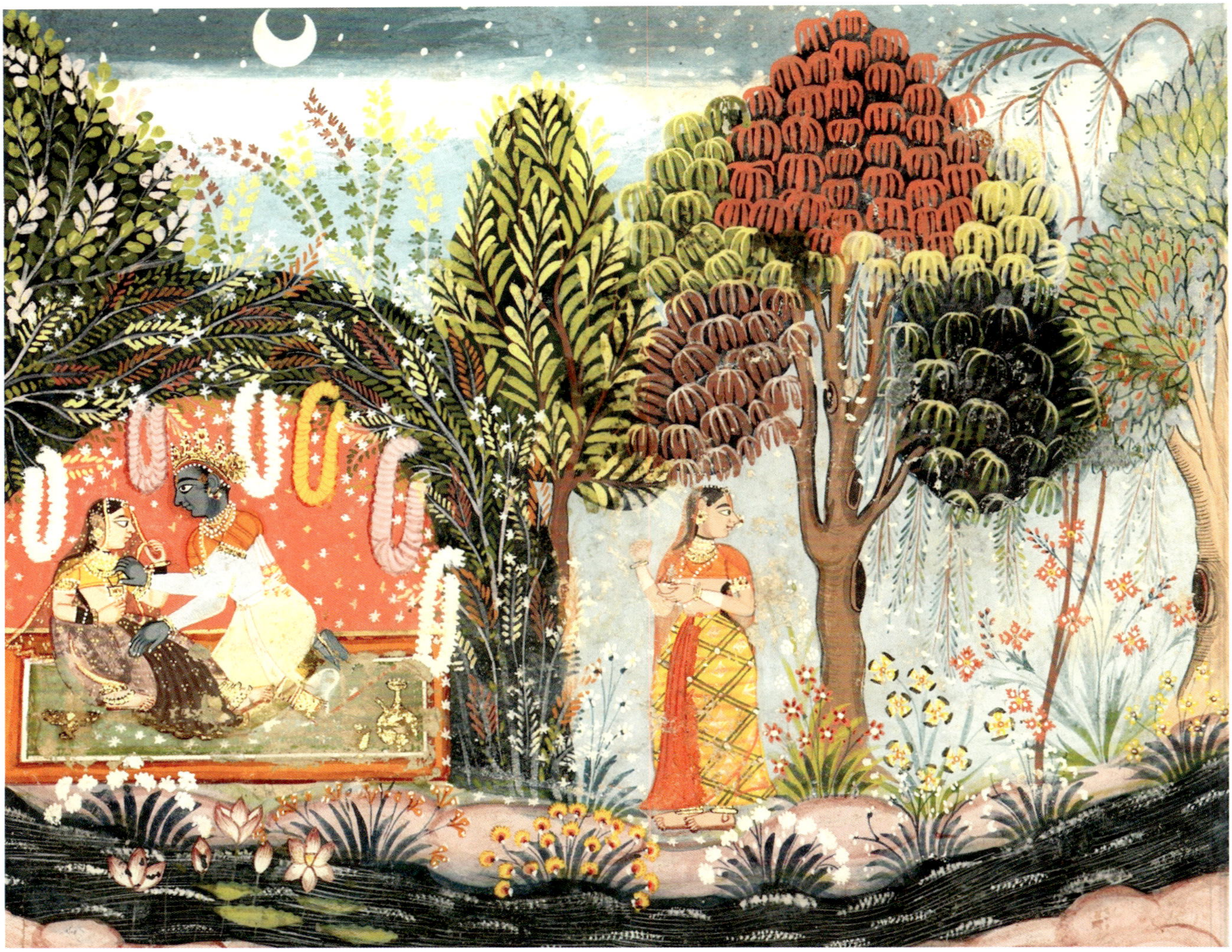

The *Upanishads* and *Aranyakas*, the more religious Puranic literature, the secular, pre-modern Sanskrit Kavya poetry of Kalidas and other greats, the Bhakti-inspired Ritikaleen poetry, all have verses suffused with the bounty of nature in which the allegorical use of flowers can be found. Flowering trees, shrubs, and blossoms themselves have been used as metaphors to describe the idealised physical beauty of men and women—the *nayak* (hero) and *nayika* (heroine)—of their fragrances adding to the mood of love, of the sensuality of perfumes.

Looking at Indian literature over time, from the ancient to the contemporary—nature, ecology and flowers are a recurring leitmotif in the writings of poets and prose writers. The medieval is replete with the sensuous, where flowers become metaphors for devotion, as an offering for the divine and a symbol of the divine. The flowers take on a more symbolic meaning, where allegory is rampant.

Indian modernity moves the flower into a more individualistic space where artists and writers treat flowers as part of personal histories, while flowers become cultural signifiers and markers of identity in the contemporary. What is really interesting is the transformation of the flower in the cultural space, where the flower changes with the transition of time.

Above: 'Krishna and Radha in a bower', *Gita Govinda*, ink and opaque watercolour on paper, Mewar, Rajasthan, c. 1665, coll. The Metropolitan Museum of Art, New York

Facing page: 'Krishna revels with the gopis', *Gita Govinda*, c. 1630–40, Malwa, Madhya Pradesh, coll. The Metropolitan Museum of Art, New York

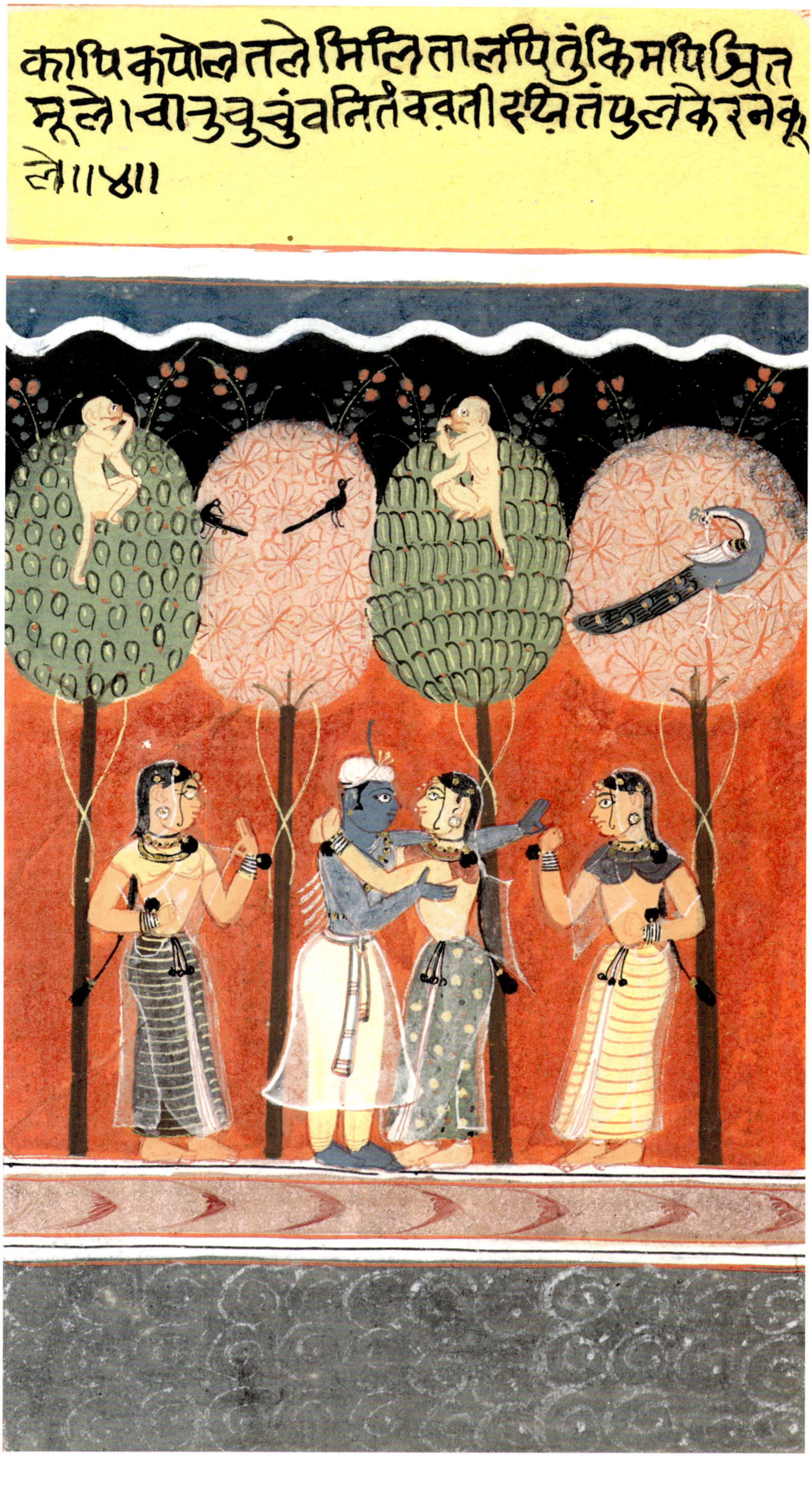
कापि कपोलतले मिलिता लपितुं किमपि श्रुत
मूले। चारु चुचुंब नितंबवती दयितं पुलकैरनुकू
ले ॥४॥

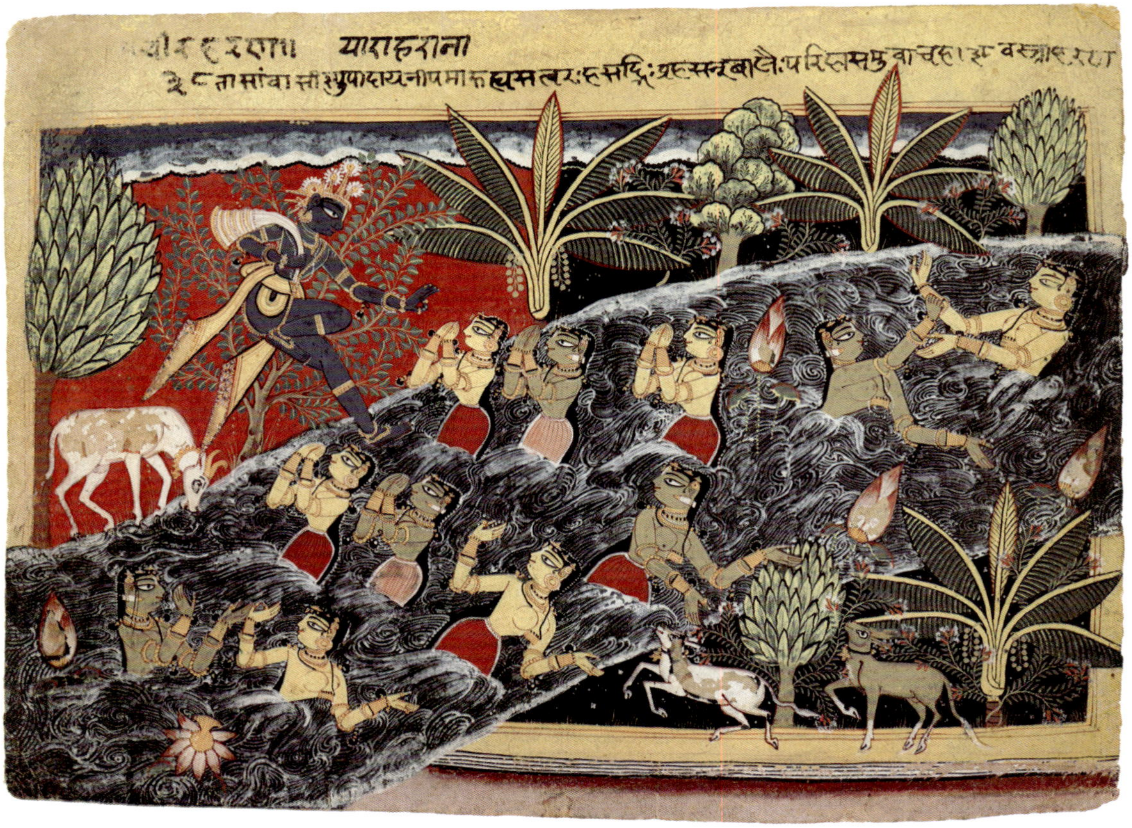

Above: 'The gopis plead with Krishna to return their clothing', *Isarda Bhagavata Purana*, opaque watercolour and ink on paper, North India, c. 1560–65, coll. The Metropolitan Museum of Art, New York

SRIMAD BHAGAVATA
THE SEEDS OF CREATIVITY

Commonly known as *Bhagavata Purana*, the *Srimad Bhagavata Maha Purana* is one of the seminal texts of Hinduism, with its focus on the bhakti or veneration of the Supreme God Vishnu (Narayana) and his incarnation Lord Krishna. Ved Vyasa is said to have authored this text, which was originally an oral work. However, it achieves final form by undergoing multiple departures from earlier texts. Although it is difficult to assign a specific date to its composition, its evolution continued across 500 BC to the 10th century CE.

It can be said that the *Bhagavata Purana* is the earliest text which introduces the flower motif to the Indian literary tradition. According to Vedic scriptures, the body of Krishna is symbolised as a bluish new-grown lotus flower. All the cantos describing Krishna are replete with the imagery of the

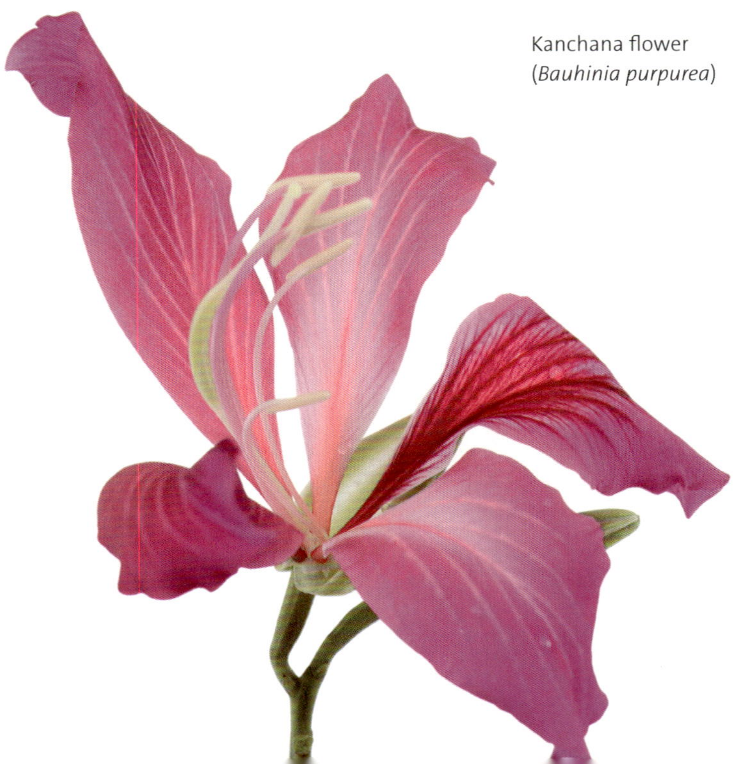

Kanchana flower (*Bauhinia purpurea*)

Right: 'Draupadi showing Bhima the sougandhika flowers', detail from the *Mahabharata*

Below: A sprig of tulsi

lotus flower, which represents purity and non attachment in the cultural metaphor of India. The *Bhagavata Purana* even has certain verses reflecting directly on the lotus. The *Mahanarayana Upanishad* says, 'In the citadel of the body, there is the small sinless and pure lotus of the heart which is the residence of the Supreme.' The lotus represents the spiritual heart, not the tangible or corporal heart. It is said that Brahma was born from the lotus growing out of the navel of Vishnu.

> Thus the demigods saw the wonderfully beautiful region known as Alaka in the forest known as Saugandhika, which means 'full of fragrance.' The forest is known as Saugandhika because of its abundance of lotus flowers.[2]

Tulsi or basil is one of Vishnu's favourite flowering herbs. In volume 10 of the Mahabharata, the tulsi is mentioned as being most important. The fruits of basil, the petals of the kanchana flower and goat meat are also said to be eternal.[3]

Another episode from the *Mahabharata* narrates the story of how, on the request of his wife Draupadi, Bhima braved a lake guarded by Yakshas to return to Draupadi with some maddeningly fragrant flowers she had been yearning for.

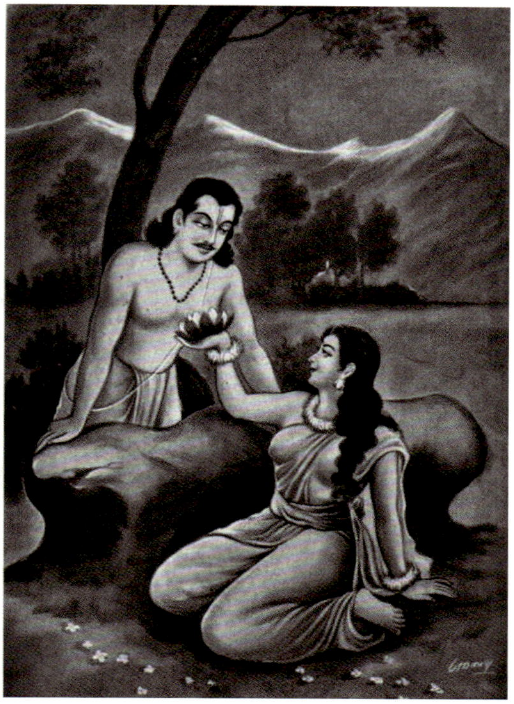

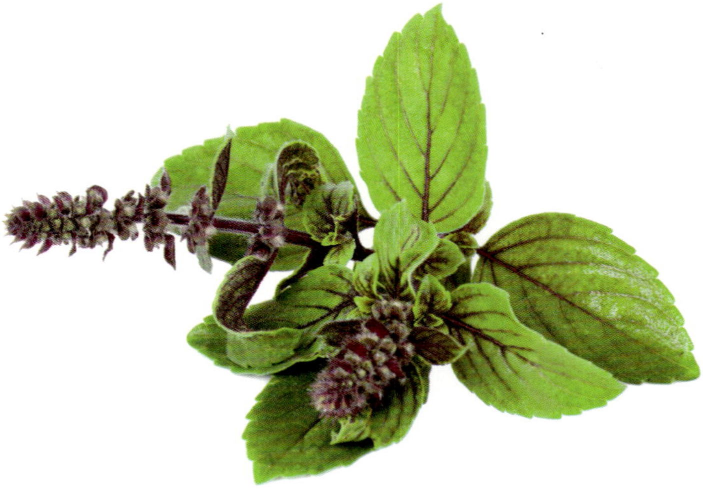

> In no time Bhima reached the lake. It was full of innumerable thousand-petalled lotuses. The air was saturated with the wonderful fragrance of the flowers. The lake was guarded by the Yakshas, the servants of Kubera. The guards asked Bhima, 'Who are you? What do you want?'
>
> 'I am Bhima, the Pandava prince. My wife Draupadi wants these thousand petalled flowers. I have come here searching these, and would pluck many.'
>
> 'But you cannot do so without the permission of our lord Kubera, the god of wealth.'
>
> 'But why should I beg anybody's favour? I am a Kshatriya and I intend to pluck these sugandhika flowers. None can stop me.'[4]

Throughout we see that flower imagery is abundant in the descriptions of sacred spaces. The tenth book of the *Bhagavata Purana*, dedicated to Krishna, talks about him in various stages and conditions of human life. The intense and personal *bhakti* described in the text is directed toward Krishna as Vishnu in human form. It reinforces the notion of *moksha* by establishing and nurturing a

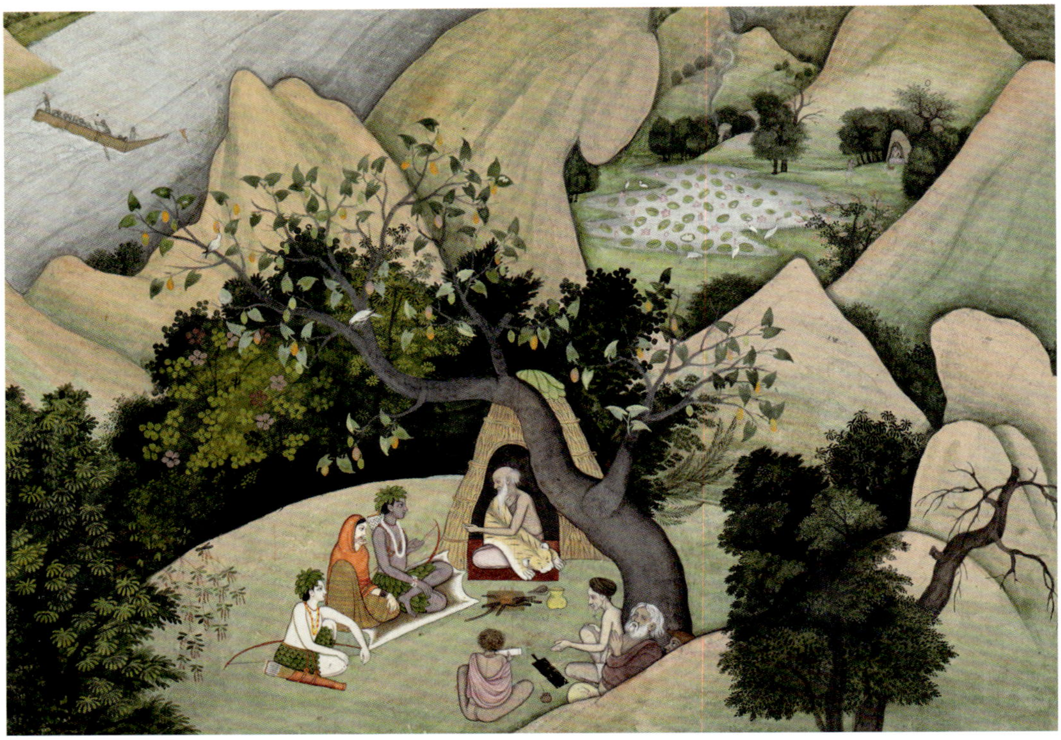

Left: 'Rama, Sita, and Lakshmana at the Hermitage of Bharadvaja', *Ramayana* series, opaque watercolor and ink on paper, Kangra, Himachal Pradesh, c. 1780

Below: 'Rama and Sita in the forest', Kangra School, 1780

relationship with Krishna, as an avatar of Vishnu. It asks for meditating on the *leela* of Krishna. The roots of the Indian tradition of dance and theatre can be first traced to the *ras-leela* where Krishna woos his *gopis* (milkmaids) with music and song.

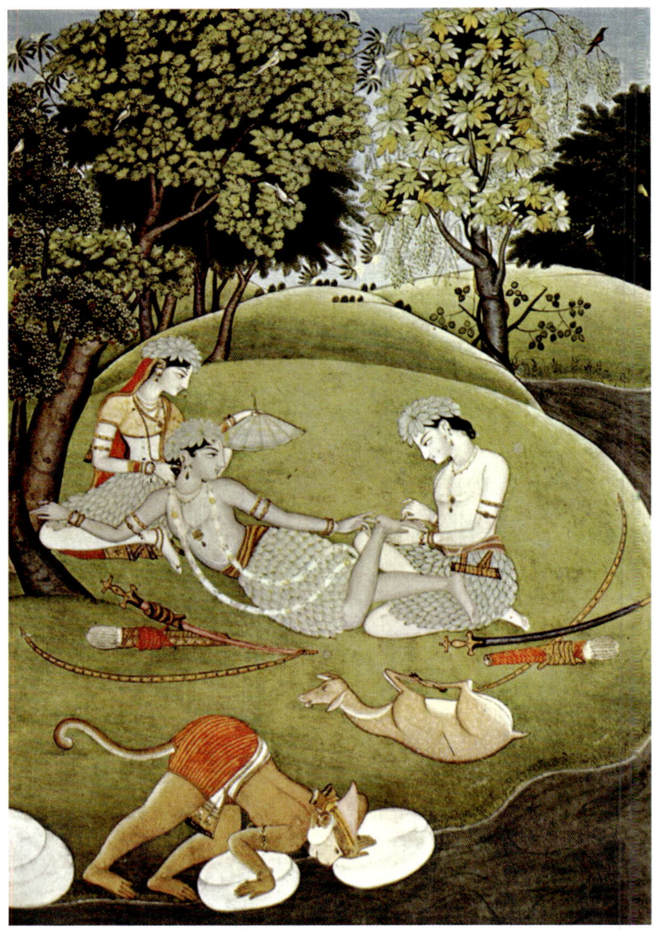

EPIC FLORAL METAPHORS

The *Ramayana* comprises 24,000 verses. This fills the pages of seven books (*kandas*) and is spread across 500 cantos (sargas). One of the *kandas* is the *Aranya Kanda* or Book of the Forest. The chapter titled *Kishkindha Kanda* delves into an analysis of the flora and fauna of the region in present-day Karnataka. Since time immemorial, India is considered to be the repository of Vedic knowledge. The learned Vedic sages gave ample knowledge about the flowers and their beneficial properties. It is important to note that the close association of man with nature or flora is also symbolic of his greater proximity with God. The purity of God

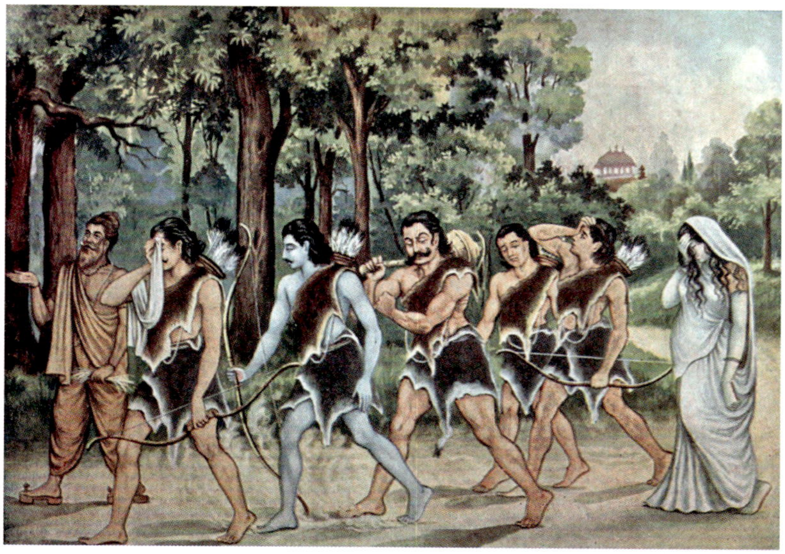

'The exile of the Pandavas', detail from the *Mahabharata*

is translated to the benevolence of nature. The *Ramayana*, while talking about filial relations, also lays emphasis on the magnanimity of Mother Nature.

Lord Rama is described with the lyrical metaphors of flowers, especially the lotus, in the epic:

> His feet were similar to the new red lotus (in full bloom), shine of his nails was removing fear out of the hearts of devotees... His eyes were large like lotus. His neck was blue and he was full of beauty. The moon of *Dwitiya* was glorifying divinity on their forehead.[5]

There are vivid descriptions of flora throughout the epic. For instance, in the shlokas 12-18, the sage guiding Rama and Laxmana to a dangerous forest on the other bank of River Ganges relates anecdotes about the provinces of Malada and Karusha, which are regularly intercepted by the description of the trees and plants of the region.

> In the East of Pumpa, the trees on Rishyamuk Mountains are full of graceful flowers
> En route Sahya and Malay Mountains, rivers and waterfalls were present.
> Varied delightful tree with the lotus flowers.[6]

When Rama succeeds in winning Sita's hand, she garlands him in the ceremony known as *jaimala*. This ritual symbolises two souls coming together as one. A marriage in India is considered to be incomplete if the bride and groom do not exchange flower garlands. Flowers represent happiness, prosperity and beauty, and the thread they are strung on, provides security to the feelings and emotions. Similarly, the *pushpa varsha*, or celestial flower shower, that took place during the wedding ceremony of Ram and Sita, has come to represent divine sanctity and happiness for a union.

The *Mahabharata* is believed to have been written by the sage Ved Vyasa. Among the chief works and tales that also comprise the individual texts, are the Bhagavad Gita, Damayanti's tale, a brief telling of *Ramayana* and the Rishyasringa. In the *Mahabharata*, the *Aranyaka Parva* recounts the 12-year exile the Pandavas spent in the forest or *aranya*. The shami tree (*Prosopis cineraria*) is where the Pandavas keep their divine weapons that they later claim on Vijayadashami day and go to achieve victory on the battlefield. Until the present day, shami leaves are exchanged to wish success and triumph.

The description of flora is also deployed to heighten the effect of tragedy. The severely hurt Karna is thus depicted in the pages of the *Mahabharata*:

> Pierced with those arrows equipped with heads like the calf's tooth, Adiratha's son of broad chest looked resplendent like an *Asoka* or *Palasa* or *Salmali* decked with its flowery load or a mountain overgrown with a forest of Sandal trees. Indeed with those numerous arrows sticking to his body, Karna, O monarch, looked resplendent like the prince of mountains with its top and glens overgrown with trees or decked with flowering *Karnikaras* or *Kimsuka* flowers.[7]

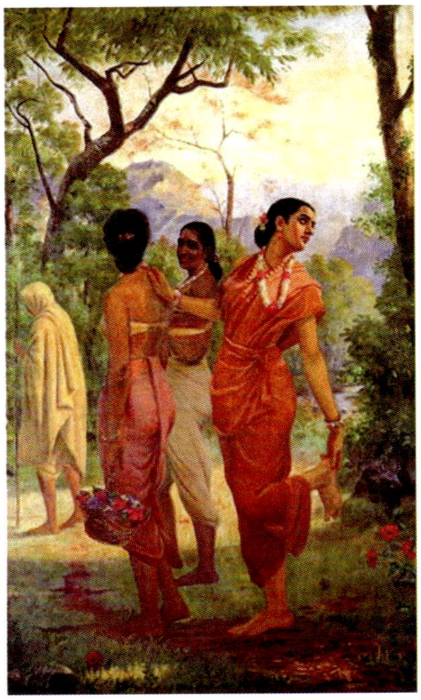

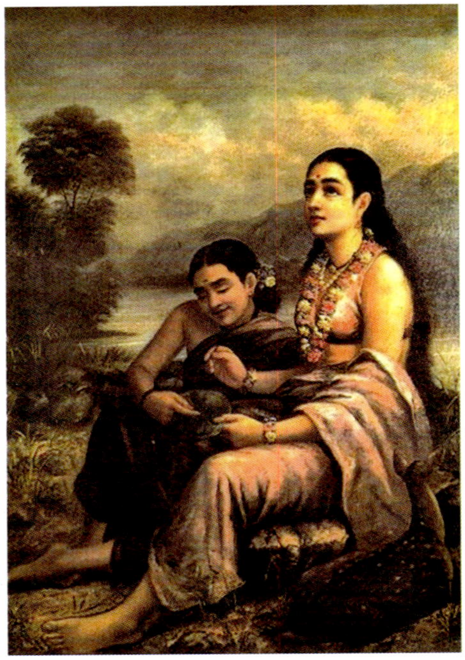

KALIDAS
THE KING OF POETRY

The epic kavya form of literature fully evolved in the time of poet and dramatist Kalidas (5th century CE) who is the author of a number of exceptional literary works known for their artistic and stylistic approach. Considered the king of all poets, Kalidas in his writings, the *Ritusamhara* and *Abhigyanashakuntalam* in particular, evocatively describes love and courtship. The imagery of flowers witnesses a shift within the works of Kalidas, from being a part of spiritual and pastoral beauty to a metaphor used in describing feminine charms and beauty.

The *Ritusamhara* marks the zenith of erotic literature in the classical period. This long poem is divided into six books, corresponding to the six seasons. The youth who has just awakened to love finds nature and its flower imagery as a powerful stimulus. In the *Ritusamhara*, apart from a vivid account of the seasons, the sentiments, which the changing seasons evoke in lovers, are described with great feeling.

The importance of *ritu* or the seasons in the history of Sanskrit poetry is unique. It brings with it the new Romantic Movement which gives the primary place to a poet's emotional experience. Sense, delight and love command the universal world of love. The flower garland becomes the symbol of eternal love and the lotus flower is harmonious with the tantric yoni. Kalidasa's *Ritusamhara* is the epitome of the same literary embellishment.

> Hair cascading down to hips,
> Fragrant flowers nestling behind the ears
> Pearl strings fondling the breasts...
> Women set the heart of their lovers on fire.[8]

In his masterpiece *Meghaduta*, Kalidas writes evocatively:

> Women twine round their coiled hair
> Wreaths woven of fresh Kadamb flowers,
> Kesara buds and ketaki fronds
> And place the Arjuna's blossoming sprays
> As pendants over the ears,
> Arranging them in many pleasing shapes.
>
> The faces of flower-gathering maidens, who
> Each time they wipe the sweat off their cheeks, bruise

Far left: 'Shakuntala looking back to glimpse Dushyanta', Raja Ravi Varma, oil on canvas, 1898

Above left: *Shakuntala patralekhan* (Shakuntala writing a letter to Dushyanta, with her friends)', Raja Ravi Varma, oil on canvas, c. 1800s

Facing page: 'Shakuntala and her friends', Raja Ravi Varma, chromolithograph, c. 1900

Below: Kadamb flower (*Neolamarckia cadamba*)

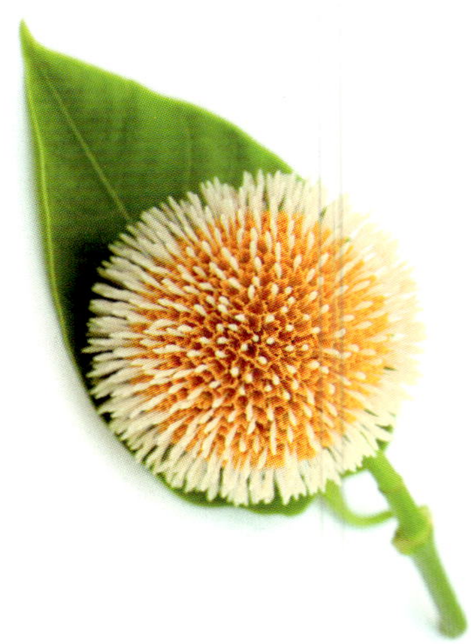

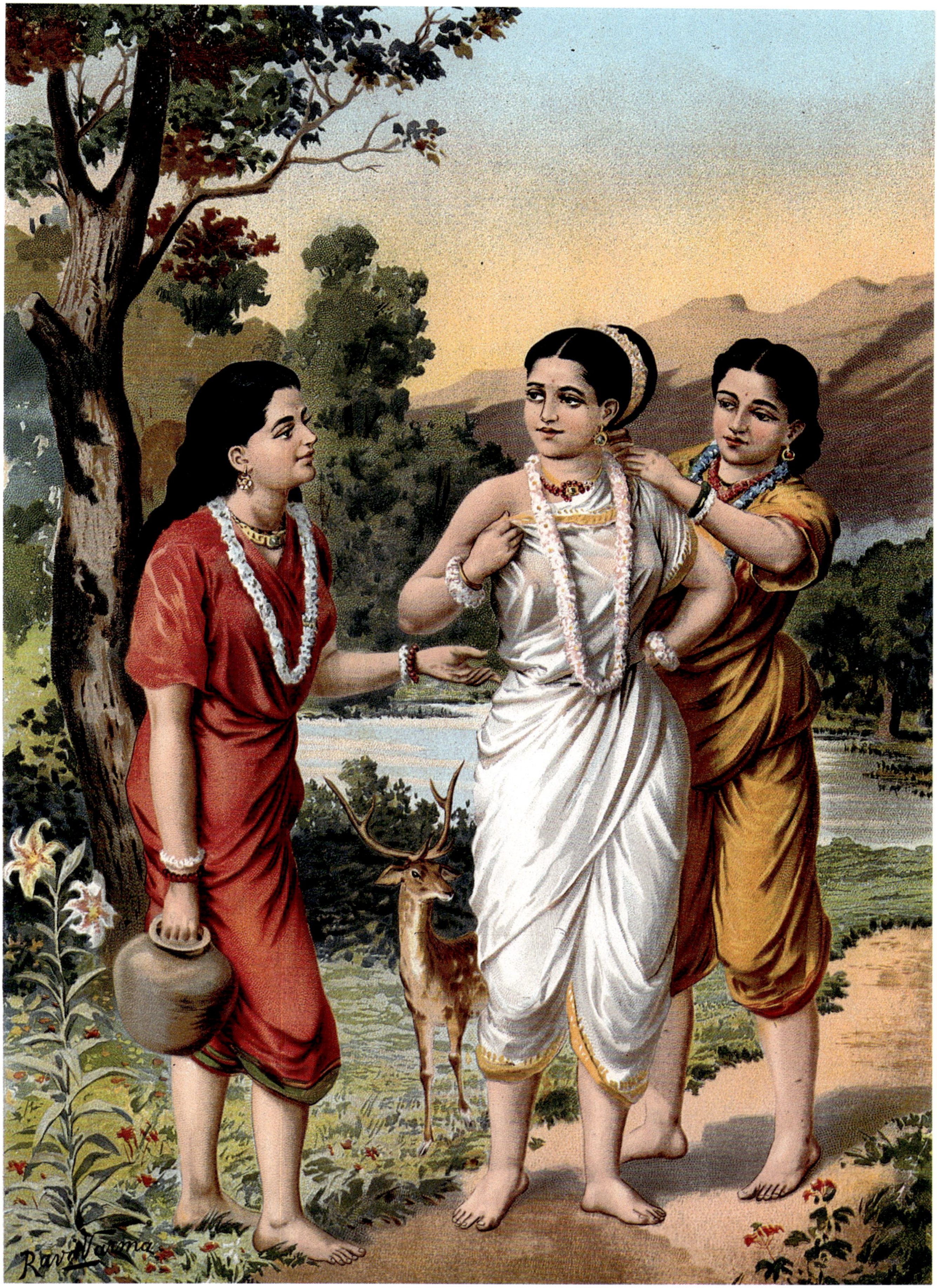
Ravi Varma

The wilting lotuses hung at their ears.[9]

Kalidasa's *Abhigyanashakuntalam* is his best-known work, in which he once again emphasises the beauty of a woman, describing her with flower imagery:

> ...Her faultless beauty is a flower not (yet) smelt, a tender sprout not plucked with nails, a jewel not (yet) perforated fresh honey whose flavour is (yet) untasted, and the fruit as it were, of accumulated (religious) merit not (yet) cut or divided. I don't know whom the providence will approach in this world, as its enjoyer.[10]

THE HEART OF DEVOTION

The Ritikaleen School of Bhakti poetry flourished from the 5th to 15th centuries and is part of the Indian *advaita* philosophy. Most poets of this genre write in the feminine voice, where the devotee is synonymous with the *atma* (soul) and is longing for union with the *parmatma* (God). The verses of the Ritikaleen style describe a cherished moment of love, in the first person by a beloved who has been torn from their loved. They then employ memory to recall their beloved. The fragrance of musk and sandalwood takes reign alongside night blossoming jasmine and lotus pollen. The sense of touch is also vividly expressed with the arms around the neck likened to vines, lips wet from wine and the lotus as a place where ardour can be expressed. Erotic descriptions of women have been used evocatively through the play of words, providing poetic expression of the highest order. One finds such examples in the following verses:

> Her yoni resembles a fresh lotus bud and her love seed [kama salila] is perfumed like a lily that has just burst into bloom.[11]

This piece references the Padmini or lotus-woman, as per the four-fold classification of women according to the medieval erotic text *Smarapradipika*.

One of the most enduring Bhakti poets, Jayadeva is nothing short of a legend among medieval Indian poets. Jayadeva is said to have lived during the 12th century. He was a learned Sanskrit scholar, giving up worldly life to become a wandering ascetic and eventually marrying a temple dancing girl. As the story goes, while she danced, he composed, thus it was a harmonious marriage of their temperaments.

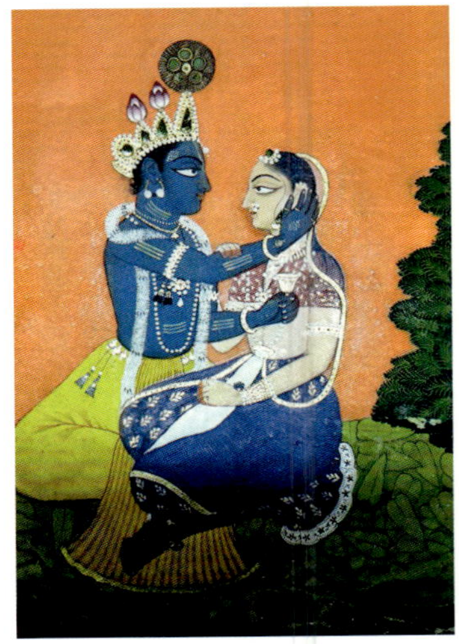

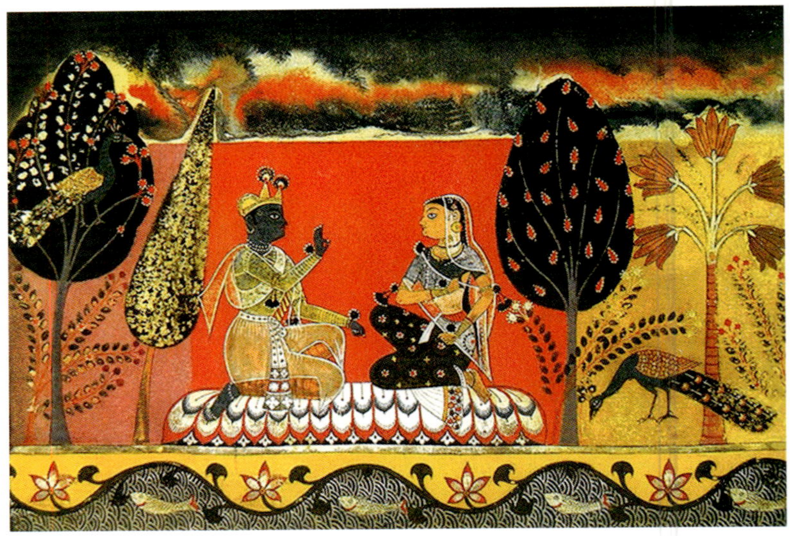

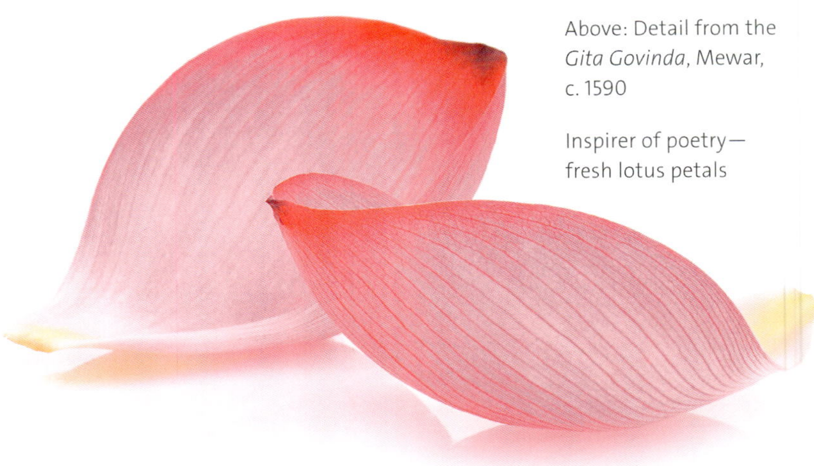

Top: 'Radha and Krishna', Manaku of Guler, from the *Gita Govinda* series, 18th century

Above: Detail from the *Gita Govinda*, Mewar, c. 1590

Inspirer of poetry—fresh lotus petals

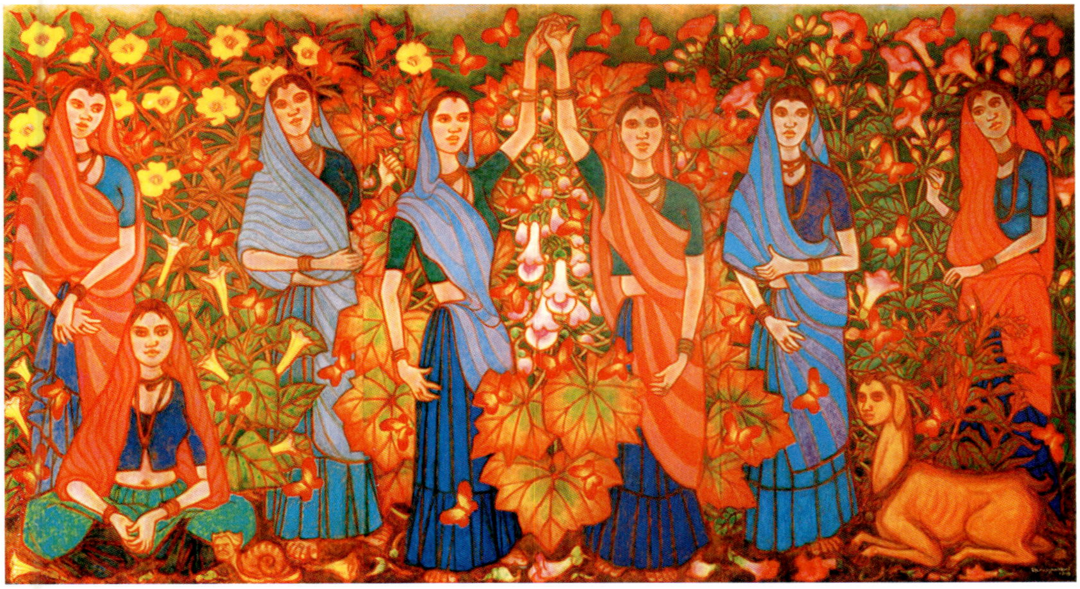

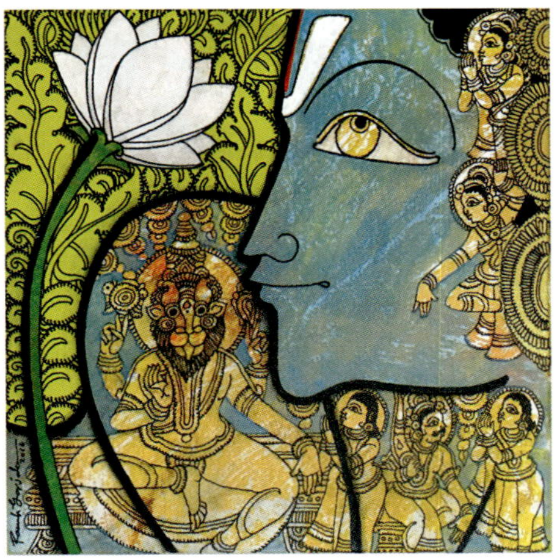

Top: 'The Monsoon Flowers', A. Ramachandran

Above: Ramesh Gorjala, mixed medium on canvas, 2016

The chef-d'oeuvre of Jayadeva is considered the world over to be the *Gita Govinda,* a simmering epic written about the deeply sensual bond between Radha and Krishna. Jayadeva is notable for transforming Radha into one of medievial India's true heroines; before this her presence in India's literary firmament was nearly non-existent. Jayadeva's love poems are replete with the erotic mood, a playful sensuality and intense devotion that have led to his works still being eulogised today.

His liberal use of flowers as a simile only underscore the delicateness of his *nayika's* beauty, serving to heighten the *shringara* rasa or erotic mood permeating his lush, lyrical poetics.

Your moist lips glow,
Like crimson autumn blossoms;
The skin of your cheek,
Is a honey coloured flower,
Fierce Radha, your eyes glower
Like gleaming dark lotuses;
Your nose is a sesame flower;
Your teeth are white jasmine.
Love's flower arms conquer worlds
By worshipping your face.[12]

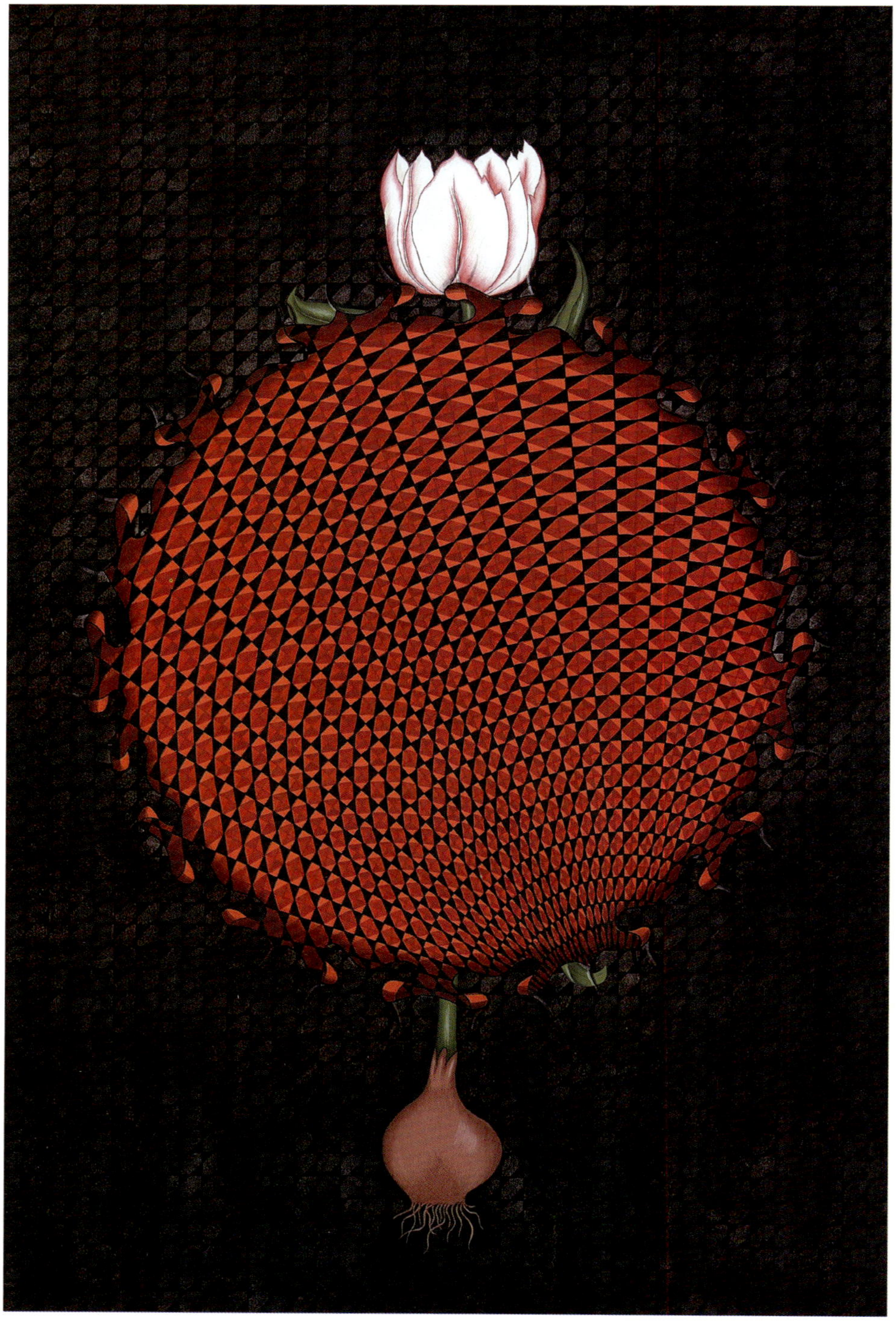

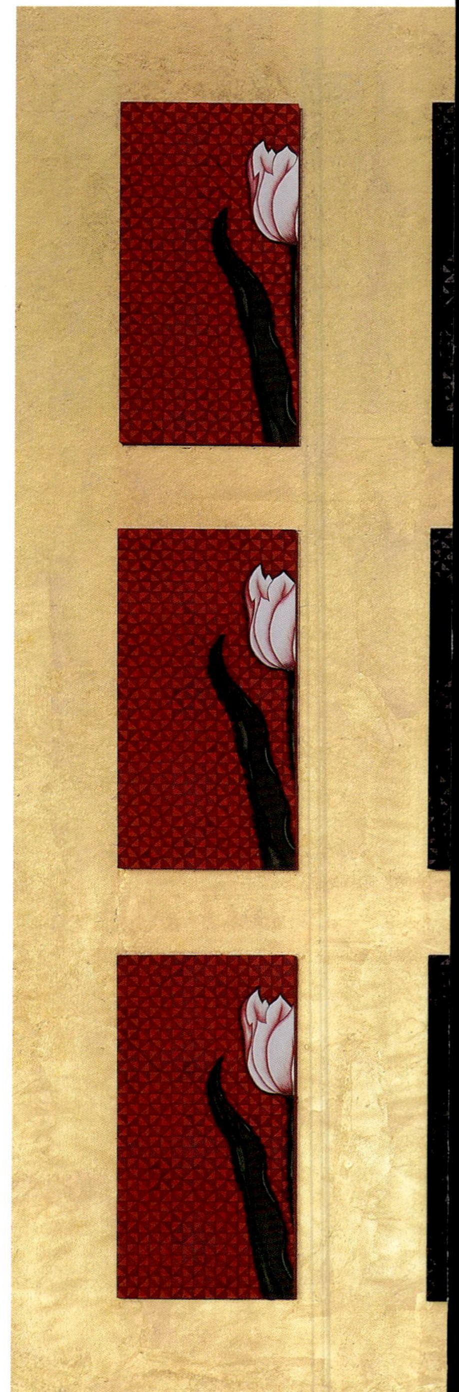

'All is gray when the black is washed away,'
Aisha Khalid, gouache on paper board, 2015

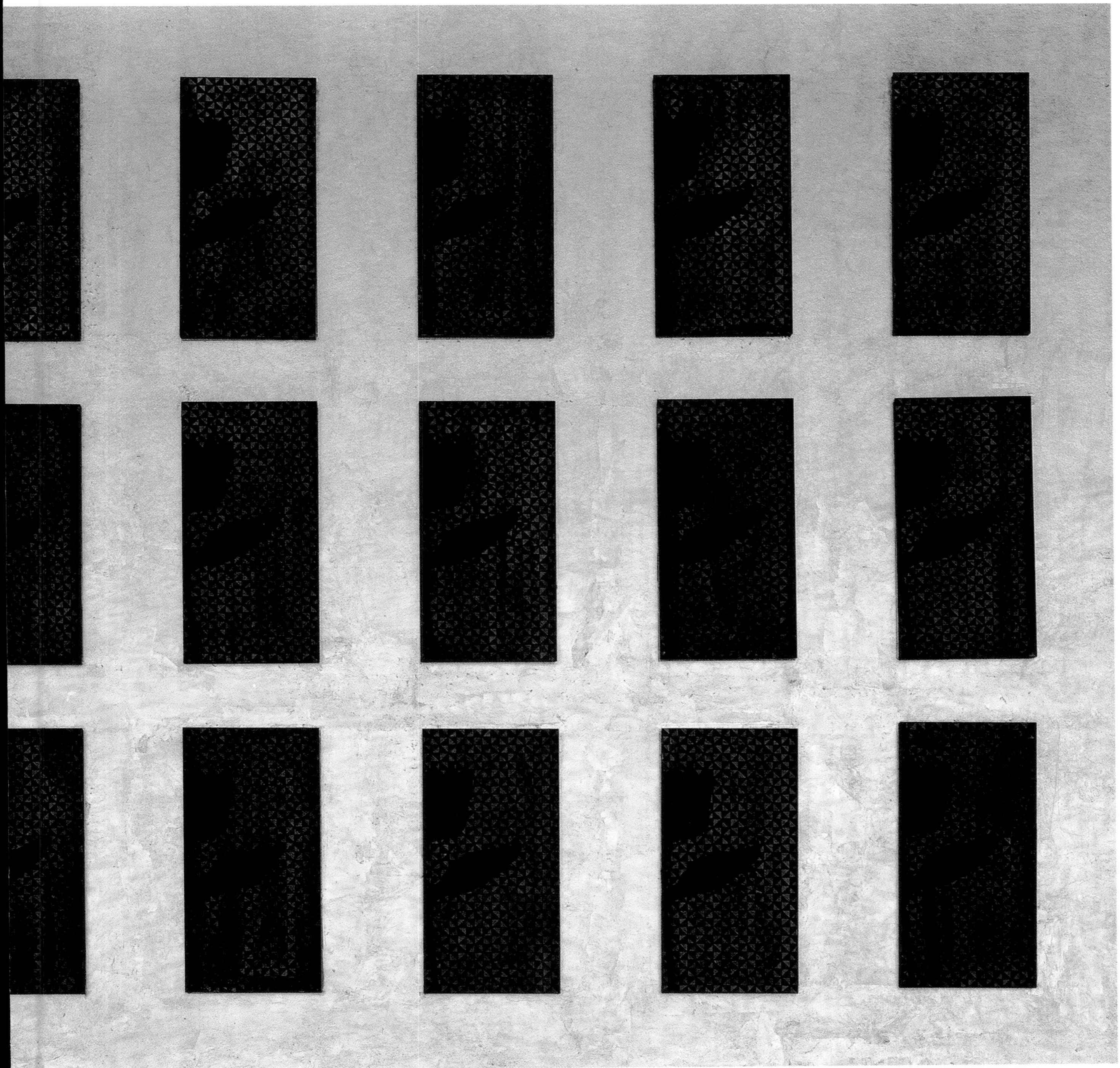

THE MODERN VOICE

The poetry written during the freedom struggle and post-independence times, especially in regional languages, carries the legacy of flower imagery. The binary of unity and separation have populated the Indian literary tradition. If flowers are synonymous with fragrance and the omnipresence of goodness, then the poet is a manifestation of the same in testing times. Flowers are a constant recurring motif in the voices of modernity through the 20th and 21st centuries, in India, from Tagore to Tarannum Riyaz.

Rabindranath Tagore (1861-1941), poet, dramatist and essayist, has time and again brought in flowers as a part of his writing.

FLOWER

Pluck this little flower and take it, delay not!
I fear lest it
droop and drop into the dust.
I may not find a place in thy garland, but honour it with a touch of pain from thy hand and pluck it. I fear lest the day end before I am aware, and the time of offering go by.
Though its colour be not deep and its smell be faint, use this flower in thy service and pluck it while there is time.[13]

THE FLOWER SCHOOL

When storm-clouds rumble in the sky and June showers come down.
The moist east wind comes marching over the heath to blow its bagpipes among the bamboos.
Then crowds of flowers come out of a sudden, from nobody knows where, and dance upon the grass in wild glee.
Mother, I really think the flowers go to school underground.
They do their lessons with doors shut, and if they want to

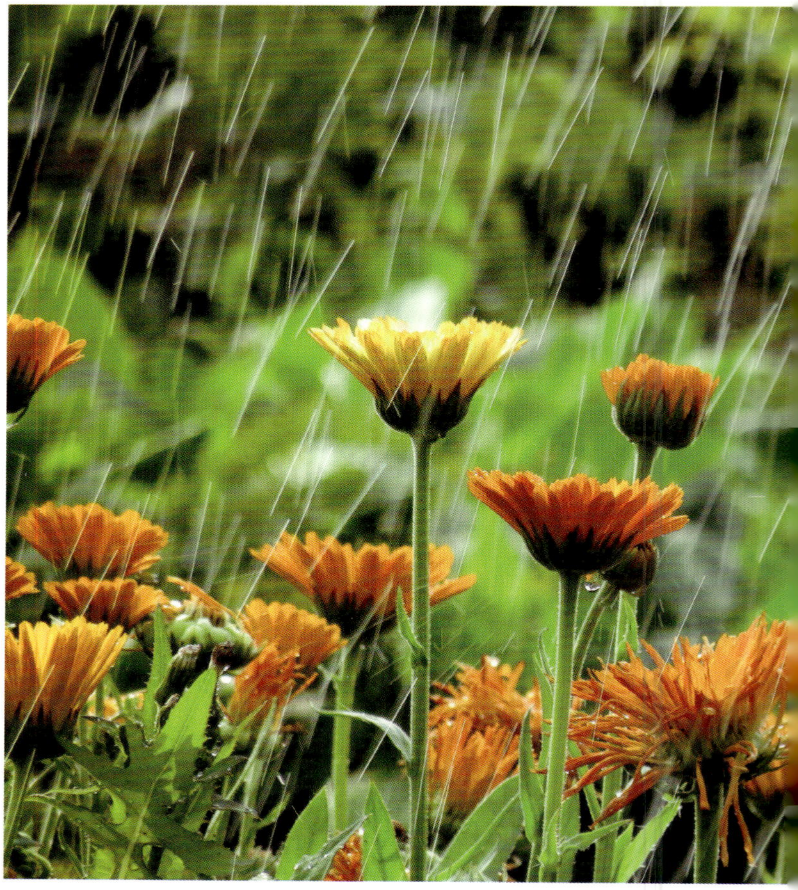

come out to play before it is time, their master makes them stand in a corner.

When the rain comes they have their holidays.
Branches clash together in the forest, and the leaves rustle
in the wild wind, the thunder-clouds clap their giant hands and the flower children rush out in dresses of pink and yellow and white.
Do you know, mother, their home is in the sky, where the stars are.

Haven't you seen how eager they are to get there? Don't you
know why they are in such a hurry?
Of course, I can guess to whom they raise their arms; they
have their mother as I have my own.[14]

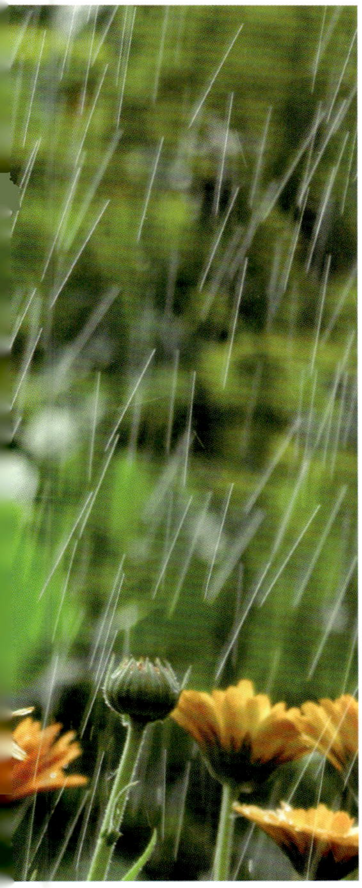

A prolific Indian writer of the modern period, **Tribhuvandas Luhar**, known by his pen name, **Sundaram** (1908–1991) made great strides in the realm of Gujarati literature, his contributions ranging from poetry to short stories to essays, literary criticism and more. In his poem 'As a Flower I Come', the flower is personified to express a fierce yet gentle devotion, at turns addressing a child, a maiden, a young man and finally God. It seems to promise an eternal passion, through the literary motif of the flower.

> And you might be the enlightened, the Blessed One, Superhuman,
> I will rest then at your feet, a heap of jasmine sweet and white,
> One with the dust of your feet
> I shall load the dust with my fragrance rich,
> And when your soul with the Supreme in communion sits,
> I'll guard your gates, a guardian pure white,
> Armed with the bow of purity.
> I'll come as a flower to you.[15]

Above: Garden flowers in the rain

Below: A heap of fragrant jasmine

A staunch nationalist, **Makhanlal Chaturvedi** (1889–1968) spearheaded the Rashtriya Dhara, a movement in poetry that encouraged passionate nationalism, originating from the Chayavad movement. Aptly writing in the national language, his poems deal with social issues as well as romantic love, nationalism and passion. This is evidenced in the poem, *A Flower's Wishes*. Here, the flower is given a voice through which it shuns idle beauty and grandeur, instead craving to be in service to those who serve the country. In a charged age when the nation was struggling to throw off the yoke of the British and rise up proud and independent, his poems went a long way in inspiring not only more similar poetry but also in inspiring fierce patriotism.

It is truly astonishing to see the many diverse ways in which the metaphor of a flower is employed to evoke varying moods and themes in poetry.

> I crave not to be woven in the wreaths of celestial maidens;
> I crave not to allure the beloved arrayed in the lover's garland;
> I crave not to be placed, O God, on the coffins of monarchs;
> I crave not to be strewn on the deities' heads and swagger on my fortune.
> Pluck me, O Gardener!
> Throw me on the path
> On which go the many gallants
> To offer their heads for their motherland.[16]

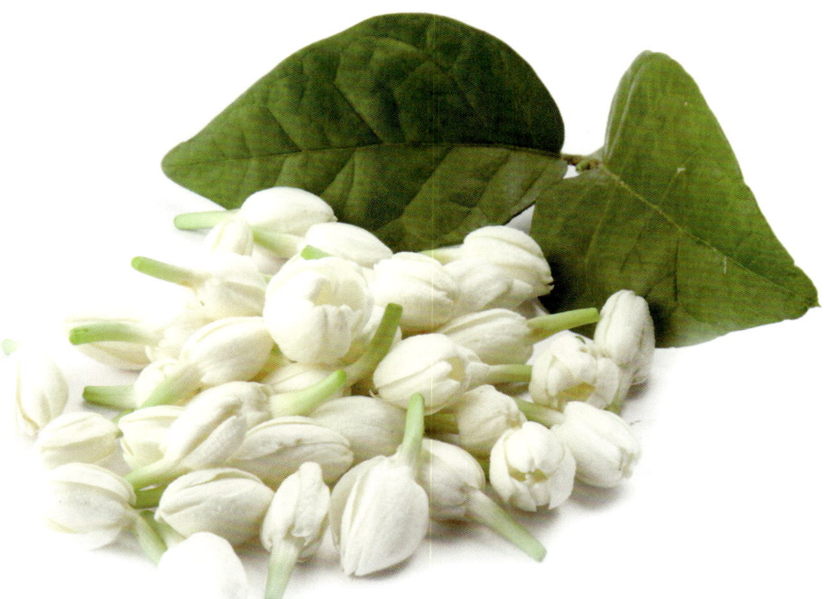

The essayist, translator and critic **A.N. Moorthy Rao** (1900–2003) is a significant name in the Kannada literary pantheon. His writing is deft and skillful. Once the director of the Department of Culture in Karnataka, he is remembered especially for his sensitive essays that incorporate lyrical writing. In his essay 'On Flowers', he dwells on various flowers and their historical and cultural significance, with regard to the Indian context. Below is an extract about the many charms of the champak blossom:

> Of these three flowers, the champak—it may be the light yellow variety of the plains or the red-gold variety of the hills—is the most conceited. In the sweetness of aroma it effortlessly out-distances all other flowers. The strength of that sweetness is its virtue and its defect. When walking near a flower-garden the fragrace of the mallige or the jaji gives you a friendly nod; and then you pass on. It is like music from far away that comes floating on the breeze, and when the breeze passes, it heard no more. But the fragrance of the champak is aggressive; it pursues, assails and conquers you. You almost hear the drumbeat of its victory. Defeat can, however, be sweet at the hands of some enemies.[17]

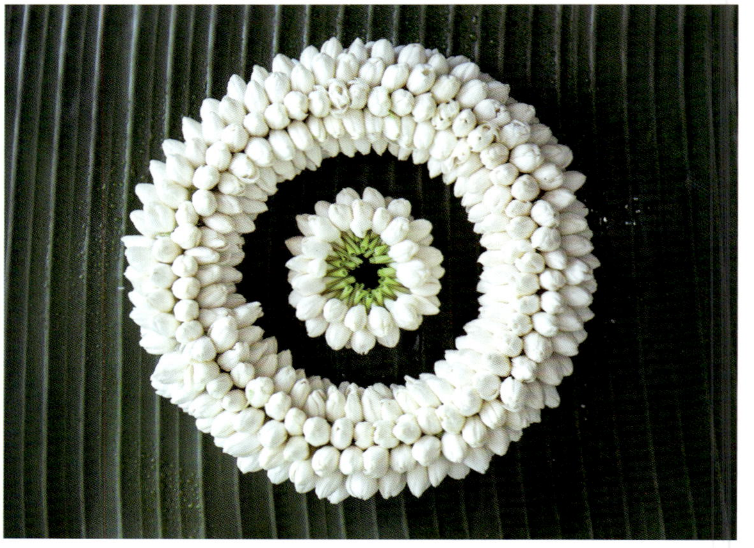

Faiz Ahmad Faiz (1911–1984) was one of the most famous Urdu poets and an influential left-wing intellectual and revolutionary poet. He has contributed to the literary development of Pakistan's arts, literature and drama significantly. The poet experienced the beauty of the Indian subcontinent and many of his works carry in them a hope for a better world. The imagery of nature is embedded in his poems; while in some cases it brings out a ray of optimism in others, such as in the following poem, they signify a powerful yearning, a longing for a love lost. Flowers have been a popular motif in the Persian literary tradition; similarly, Urdu literature carries forward the legacy.

Above: Garland of Jasmine flowers on a banana leaf

Below: Yellow champak flowers

THE ASHES OF PARTING
AND THE FLOWERS OF LOVE

Today once again on the thread of my pain
I have strung the flowers of your memory.
Having picked them in the desert of exile
These flowers, months and years of our intimacy,
I want to lay them on your doorstep
Offering to the shrine of your memory,
Bound in the silver lace of yearning,
The ashes of parting and flowers of love.[18]

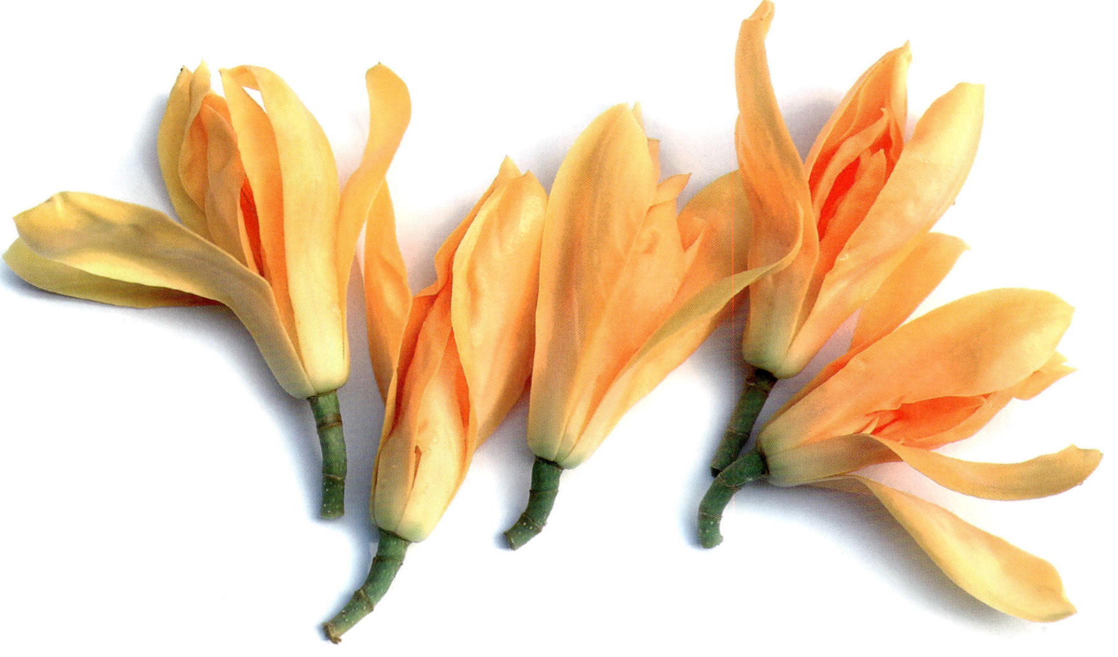

Tarannum Riyaz is an award winning modern poet, writer, essayist and critic. Residing in Delhi her poems explore a familiar theme that permeates Urdu poetry: the complexities of love and longing, making these age-old feelings find a contemporary space. It is of note, however, that in her explorations the voice she employs is markedly female, a departure from the male poetic voice that reigned in the sphere of Urdu poetry. Her work has featured in a number of Urdu as well as Hindi magazines across both India and Pakistan. In her poem *The Flowers of Love,* she writes:

> If you should leave me now and go away
> the world will not come to an end
> the gift of life is given to us but once
> I will not pine and sigh
> go, be unfaithful if you will
> I shall seek consolation with someone else
> in the spring-scented garden of life
> I too shall pick the flowers of love.[19]

ENDNOTES

1. Edward Moor, *The Hindu Pantheon.* (London: J Johnson, 1810), p. 210.
2. A.C. Bhaktivedanta Swami, *Srimad-Bhagavatam.* Fourth Canto: The Creation of the Fourth Order (Mumbai: The Bhaktivedanta Book Trust, 1974).
3. Bibek Debroy, *The Mahabharata.* Volume 10 (New Delhi: Penguin India, 2015).
4. C. Rajagopalachari, *Mahabharata* (New Delhi: Diamond Pocket Books. 1958), p. 59.
5. William Buck, *Ramayana* (New Delhi: Motilal Banarsidass Publishers, 2nd edition. 2000), p.17.
6. Ibid, p. 215
7. Shakti M. Gupta, *Plant Myths and Traditions in India* (New Delhi: Munshiram Manoharlal Publishers 2001). p 7.
8. Rajan, Chandra, ed. and trans., *Kalidasa: The Loom of Time* (New York: Penguin Books, 1989)
9. Ibid.
10. S. Viswanathan, ed., *Abhijñana S´akuntala: With English notes and translation by C. Sankara Rama Sastri* (Mylapore, Madras: Sri Balamanorama Press), p. 28.
11. Sandhya Mulchandani, *The Five Arrows of Kama: The Art of Love, Sex and Desire* (New Delhi: Penguin Books India, 2010), p 38.
12. Jayadeva, *The Gitagovinda of Jayadeva: Love Songs of the Dark Lord, The Tenth Part,* trans. Barbara Stoler Miller (Columbia University Press, 1997), p 114.
13. http://tagoreweb.in (Accessed on 07.07.2016).
14. http://tagoreweb.in/ (Accessed on 07.07.2016).
15. K.M. George, ed., *Modern Indian Literature: An Anthology – Volume One* (New Delhi: Sahitya Akademi. 1994), p. 569.
16. Makhanlal Chaturvedi, *Modern Indian Literature: An Anthology – Volume One,* trans. Shaileshwar Sati Prasad (New Delhi: Sahitya Akademi. 1994). p. 583.
17. K.M. George, ed., *Modern Indian Literature: An Anthology – Volume One* (Sahitya Akademi. New Delhi, 1994), p. 318.
18. Faiz Ahmed Faiz, *100 Poems by Faiz Ahmed Faiz, 1911-1984.* (New Delhi: Abhinav Publications, 2002), poem 58, p. 205.
19. Tarannum Riyaz, 'The Flowers of Love', trans, Jaipal Nangia. *Indian Literature* Vol XLVII No. 3 (2003: 82.)

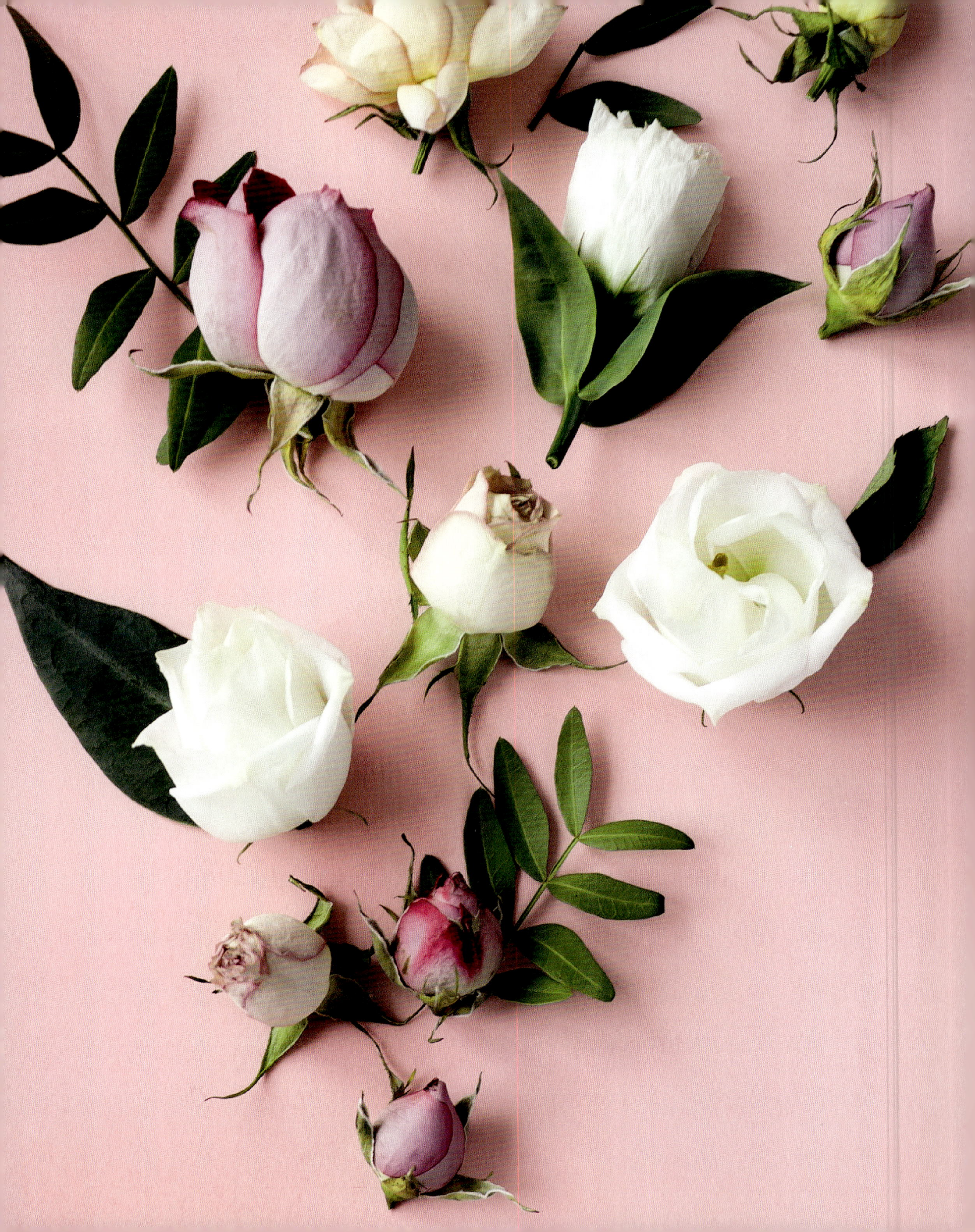

The Sensual and the Fragrant

Perfumes are the feelings of flowers, and as the human heart,
imagining itself alone and unwatched, feels most deeply in the night-
time, so seems it as if the flowers, in musing modesty, await the mantling
eventide ere they give themselves up wholly to feeling, and breathe forth
their sweetest odours. Flow forth, ye perfumes of my heart, and seek
beyond these mountains the dear one of my dreams!

—Heinrich Heine[1]

As Heine stated in his beautiful poem, the feeling of the flower is its perfume. This delicate analogy fits the Indian cultural métier like a glove. For each flower with its distinctive fragrance has a singular relationship to the many Gods of the Hindu pantheon. Sometimes its very aroma becomes a physical site of not simply pleasure, but also of hygiene and even medicine as part of aromatherapy.

Embellished with flowers, armed with a bow of sugarcane and a line of honeybees and floral arrowheads, Kamadeva the god of love, is often portrayed looking resplendent in vibrant red and green hues much like the parrot/Shukha, the messenger of love and the vehicle of Kamadeva. The beautiful Rati and Priti are his consorts. The champa flower is one of the five flower-darts/love-darts that Kamadeva possesses. The champa tree is popularly grown in many temple precincts. It is also considered sacred to another amorous god, Lord Krishna.

Facing page: An array of fragrant roses

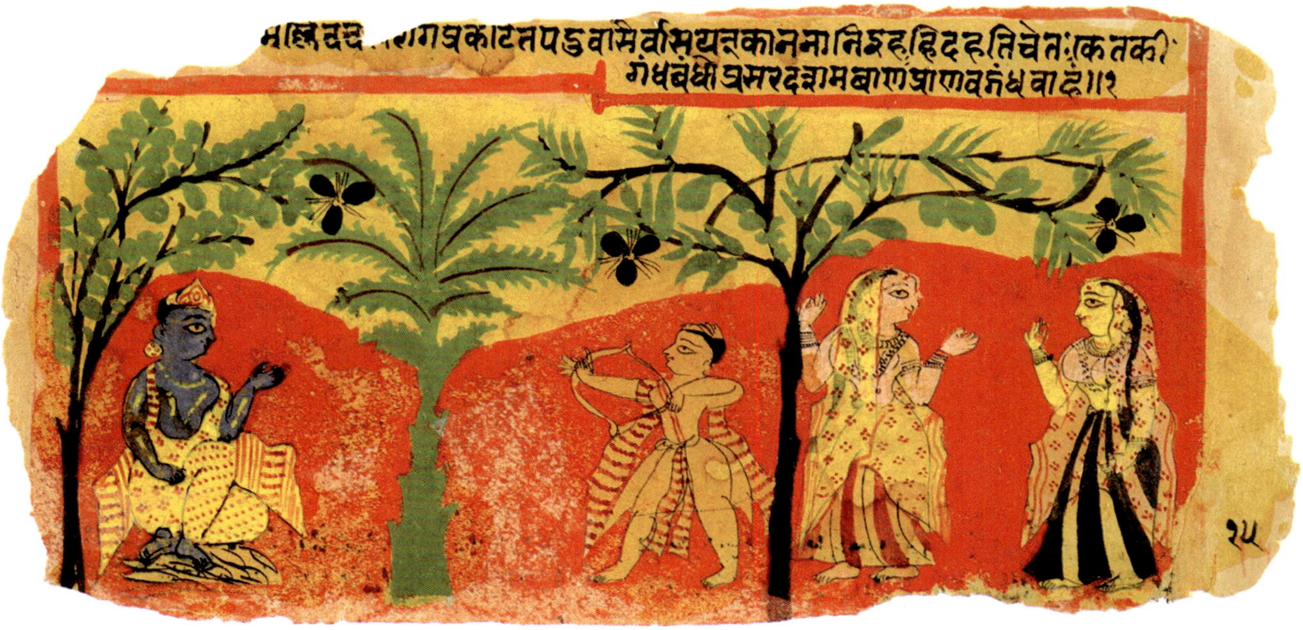

Kamadeva, God of Desire, shooting an arrow at Krishna, *Gita Govinda*, opaque watercolour and ink on paper, Gujarat, c. 1585

Men react differently to floral aromas, as in the case of King Vasu, who is aroused by the smell of springtime. According to legend, a fish, Adrika, who was actually a cursed apsara, swallows the semen ejaculated by King Vasu when he smelt the flowers, and became pregnant. The girl child Satyavati, born to Adrika, starts her life smelling of fish and is also known as Matsyagandha. She is later blessed by Sage Parashara to smell alluringly fragrant and is then known as Kasturigandha. Her exceptional fragrance enchants King Shantanu, the father of Bishma. Shantanu marries Satyavati, accepting the pre-condition that her children would inherit the throne, while Bhishma takes a vow of life-long celibacy. This is a manifestation of floral fragrances in Sanskrit literature.

An exceedingly enjoyable fragrance is described in the *Mahabharata*, and in this case, it emanates from quite a rare, heavenly flower, that one can come by only after great trials. More than the diffusiveness of the perfume, what is emphasised is the desire the fragrant object creates—a desire to possess the object and experience more of the perfume. This romantic allegory is found in the '*Aranyakaparvan*' section of the *Mahabharata*. While the Pandavas are travelling and Arjuna goes to acquire some divine weapons from Indra, the sage Lomasa tells the other four brothers and their wife Draupadi that they should undertake a pilgrimage to the sacred bathing places. While on this pilgrimage, they wander through the fragrant woods to the hermitage of Nara and Naryana near Mount Kailasha. Here, Draupadi comes across an unusually fragrant flower that she desires.

The Pandavas find the source of the flower in the Saugandhika forest on Mount Gandhamadana, which literally means 'intoxicating with perfume'. This particular flower grows in the pleasure garden of the god Kubera. It is a golden, perfumed water lily. Bhima goes to fetch more of the water lilies and in his long and arduous journey, his father Vayu, the wind god, cools his tired body and also provides a scent trail for Bhima to follow and find the lily with the amazing perfume. Thus it is the wind which is instrumental in bringing the flowers to Draupadi. In Bhima's attempt to please Draupadi, we see the wonderful meeting of wind, perfume, desire and union.

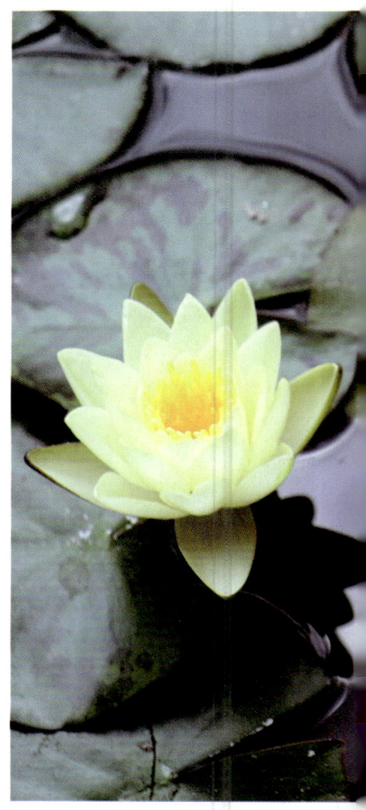

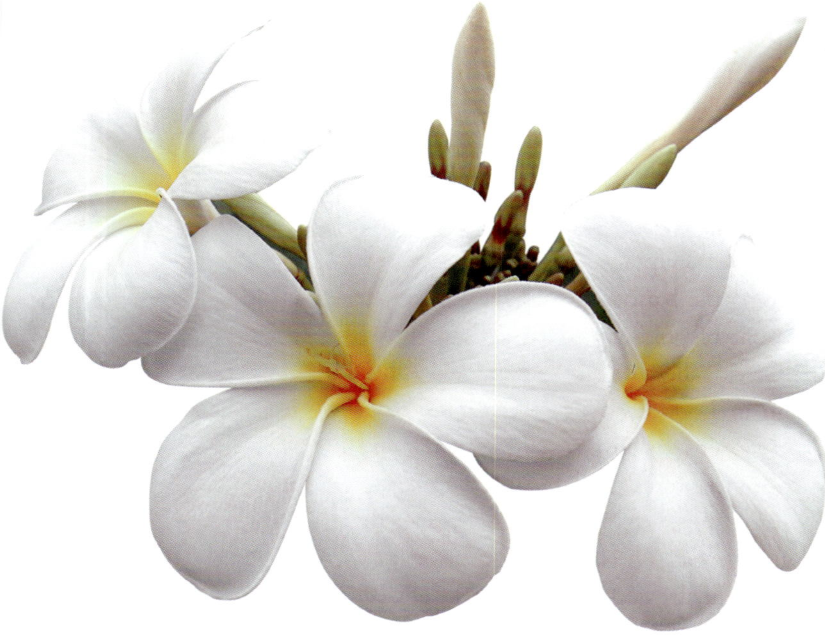

Above: Champa flower

Below: Yellow water lily

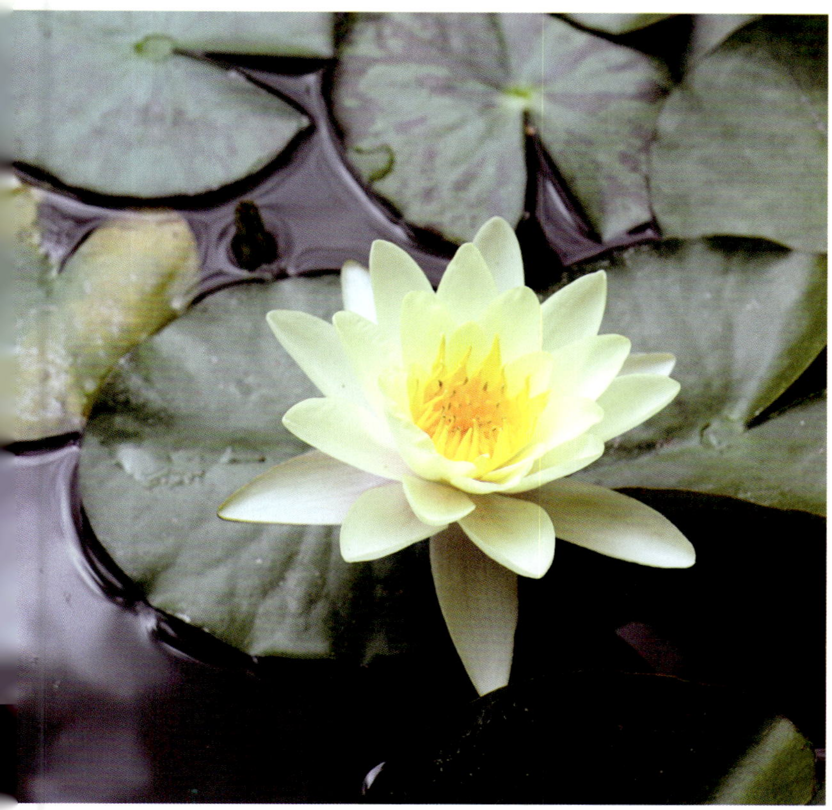

In that place, the wind blew, with a good smell and pleasant feel,
Gladdening all the Pandavas together with Krsna and bulls of sages.
Then, unexpectedly, a clear wind from the north-east carried
A thousand-petalled, sun-like, divine lotus.
Pancali saw that lotus, of divine perfume, and pleasant,
Brought by the wind, pure, fallen on the ground.
The beautiful one, having found that beautiful, excellent, white water-lily,
Being extremely joyful, oh king, then said to Bhimasena:
'See, O Bhima, the divine, very radiant, most excellent flower,
Endowed with perfume and form, delight of my mind.
But I will offer this one to King Dharma, oh enemy burner,
For love of me, you should take this one back to the Kamyaka hermitage.
If I am dear to you, oh Partha, fetch many of these,
I want to take them back to Kamyaka hermitage.'

Facing that very wind whence the flower came,
Desirous to fetch other flowers, swiftly he went.

He, of great vigour, smelling that unfettered perfume produced of flowers of all seasons,
Unfettered in the forest, like an intoxicated elephant
His body hair was greatly thrilled; his tiredness removed by his father,
With the wind on the Gandhamadana mountain, cool with the touch of his father.[2]

THE ESSENCE OF FLOWERS

The aroma of flowers, *sugandh* (fragrance), precedes their physical manifestation. Because of their delicate aromas, flowers have qualities that are distinct from other plant parts that are used as aromatics, like sandalwood and camphor. There are some finely nuanced differences between flowers which come from trees, shrubs or herbs. Depending on where flowers are grown or sourced, their aromas can be inhaled directly, used as aromatic oils or made into perfumes. Flowers have a limited shelf life, bound by time and space. And because flowers are usually seasonal, they are associated with typical aspects of that season. The other feature that characterises flowers is how quickly they fade, which has become a metaphor for impermanence. They cannot be re-traded as easily as, say, sandalwood.

The aroma of fragrant flowers is heady and sensual, and was used to add to the erotic mood of the boudoir by courtesans in the bygone eras. There is a difference in the treatment of scents in Indian and Western literature. In Indian narratives, relishing a diffused aroma is never enough—the person smelling the scent wants to unite with its source and experience the rest of that object. These affective and attractant powers of smell were consciously exploited in traditional Indian culture. As a result, the search for powerful aromatic materials and the means of using them produced a sophisticated and cosmopolitan culture that appreciated and valued perfume.

AGE-OLD AFFAIR WITH FRAGRANCE

The custom of fragrance was prevalent right from the Indus Valley Civilisation, through the Mauryan and Gupta periods to the present day. During the Kushan periods, too, there are descriptions of gold-embedded perfume boxes. In the Greek gallery of the Victoria and Albert Museum in London, there is a pair of stone studded earrings designed to hold scented liquid. The fragrance would waft out of the earrings with every move of the neck, scenting the surrounding area. The illustrious poet Kalidas, in his seminal works *Abhigyanashakuntalam* and *Meghaduta*, describes how fragrances were employed by the *nayika*. She would scent her hair and dresses using fragrances made with sandalwood, jatamasi or kasturi.

Fragrance and perfume are mentioned in the ancient Indian texts and there are descriptions of making the floral essences. In Sanskrit it was known as *sugandhdravya* or *gandhika* and later in the Indo-Islamic period, which predates the Mughals, it came to be known as *attar* or *itr*. The king usually

Facing page, above: Rose water is a popular floral perfume

Facing page, below: White sandalwood flower

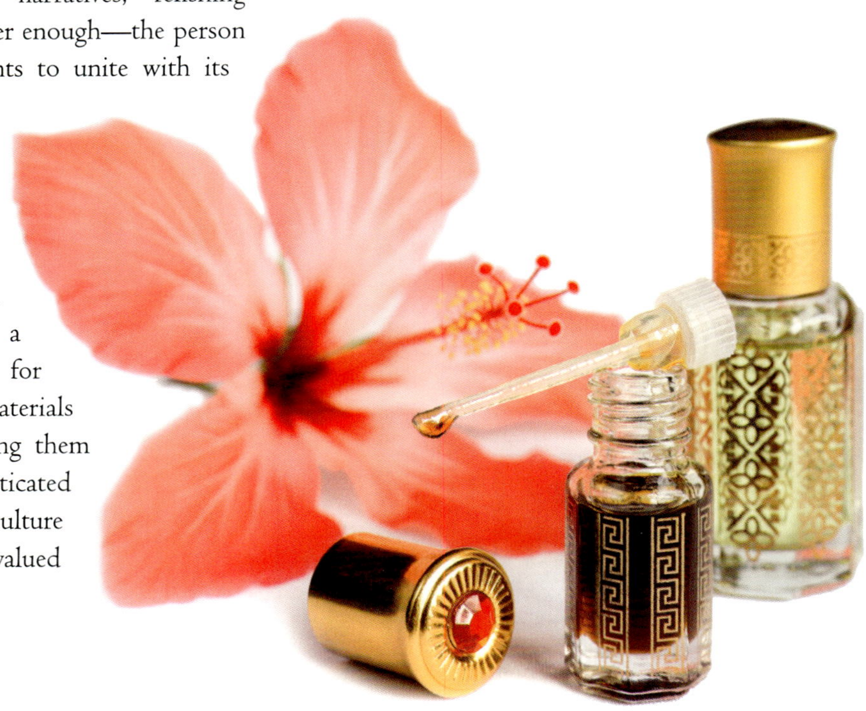

Below: Attar bottles

began the day in the morning with fragrance. According to the *Agni Purana*, the kings bathed with over 150 varieties of fragrances. The consumption of fragrances was also popular in the royal harems. It was mainly women, called *gandhkarika*, who prepared the fragrances. Indian *attar* was exported to Iran, Egypt, Arabia, Turkey and Afghanistan.

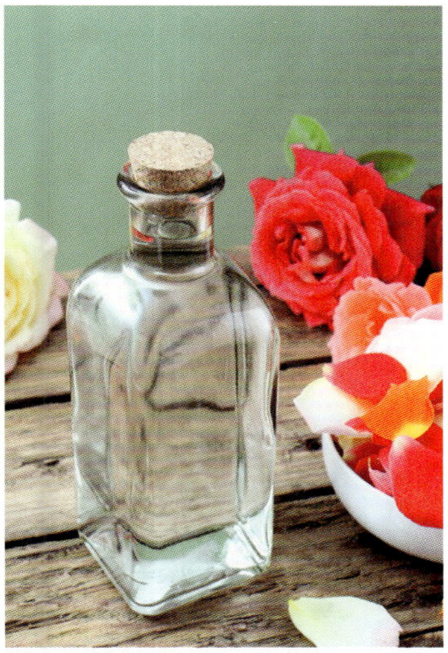

Attar, is a kind of diluted oil. The word has its roots in the Persian 'atr', which means 'aroma'. It is one of the oldest existing perfumes. They are made from natural ingredients from plants and flowers that are distilled at a temperature, where the vapours are condensed. Sandalwood, herbs and spices are some of the other raw materials that go into the creation of *attar*.

In ancient India, copper distillation containers were used to extract the *sugandhdravya*. The perfumes were collected into jars, remnants of which have been found across Asian and Arabian countries. Perfumers voyaged far and wide to trade as well as to procure intriguing ingredients. Hindu literature contains some of the early formulations to create the first perfumes from a melange of herbs. Ayurvedic texts such as the *Charaka Samhita*, and later, the *Kama Sutra*, speak of the creation of fragrant oils.

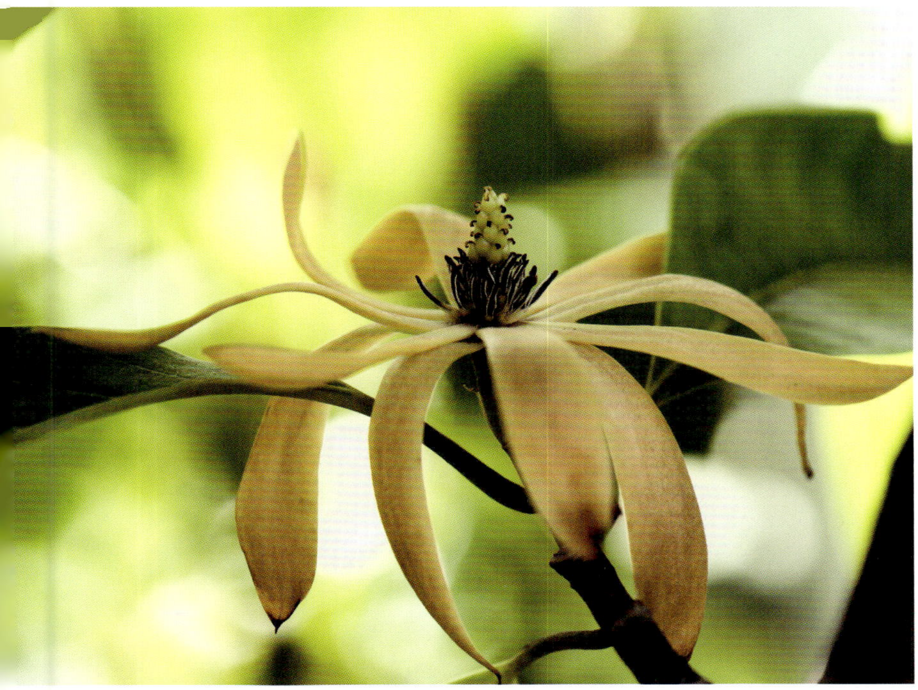

In ancient India, people would luxuriate in baths of perfumed water. A variety of petals in various combinations would be employed as cosmetics. Rose water is one of the favourites that still reigns supreme across Asia and the world. Attars were also a great indulgence of the Mughal emperors, who acquired them from the Far East. Urdu masters like Mirza Ghalib have waxed poetic about it. Uttar Pradesh is well known as a centre of *attar* production, Kanpur and Lucknow being notable cities where traditional attar makers plied their trade. The traditional way of extracting the fragrances, however, is slowly being lost to time.

ATTAR, A MUGHAL OBSESSION

The obsession for floral fragrances reached its peak during the Mughal period. Nur Jahan, wife of the Emperor Jahangir, was a connoisseur of attar—she had '*Itre Heena*' made for herself. Five *sers* (roughly 4.6 kg) of attar could be produced by combining the essences of 30 varieties of roses and five varieties of *chandan* (sandalwood) oils. She used it so lavishly that this quantity was not enough to last her a fortnight.

The Mughals had settled the *gandhis* (experts in making attar) in the kingdom, sanctioned them zamindaris and commissioned them to make the different varieties of attar.

Roses were cultivated extensively in the Malwa region bordering Madhya Pradesh and Rajasthan even today, and different species were exported to foreign countries and shipped across India.

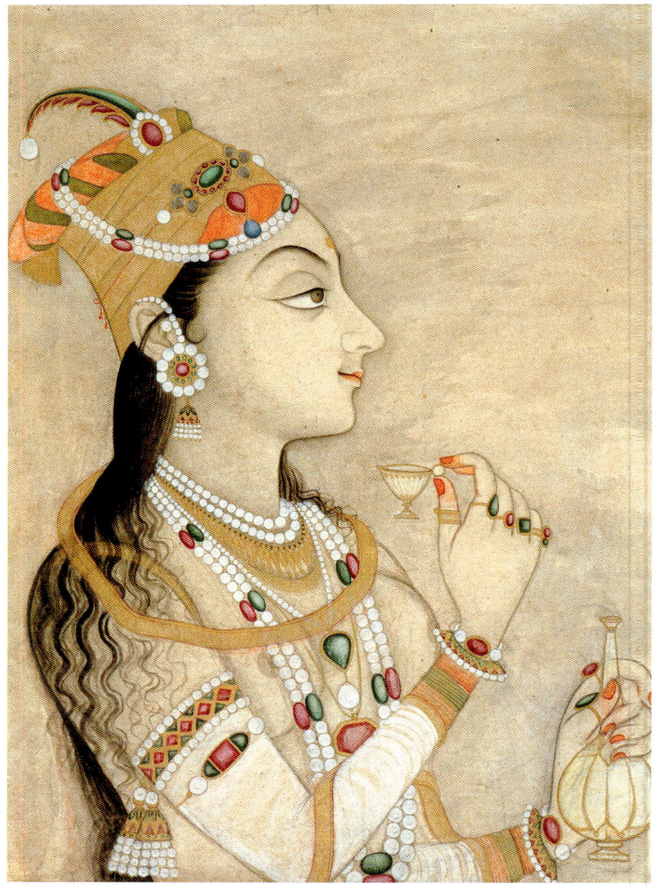

FLOWER POWER

JASMINE

Belonging to the genus known as *Jasminum,* jasmine has over 200 species of scented white flowers found in many parts of the world. In India the common species is *jasminum sambac,* known as *mogra* in the north, and *mallige* in the south. Jasmine perfume has been called 'Queen of the Night' and 'Moonlight of the Grove', and is beloved by women from time immemorial for its sensual and lovely scent. Jasmine oil is considered invigorating and alleviates nervous exhaustion, anxiety and apathy. As a perfume it is said to arouse stimulating emotions. Oil of jasmine is a relaxant and is used in massage oils and creams as it warms up. It helps to relieve dry, sensitive and itchy skin providing relief from irritation.

ROSE

Herbalists and followers of Ayurveda make use of rose essential oil in order to treat a number of afflictions. Its fragrance has always been thought of as soothing, no wonder it was used to help the mind keep calm during ancient times. Other benefits include helping maintain a healthy vascular system and eliminating toxins. Cosmetically speaking, rosewater is a tonic for a number of beauty problems, including treating the delicate complexion of the under-eye region as well as treating large pores and toning.

HIBISCUS

Known as *jabakusuma,* the hibiscus is in wide use across India. The extract from the flower was once used in making perfumes. Its refreshing properties made it an ideal ingredient in balms. Its cooling effects means it was also used in hair oils and in treating afflictions such as dandruff as well as to dye hair. An effective astringent, it is said to open up choked pores and in increasing blood flow.

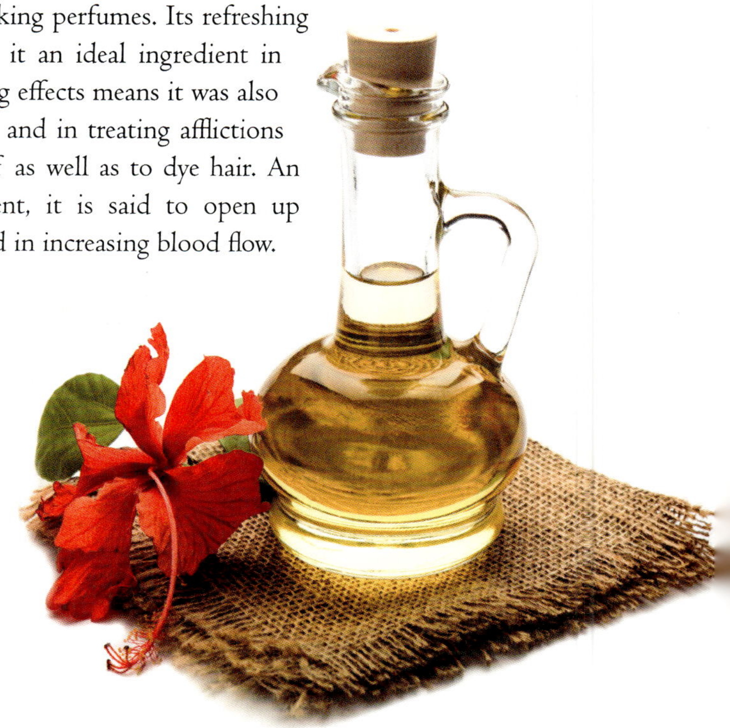

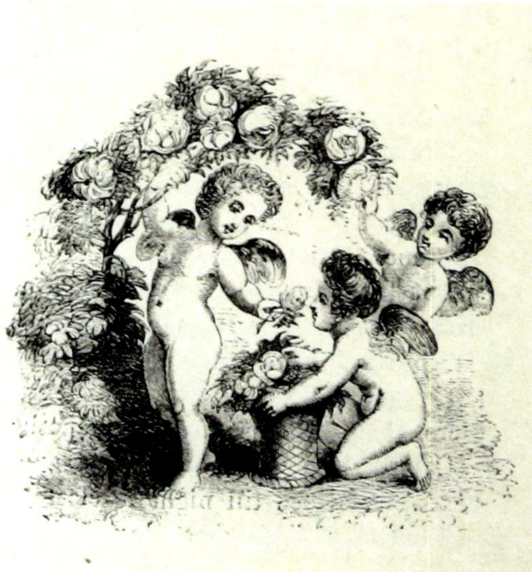

GUL REA ZEE

In New York City in the early 20th century, one of the most sought after perfumes was known as Gul Rea Zee. It was a Greek fragrance its name based on the Persian festival called the Scattering of the Roses. This festival originally took place when Lalla Rookh, the daughter of Aurangzeb, left Delhi for Kashmir for her marriage with the young king of Bukhara. As Thomas Moore described the scene in his Oriental romance, *Lalla Rookh*, 'beautiful children went strewing the most delicious flowers around...till every part of the city was as fragrant as if a caravan of musk...had passed through it.[3]'

Image: 'Moore's Irish Melodies, Lalla Rookh, National Airs, Legendary Ballads, Songs, etc., with a memoir by J. F. Waller. Illustrations, etc', p. 462.

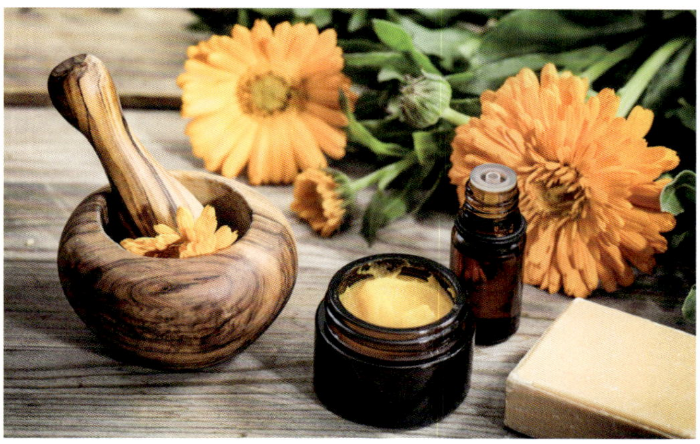

Above: Essential oils and ointments made from calendula

Facing page, above: Portrait of the Mughal Empress Nur Jahan, opaque watercolour and gold on paper, Kishangarh, Rajasthan, c. 1725-1750

Facing page, below: Hibiscus oil has many aromatic uses

LOTUS

Poets and writers have waxed eloquent about the lotus flower. Its nature as an anti-allergic, astringent and anti-spasmodic substance render it ideal for medical purposes as well as in cosmetics. It can be used to treat sun damage to the skin, as well.

MARIGOLD AND CALENDULA

Both these yellow blossoms have essential oils that are therapeutic, tonic and astringent. The oils can be employed as natural remedies against dandruff and skin problems such as enlarged pores and excess sebum.

FRAGRANCES IN TODAY'S WORLD

The daily manipulation of our olfactory universe and the olfactory identities of our bodies is a recent phenomenon. In the West, the sense of smell was not looked upon as essential for long centuries. It wasn't until the Cartesian Enlightenment that the cleaning up of smells was accompanied by an increased medical concern where getting rid of odours was seen as a civilising process. As part of global cosmopolitanism, the multi-million dollar fragrance industry, the growing interest in aromatherapy and numerous advertisements offer a variety of odour-killers that powerfully focus our attention on scent and stench. Today, even candles are perfumed and everything—from soaps to talcum powder—is loaded with aromas. Unfortunately, most essences today are chemically created in labs, to make them affordable for the masses. However, the discerning user who can afford to buy genuine perfumes created from flowers and natural products still continues to do so.

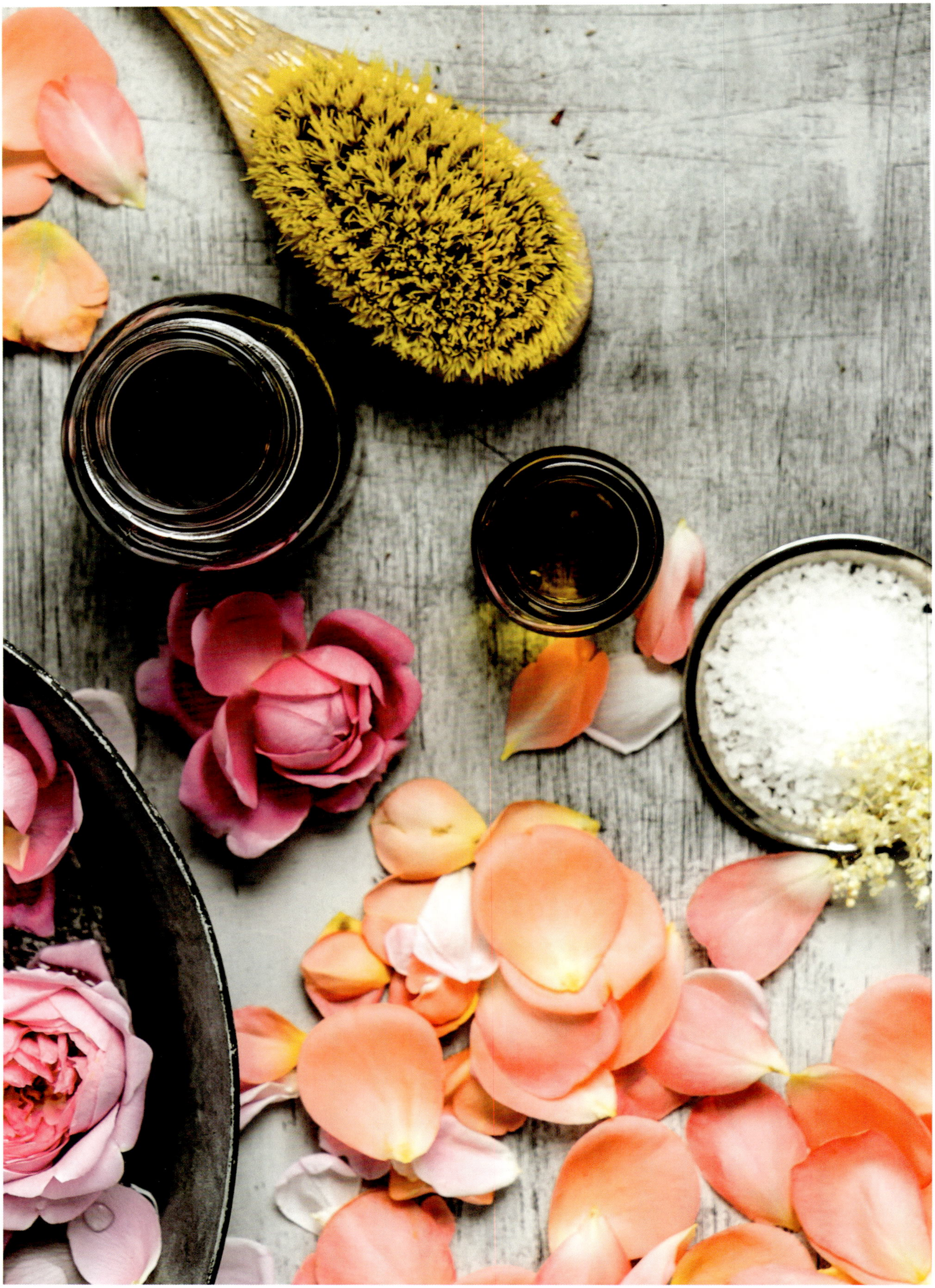

Facing page: Spa treatment with flowers

Natural substances derived from plants have been used as fragrance, as well as for embellishment, for thousands of years, across civilisations. Modern perfumers have taken their cues from these age-old practices, employing natural substances in their creations. We are seeing a return to natural perfumes, and an eschewing of chemicals and synthetics. Many top-end Indian companies sell original natural products, and they are known as 'ayurvedic' rather than 'cosmetic'. Shahnaz Husain, the Indian herbal cosmetologist, is famed for her innovative use of flowers in her beauty and healing products. She was one of the first to recognise the benefits of flowers in beauty treatments and to use flower extracts across many ranges of beauty care solutions. She led the way for other companies to begin using Ayurvedic systems as a base to their beauty products.

In addition to their use as aids to natural beauty, flowers are soothing and offer a relaxing effect. They can be harnessed to combat stress from environmental and mental factors. The stress of modern living can lead to a number of beauty dilemmas; acne, hair loss, dandruff, dry skin, under-eye dark circles and many others, for which floral essences provide effective remedies. Floral fragrances can also set the mood and enhance pleasure. Aroma oils and essences invested with a potent perfume when placed in bedrooms in diffusers are being used as accessories to heighten sensuality and desire.

The essential oils extracted from flowers can aid in lessening stress and their fragrances offer a rejuvenating effect. Flowers have potent healing properties too, as is evident in aromatherapy massages that use combinations of essential oils. Aromatherapy is an alternate science whereby essential oils of flowers are used to cure headaches, reduce stress and other ills, enrich beauty treatment, and bring about a great deal of ambient fragrance through the aroma of flowers. Practitioners of nature cure also believe that a floral aroma helps the body find a natural way to cure itself and improve its immune response. Today flowers are also an essential part of the interior décor of spas and wellness centres.

One of the signature selling points of the more cosmopolitian spas in wellness centres or boutiques is the use of aromatherapy in the massage oils used. Increasingly, the curative effects of aroma therapy is being recognised, with the belief that there is a direct relationship of aroma on the brain. Even though there are no definitive scientific data and studies available, the pharmacological effects of essential oils bear a direct reference on the limbic system through the olfactory system. Essential oils and absolutes have different processes of extracting the essence of aroma either directly from the flowers, or from the delicate plant tissues. For example, steam distillation as in eucalyptus oil, or expression as in grapefruit oil, or through solvent extract.

ENDNOTES

1. Heinrich Heine, *Great Thoughts from Master Minds, Volume 2* (London: A. W. Hall, 1884), p. 88.
2. James McHugh, *Sandalwood and Carrion: Smell in Indian Religion and Culture* (New York: Oxford University Press, 2012), p. 95.
3. Thomas Moore, *The Poetical Works of Sir Thomas Moore* (Philadelphia: J. Crissy, 1835), p. 27.

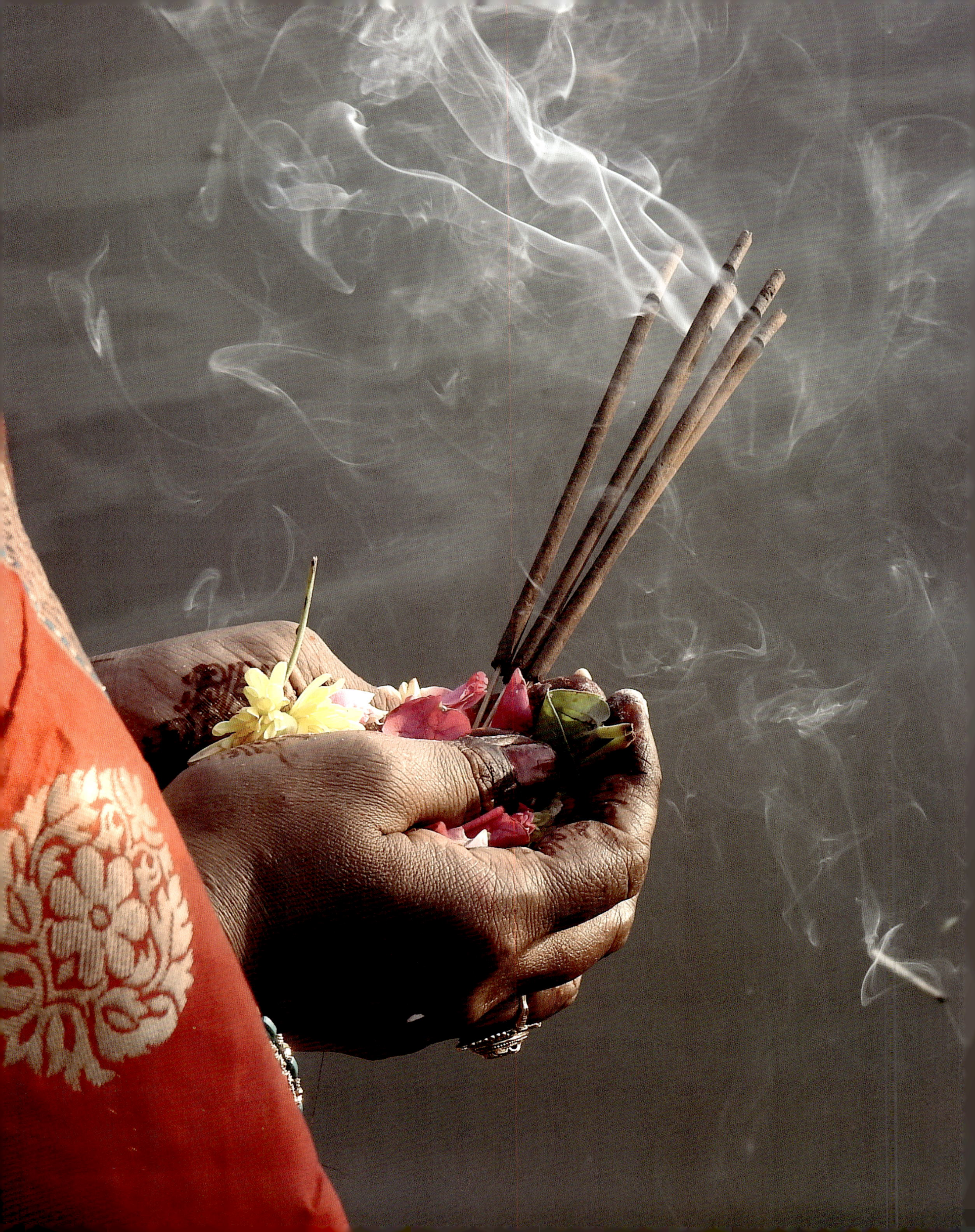

The Divine and The Sacred

patram´ pus.pam´ phalam´ toyam´
yo me bhaktya- prayacchati
tad aham´ bhakty-upahr.tam
as.na-mi prayata-tmanah
(If one offers me with love and devotion a leaf, a flower,
fruit or water, I will accept it.)
—*Srimad Bhagavatam,* Canto 10, Chapter 81, Verse 4.

Be it at a wayside shrine or inside the sanctum sanctorum of famous places of worship like the magnificent Shiva Temple at Chidambaram or the bejewelled Krishna shrine in Nathdwara, flowers are the most precious ornaments of Indian gods. What is particularly alluring is the way each god has their own way of adornment: *tulsi* leaves for Vishnu or a *bel* leaf garland for Shiva; while Ganesha, the remover of obstacles, sometimes wears a garland of *durva* grass. The Devi at the Kalighat temple in Kolkata is inundated with her favourite red *jabakusum* or hibiscus flower. In general the gods are decorated with garlands of different fragrant flowers: jasmine, marigold, and now, increasingly, *desi gulab*. Other popular flowers for garlands include *champa,* lotus, lilies, *ashoka,* oleander, *yuthika, juhi, akund, vakula, punnag* and *kadamb.*

Facing page: A Hindu woman prays and makes offerings to the Sun God, in a ritual performed while standing in water, during the Chhath Puja festival

MYTHIC QUOTIENT

Mythology is not a lie, mythology is poetry, it is metaphorical. It has been well said that mythology is the penultimate truth…penultimate

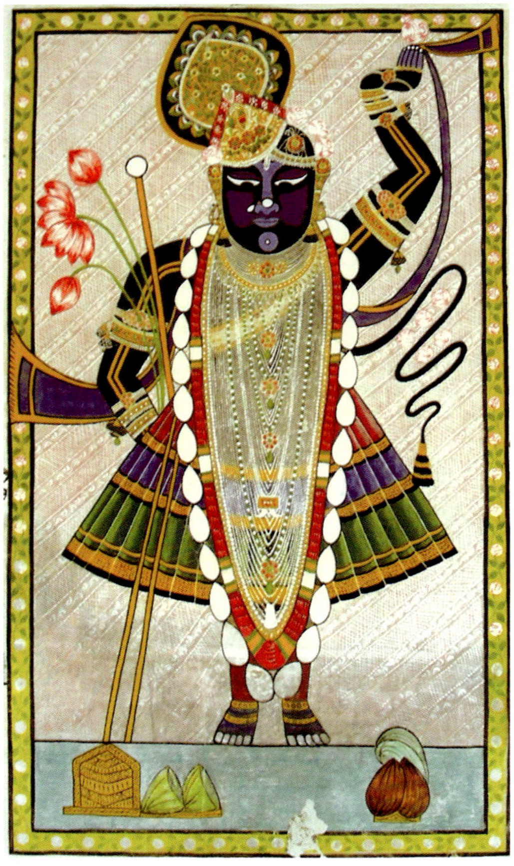

because the ultimate cannot be put into words. It is beyond words. Beyond images, beyond that bounding rim of the Buddhist wheel of Becoming. Mythology pitches the mind beyond that rim, to what can be known but not told.[1]

In a parallel trajectory there is an extremely close connection between Indian flowers and their mythical orientation. Sixteen flowers akin to the *solah sringara* or modes of adornment have been strung together to form a garland of the sacred, the sensual and the fragrant.

Cherry-picking from a host of flowers, the more frequently acknowledged, referred to and used are:

ASHOKA

Ashoka is a Sanskrit word meaning 'without grief.' In the *Ramayana*, after Sita was abducted, the space in the shade of the tree is mentioned

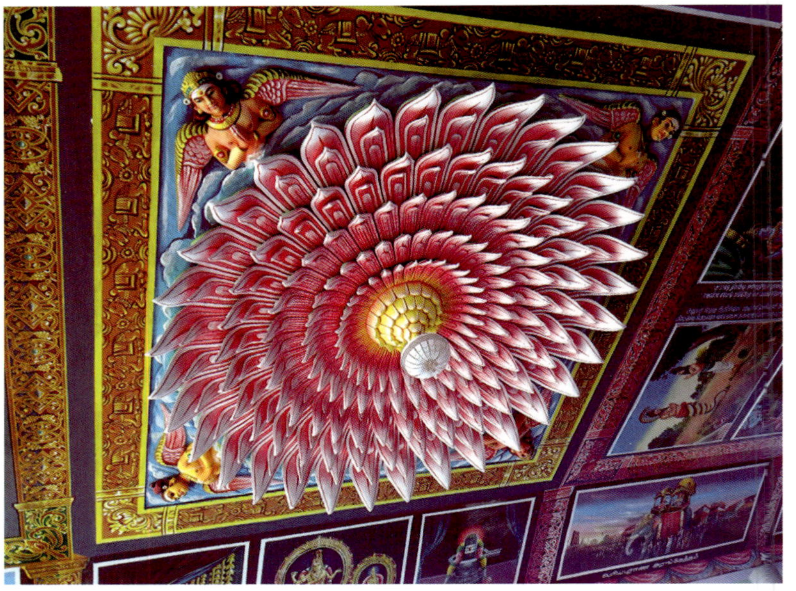

as her resting place in Ravana's garden, which is itself called the Ashoka Vatika. The *ashoka* tree blossoms with bunches of crimson flowers which bloom in early March; it is the harbinger of Spring.

PARIJATA

The *Parijata* was believed to have sprung along with the *amrit* during the *samundramanthan* or churning of the ocean. It is said that the seeds, flowers and leaves possess great medicinal qualities. It was also one of the trees in the garden of Indra, the chief of the Gods. Even though it emerged from the churning of the ocean, it seemed to instigate contentious relationships between gods (it was the root of a feud between Krishna and Indra) and their wives as well (also being at the root of another feud between two of Krishna's wives).

KADAMB

The *kadamb* is a universal favourite among the gods and it is said that Durga resides in a *kadamb* forest. The *kadamb* forest was also fabled to be the playground for Lord Krishna.

Left: Krishna mural, in the form of Srinathji of Nathdwara, City Palace, Udaipur

Above: Lotus motif on the roof of the Natarajar Temple, Chidambaram, Tamil Nadu

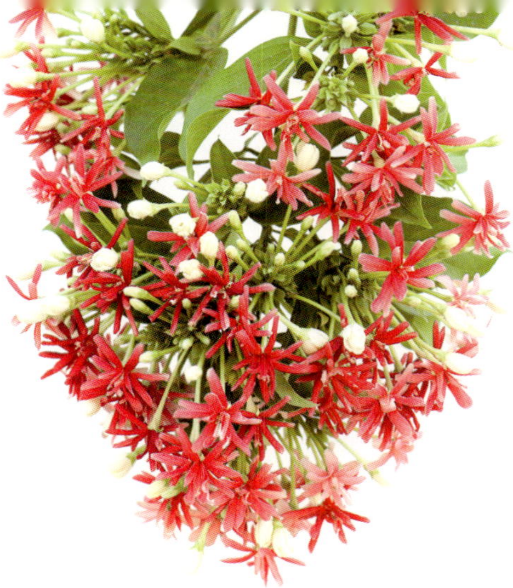

Above: Malati flowers

Below: Bakul flowers

Lower right: Akund flower

KAMAL

The *kamal* is primarily associated with Brahma, Vishnu, Lakshmi and Buddha and is a symbol of spiritual liberation. It is emblematic of beauty, purity, enlightenment and holds pride of place in Indian and South Asian philosophy.

BAKUL

It is often believed Lord Krishna used to seduce the milkmaids of Vrindavan while playing his flute under the *Bakul* tree on the banks of the River Jamuna. It is said that the tree blossoms when sprinkled with nectar from the mouths of lovely women.

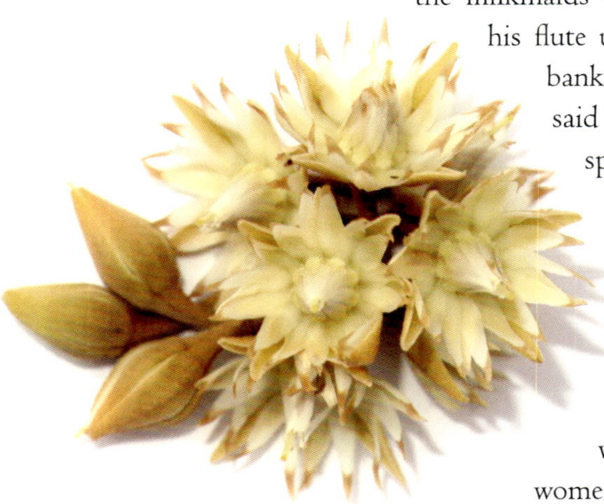

MALATI

Malati is an extremely fragrant vine. Ancient Indian texts abound with stories about beautiful women whose braids were adorned with *malati* flowers. Under the *malati* vines, lovers would meet.

KUMUD

This white waterlily which blooms only at night seems to find an enemy in the sun. The *kumud* is a much-used comparison for the face of a beautiful woman. A pond brimming with this white waterlily is also favourably compared to a woman's loveliness.

KETAKI OR KEVADA

One aromatic bloom which is surprisingly not used in worship rituals for Lord Shiva, as per the *Shiva Purana* is the *ketaki.* According to a Puranic legend, it was cursed by Lord Shiva for bearing a false witness of Lord Brahma. It is sometimes, however, offered in prayer to goddesses.

NEEL KAMAL

It is also known as the Krishna *kamal* for the Dark Lord's complexion is compared to this blue-hued water lily. A touching story from the epic *Ramayana* recounts that before fighting the demon Ravana, Lord Rama sought the blessings of Devi Durga by seeking to make an offering of a 100 of these blue flowers. Failing to procure a 100 and only having 99, he sacrificed one of his eyes, said to resemble a blue lotus, as well. This was immensely pleasing to the goddess, who blessed Lord Rama.

AKUND

Also known as the crown flower, it has a sweet, agreeable fragrance. Sacred to Shiva, it is one of the five love-darts of Kamadeva, the God of love. It also possesses medicinal properties.

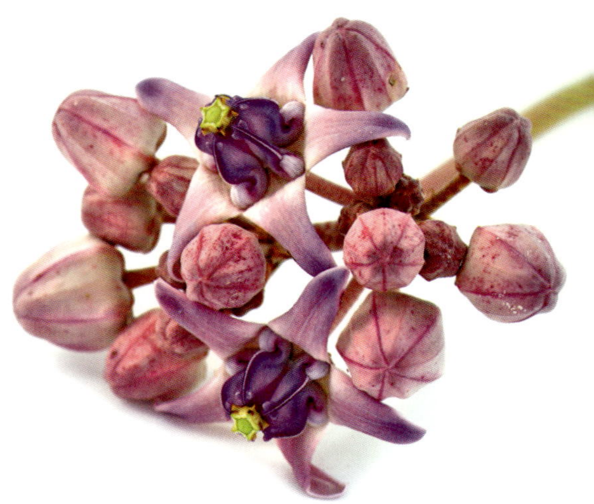

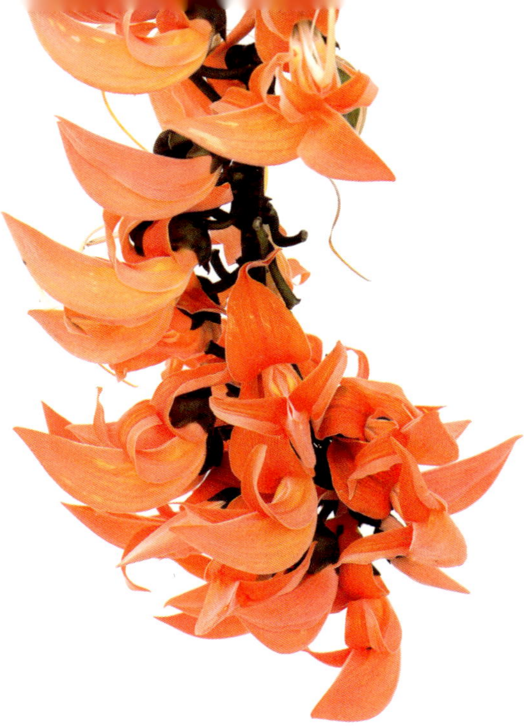

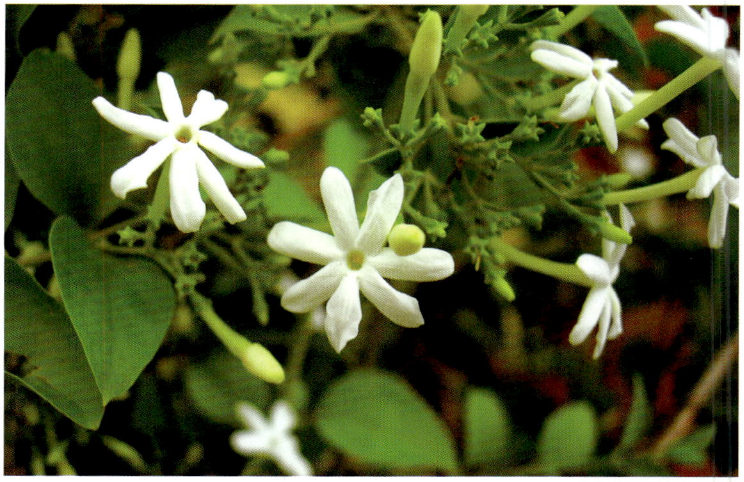

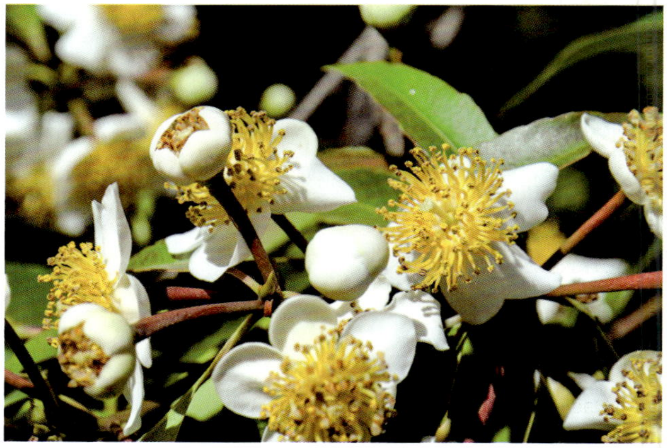

Top left: Palash flowers

Top right: Yuthika flowers

Above: Punnag flowers

PALASH

The Flame of the Forest is said to resemble the parrot's beak. This is why in Sanskrit it is referred to as the *kimshuka* meaning, 'Are you a bird?' The twig of the *palash* tree is also held by young Brahman boys during their thread ceremonies. It is also used to feed the sacred flames during the *havan* ceremonies.

PUNNAG

The *punnag* serves a dual purpose; it is employed both in worship as well as in *sringara*/ adornment. When *punnag* flowers blossom they are believed to indicate abundance. There is a myth that relates that Parvati also worships the *linga* under a *Punnag* tree in the form of a peahen after being banished by Shiva when she became distracted by a beautiful peacock.

JABAKUSUM

The resplendent red hibiscus flower or China rose is offered to the goddess during Durga pooja. The flame-coloured flower, believed to resemble the rising sun, is said to be a match for the ruddy complexion of the fierce goddess.

YUTHIKA

The *yuthika* is a type of jasmine vine. Many songs in praise of Lord Krishna employ the *yuthika* as a reference. It holds especial import in the worship rituals and everyday lives of the people of South India. It is used to make offerings to the Lord Vishnu and is also worn in the hair of Goddess Lalithambika.

BANDHOOK

The midday flower has an alluring radiant orange-red colour. In the *Suraya Ashtakam*, the complexion of the sun god is alluded to be 'as red as the *bandhook* flower.'

KOVIDARA

In the *Ramayana*, Bharat's chariot's standard bore the ensign of the *kovidara*, or orchid. Another name for the bloom is the blood flower or *raktapushpa*.

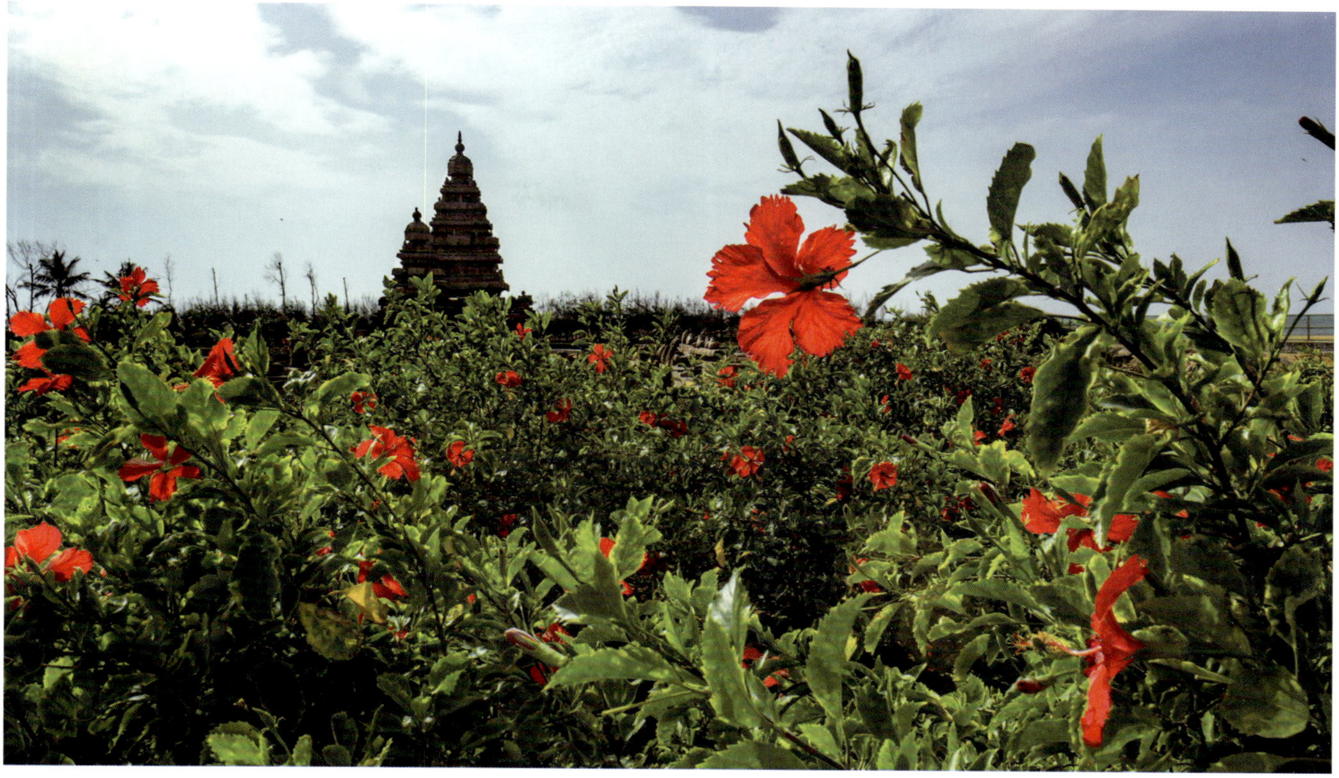

View of the shore temple, Mahabalipuram, Tamil Nadu, from its garden with hibiscus flowers

THE ACT OF WORSHIP

In Tamil Nadu, each temple had its own *nandavanam* or flower garden where the kings had appointed people—*sattharars*—to make garlands for the deities. Only the garlands made by the brahmachari or unmarried *sattharars* could be offered to the deities. Other traditional rules had to be adhered to—the flowers had to be plucked before sunrise, by the freshly bathed *sattharars*; any flowers which had fallen on the ground could not be used; the flowers could not be smelt by anyone and the garlands were to be made while chanting the holy names associated with the deities.

In Hinduism the flowers are an important part of the offering to the deities. The word *puja* (ritual worship) is literally translated as the 'flower act of worship'. Flowers are used while performing *puja* at home, in the temple and at other sacred places. In fact, rarely do we come across a *puja* in the Hindu religion without flowers. The idol or image of the deity in the temple and home is decorated with garlands of flowers and leaves. Flowers are also placed at the feet of the deity and the idol is showered with flowers. The word *puja,* meaning worship to the God, is the attainment of the rituals (*puryante sarva karmani*) and realisation of the divine powers (*jayate janam atmani*). The act of offering flowers to the gods and goddesses during the *upachara* signify complete submission of the gratifying self to the ultimate Lord.

Flowers represent nature or *prakriti.* By looking at flowers, we are directly connected to nature and feel contentment and compassion, which leads to reduced anxiety and worry. Flowers are also a symbol of our inner goodness, which we offer to God during prayer and worship. In addition, they are a gift that almost anyone can give to God, especially if the devotee is not rich enough to offer money or gems. In another sense, the symbolic offerings of a leaf, flower, fruit and water, correspond to the human body, heart, mind and tears of bliss.

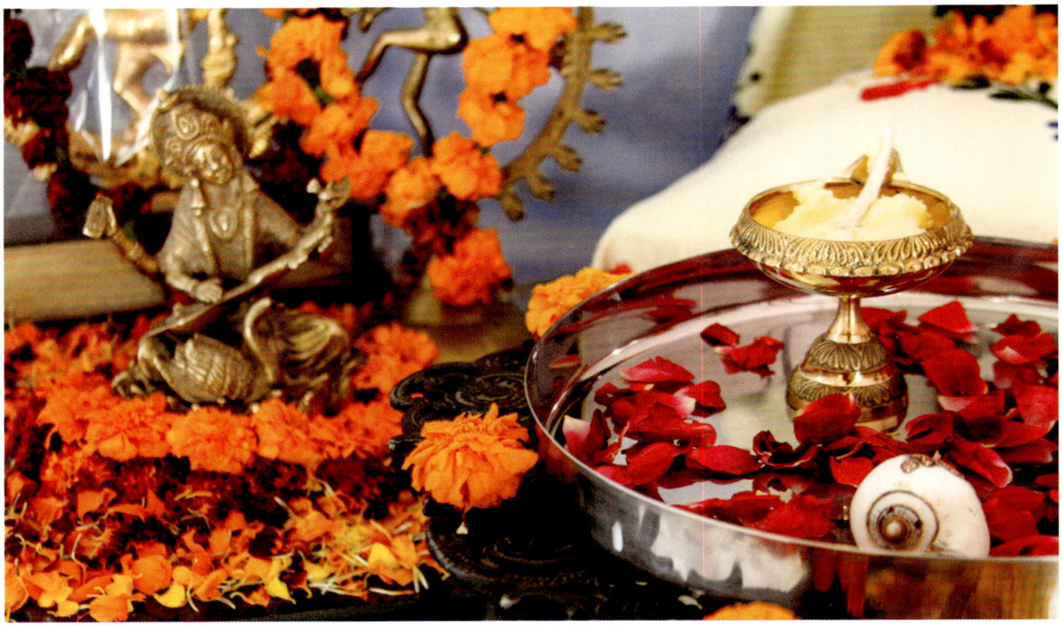

Left: Incense and floral offerings are placed before Goddess Saraswati and Lord Shiva

The flower is an important part of the *Panchopachaar Pooja* or worshipping the deity with five objects in a dish: *jal* (water) symbolising compassion; *akshat* (rice) representing prosperity; *pushp* (flowers); *dhoop-deep* (incense and lighted lamp) which spreads the light of self-effacement; *naivedya* (offering to God) conveying the sweetness and gentleness of nature.

ANCIENT TRADITIONS

The 'Anushasana parva' of the *Mahabharata* (Section XCVIII) describes the various flowers fit for offering to deities. Some are wild, while some grow in the midst of human habitation. Some flowers are from plants grown on mountains; some are from trees that are not prickly. Fragrance, beauty and taste are also the reasons for their classification. The flowers with pleasing scents should be offered to deities. If seeking wisdom, one should offer garlands of aquatic flowers like the lotus. Flowers grown in a garden (*aarama*) are supposed to be the best; flowers from the forest (*vana*) are ranked next; and flowers that are bought with money (*vikrita*) are at the lowest level. Details about the types of flowers that can be used for puja and the method in which they are to be used are mentioned in the scriptures. Some flowers are associated with a particular deity.

In the texts of antiquity, flowers are also classified according to the *guna*s (virtues or merits) they represent. For example, the *sattvika* (superior) type of flowers have mild colours and mild and sweet fragrances. These include *svetarka* (white calotropis), *nandyavartha* (*Leucas linifolia*), *sveta-padma* (white lotus), *jati* (jasmine), *mandaara* (coral tree, *Erythrina indica*), *maalathi* (*Jasminum grandiflorum*), *punnaga* (*Rottleria tinctoria*), *kurava*, and *sara-patrika* (*Anethum sowa*, a kind of rose).

The *rajasa* (middling) flowers are of slightly stronger colour and fragrance. These include flowers like *asbrhati* (solanum), *datura* (white thorn apple, *Datura alba*), *raktotpala* (red lotus), *palasa* (*Butea frondosa*), and *patala* (trumpet flowers).

The variety of flowers may not be used in daily worship in temples but may be used on festival occasions, say for decorating the temple chariot or the image carried in procession in public. The blossoms of the cotton plant (*karpasa*), *ketaki* (*Pandanua odoratissimus*), *kasa* grass (*Saccharum spontaneum*), *jaba* (China rose), *gambar* (*Gmelina arborea,* found in Eastern India), *vibhuthika* (*Terminalia bellerica*) and *bandhuka* (*Phoeniciea*) form part of this category.

Apart from these flowers, certain leaves i.e. *tulasi* (basil), *bilva* (bel) and the *dhurva* grass are used in daily worshipping. While decorating the image of the God being worshipped, one part of it must be covered with flowers, and other part with cotton or silk apparel, while the remaining part should be covered with ornaments and jewels and one part must be left clean for worshipping (*puja bhaga*). Colourful garlands are hung around the neck of the deity, reaching the ankles. Unstrung flowers are placed at the feet of the deity; the ritual being called *pushpa nyasa.*

Often a lotus is used as the locket of the garland or in between the garland of other flowers. In West Bengal, devotees offer garlands of red hibiscus to Goddess Kali and the same is worn by the tantric sages who worship this particular goddess.

Below: A flower garland adorns a Hindu deity

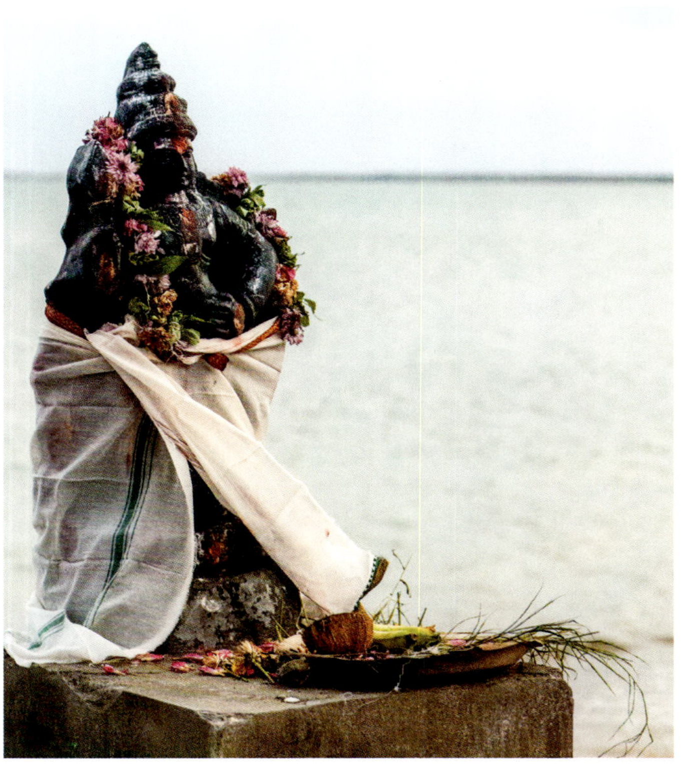

THE GOD'S FAVOURITE

The deities in Hinduism have their own favourite flowers and leaves which are used during worship and *pujas* performed at the Hindu festivals. All festivals and rituals in Hinduism begin with the Ganesh Puja. Red coloured flowers are the favourite of Ganesha. The gudahal flower (red hibiscus) is especially dear to Ganesha and *durva* grass is also an important offering to him. Another important flower offered is the milkweed flower, known as *Arka* in Hindi and *Eruku* or *Erukkum Poo* and *Jilledi* in Southern India. Tulsi leaves and flowers are not generally used in Ganesh Puja, but but are used during Vinayaka Chaturthi. Ganesh Patri are the numerous types of sacred leaves that are offered during Ganesh Puja. Interestingly, most of the *patri*s that are used in Ganesh Puja also have medicinal properties.

The tulsi flower (*manjari*) is special to Vishnu. If this flower is offered to him during worship, no other flower is needed. Tulsi leaves are used as *naivedya* (offering to God) offered to Vishnu. That is why all Vaishnavites usually have a tulsi plant in their house, as they have to use it at least twice a day.

The datura flower, which is a poisonous plant, is sacred to Shiva as the lord of life and death, and is specially offered to him during worship. This flower is also frequently used in temple architecture in south India. It is a recurring motif in brackets that support the roof of a temple *mandapam*, or balconies and other structural details.

Lakshmi Puja is one of the most important pujas performed by Hindus. It is performed to seek the blessings of Lakshmi—the goddess of wealth and prosperity. Lakshmi Puja is mainly performed during Deepavali. The Goddess Lakshmi is associated with the

MAJOR DEITIES AND THEIR RITUAL OFFERINGS

SHIVA
Flowers Purple and white dhatura; akund; nagalingam; thumpa; erikku; banyan; bilva; mandarai; rajnigandha
Leaves Koovalam/bel; amalaki

VISHNU
Flowers Tulsi; mallika; mogra; pichi; marigold; punnaga; yellow flowers; lotus
Leaves Ashwatha; amalaki

BRAHMA
Flowers Lotus

GANESH
Flowers Pink kanare; red hibiscus; milkweed; lotus; bandhook; sanku pushpam; pomegranate; ketaki; screw pine. Tulsi only during Ganesh Puja or Vinayaka Chaturth
Leaves Pomegranate; screw pine. Tulsi only during Ganesh Puja or Vinayaka Chaturthi
Grass Durva

KRISHNA
Flowers Atasi; tulsi; lotus; champaka; kadamba; nandhyavattai; vakul or maulasari; yuthika

HANUMAN
Flowers Akra; chameli

SURYA
Flowers Bandhook; Jabakusum
Tree Plantain

DEVI
Flowers Red hibiscus; China rose; red flowers
Leaves Neem

LAKSHMI
Flowers Lotus; marigold; screw pine; red jasmine; thazhampoo or ketaki
Grass Durva
Fruit Koovalam

SARASVATI
Flowers White lotus; yellow champaka

DURGA
Flowers Aparajita; hibiscus; jasmine; kadamba; neel kamal

SITA
Flowers Ashok

lotus flower. *Durva* grass is also an important offering to Lakshmi.

Tulsi leaves or flowers are the most important leaves and flowers used during pujas and prayers dedicated to Krishna. In most temples tulsi is also offered as prasad. The dark variety of the plant is also known as Krishna Tulsi. It is called holy basil in English and in most Indian languages it is known as tulsi or thulasi. Usually the local flowers that are used in pujas and prayers are used in worshipping Lord Krishna. However few exceptions being the hibiscus, milkweed and angel's trumpets.

Facing page: Lakshmi, the Goddess of Wealth on a lotus flower, Raja Ravi Varma, oleograph, 1906

Below: Durva grass

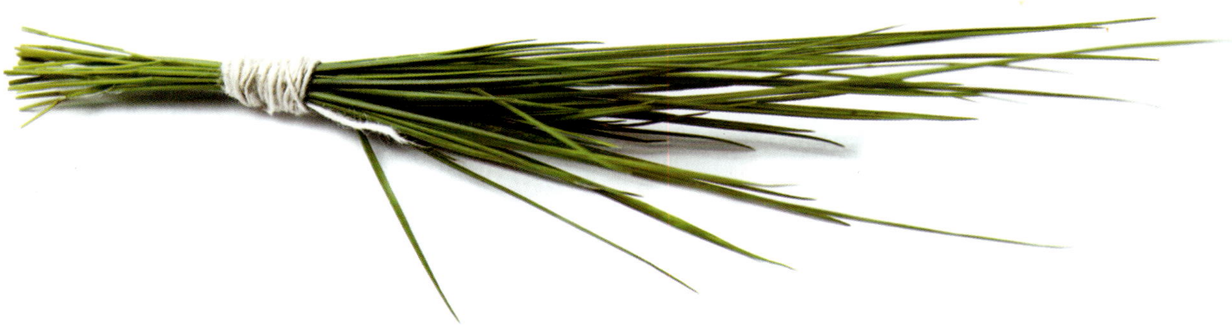

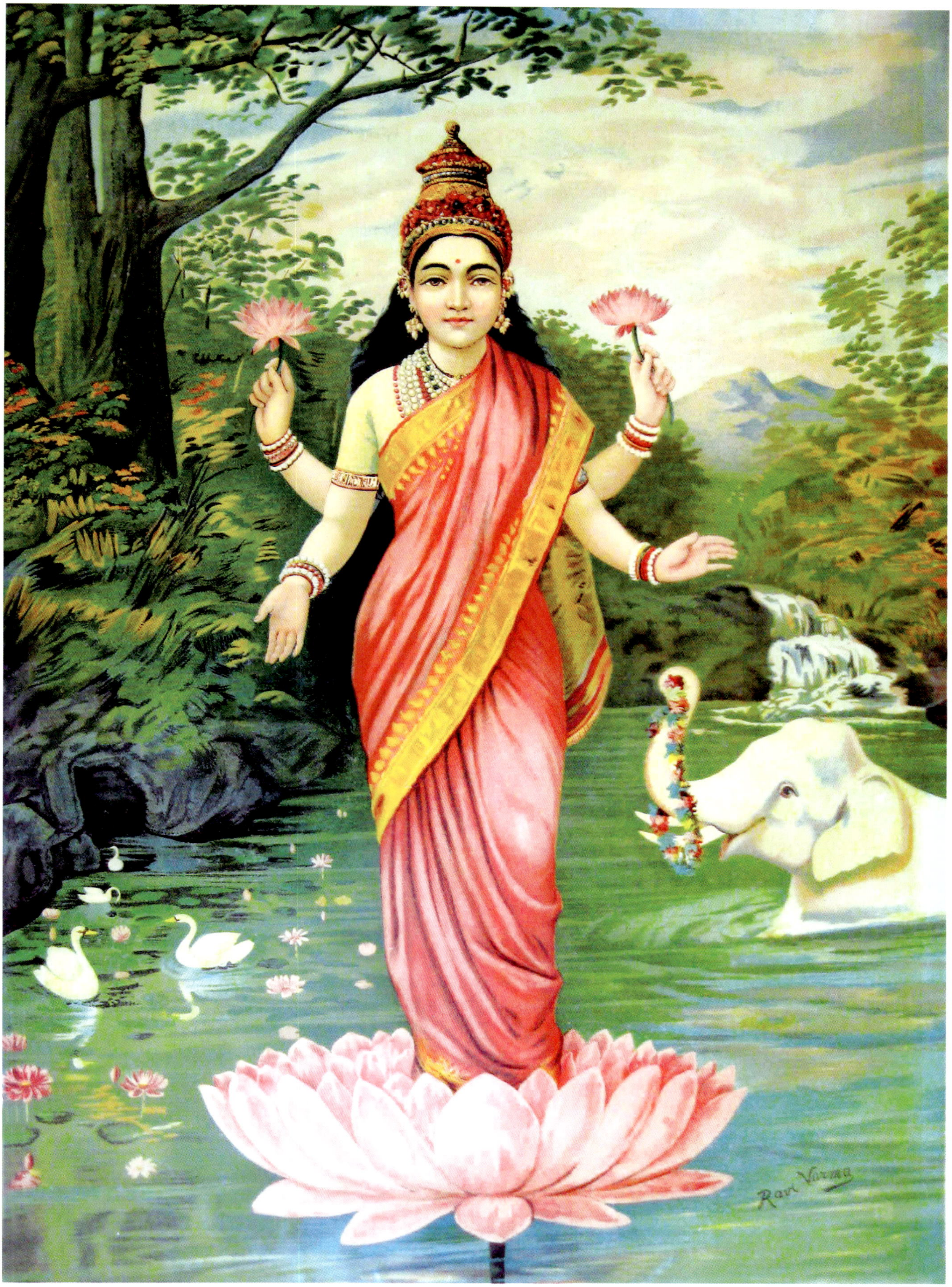
Ravi Varma

THE STORY OF MALATI

The malati flower is considered to be a special flower since the Goddess Parvati gave its seed to the Devatas in order to detach Vishnu from the sorrows of Vrinda (the daughter of the Asura Kalanemi, who is also known as Tulsi). The story goes like this: the demon-king Jalandhar is married to Tulsi. After the disappearance of the asura Jalandhar, Tulsi is worried about his whereabouts. Lord Vishnu takes the form of Tulsi's husband Jalandhar and starts living with her. Tulsi becomes suspicious of him as Vishnu does not possess any evil qualities unlike her husband Jalandhar. When she becomes aware of Vishnu's disguise, Tulsi curses him to turn into a stone and stay at her feet. Vishnu accepted her curse. However, he made Tulsi realise that she is not as pious as she considers herself to be: she never stops her husband Jalandhar from spreading evil around the earth. Soon realising her mistake, guilt-ridden Tulsi fell at the feet of Parvati as an act of penance.

After Tulsi's demise, Jalandhar's demon army attacked Lord Shiva. Shiva dipped his toe into the ocean to churn the water. From this churning, a weapon called the Sudarshan Chakra was created, which beheaded Jalandhar. Shiva sent Jalandhar's flesh and blood to the lord of death—Yamraj. However Jalandhar's soul united with Shiva. On the other side, Vishnu was saddened by the death of Tulsi. He took the ashes from the pyre and after applying them on his body, he started wandering fanatically in the universe. The obsession with which Vishnu was living, worried the rest of the Devtas and they went to Lord Shiva, and requested him to eliminate the false attachment from which Vishnu was suffering. Shiva sent Devata to Parvati, saying that she would certainly help in this regard. The deities went to Parvati and prayed to her. She was very pleased and with the assistance of Lakshmi and Saraswati gave some seeds to them. Devta sprayed those seeds on the pyre of Vrinda. Three holy plants manifested from the pyre—aamla, tulsi and malati. Tulsi and malati later attained the Vishnu Lok by virtue of their respective penance.

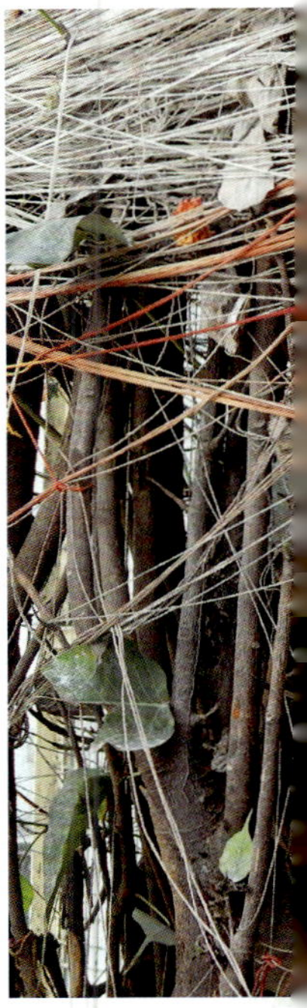

IN THE LAND OF LEGENDS

Shiva's body is always covered with ash and Shiva devotees wear an ash tilak on their forehead and hands. An interesting background story in the *Shiva Purana* narrates how Lord Shiva came to be synonymous with ash. There once lived a sage whose lineage was traced to the famous Saint Bhrigu. This sage performed deeply passionate austerities and became very powerful. He used to partake of only fruits and later only green leaves and thus got the name Parnada.

Sage Parnada continued his intense tapas and controlled all the animals and plants in the forest where he lived. Once while cutting some grass to repair his hermitage, the sage cut his middle finger. To his surprise, instead of blood, he saw the sap of a tree oozing from the wound. Now Parnada thought that he had become so pious that instead of blood, sap was flowing through his body. Pride filled his mind and he started shouting in joy that he was the most pious man in the world now.

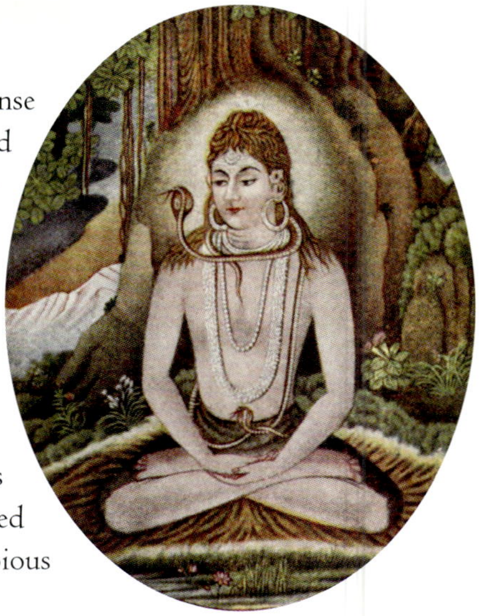

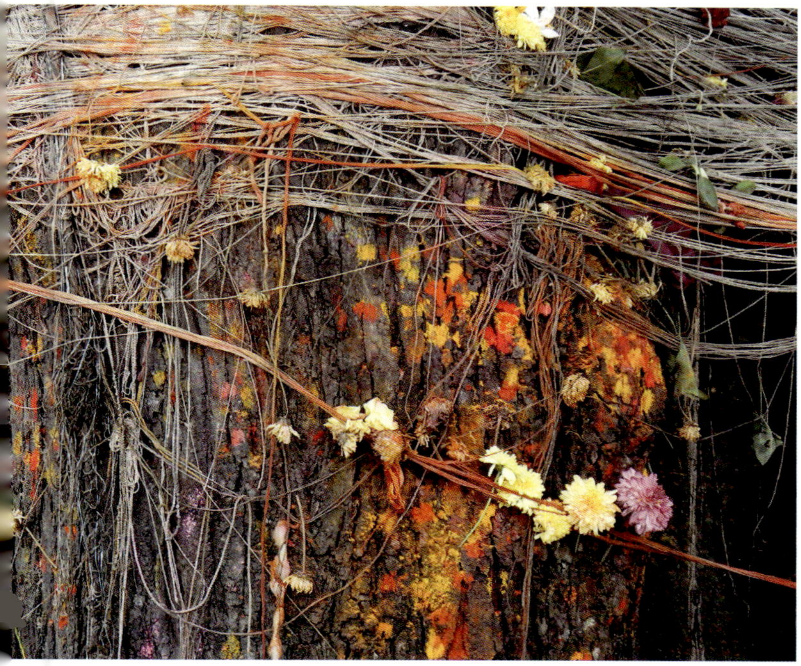

Above: A banyan tree with sacred threads tied around it, symbolising marital happiness and longevity

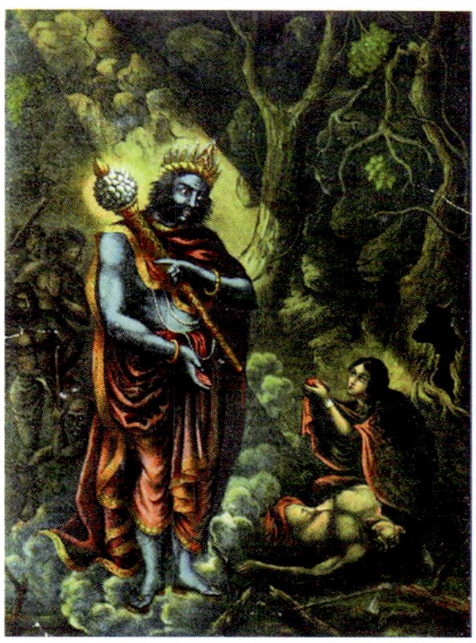

Right: Savitri's defeat of Yama, the God of death, colour lithograph, early 20th century

Facing page: Lord Shiva covered with ash

Shiva, who witnessed this event, took the guise of an old man and reached the spot. When the old man asked the reason for his uncontrollable delight, Parnada said that he had become the most pious man in the world as his blood had become similar to the sap of fruits and trees. The old man asked what there was to be so elated in this because it is just sap. However, when trees and plants are burned, they turn into ash. Only ash remains, so that is surely the higher stage. To demonstrate it, the old man sliced his finger and suddenly ash came out of it. Sage Parnada soon realised that it was Shiva standing before him and prayed to him for forgiveness for his ignorance. It is said that since then, Shiva wears ash to remind his devotees about the ultimate reality and foolishness of getting enamoured by physical beauty.

Cotton threads of various colours like red, yellow and white are tied around peepal (*Ficus religiosa*) tree trunks, especially in northern and western India. This ritual is performed especially on the Vat Savitri Puja day (May-June). During Vat Savitri Puja, the banyan or peepal tree symbolically represents Brahma, Vishnu and Shiva. The root of vat vriksha is Brahma, the stem is Vishnu and the upper part is Shiva.

The peepal tree represents the tree of life and is sacred in Hinduism as it supports life and is famous for its longevity. The peepal tree also has the property to purify air. The cotton thread is just the opposite. It represents the fragile nature of life, love, trust, faith—and all things that go on to make up a relationship. A single thread may be weak, but, when it is wound 108 times around the trunk, it becomes strong. It is no longer fragile and no longer easy to break. By walking around the tree 108 times, the wife contemplates on these matters – love can only be strengthened by trust, faith and the desire to make it work. With each step, the woman strengthens her relationship with her husband. She prays not just for her husband's long life, but an enduring relationship that will last beyond this life and into the next.

The banyan tree plays an important role in the famous story of Satyavan and Savitri. It is believed that Satyavan spent his last moments under a vat or banyan tree on the full moon day in the month of Jyeshtha. Yamraj, the Lord of death, appeared here and Savitri pleaded with him under the banyan tree. In memory of this event, women go around the banyan tree 108 times, tying threads and fasting for the health and longevity of their husbands.

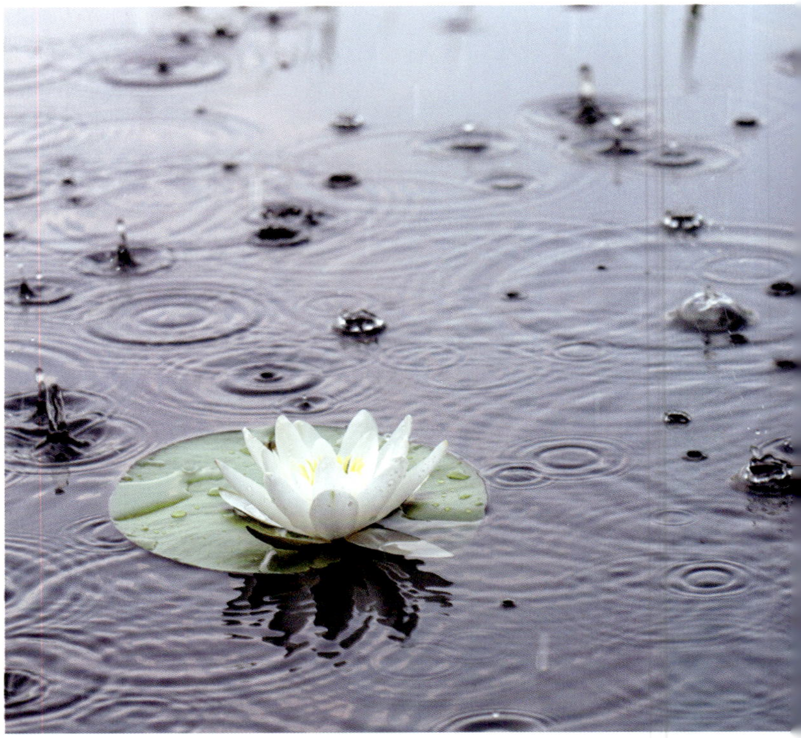

THE FLOWER SALUTATION

'Let there be a flower salutation to the one who is worshipped with a string of white lotuses, lotus buds, full-blown flowers, a bouquet of flowers of varied colours like the nine jewels and with leaves of gold, with nine red lotuses, to this one whose mount is the peacock.' While this *puspanjali* sloka is recited in Bharatanatyam, the dancer stands still, forming a cup with both hands containing imaginary flowers which she scatters at some particularly significant words. Both the words and the actions are very simple; at close study, though, they contain rich symbolism.

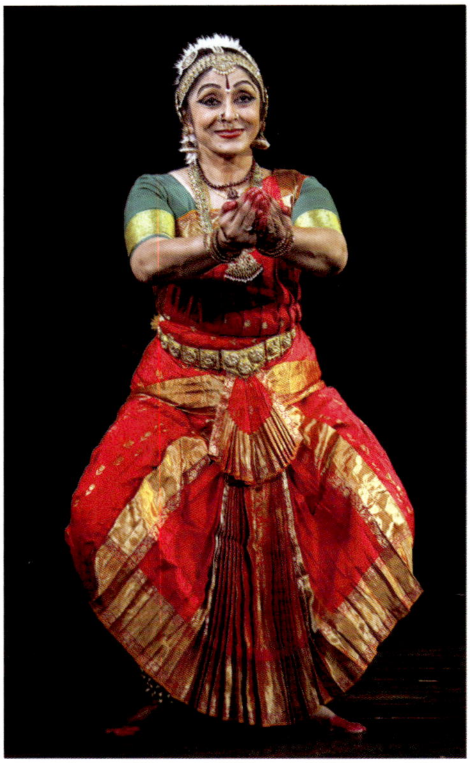

The flowers that are mentioned vary in colour: the kalhara flower is white (*Nymphea* or water lily), the bouquet of flowers shows nine colours, while the last flower, the hallaka, is the red lotus. The symbolism of the red and white flowers is more evident than that of those with the colours of the nine gems. The number nine is often connected with the worship of the Devi, and, generally, gems are considered to exert great influence on health, prosperity and wisdom and therefore they are often used in astrology.

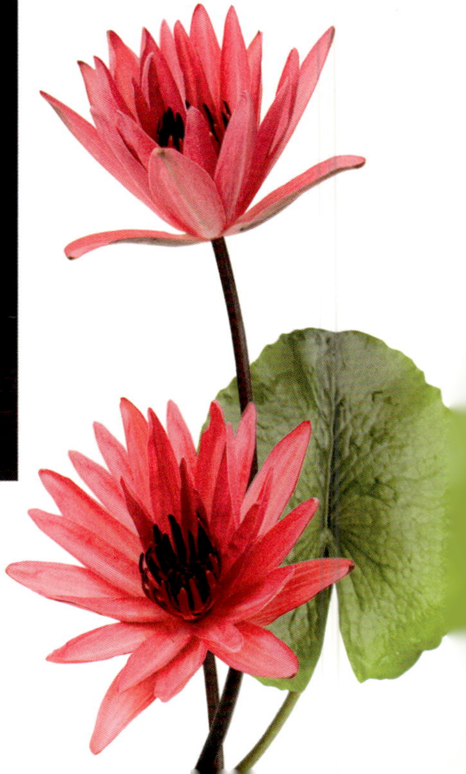

In the beautifully illustrated two-volume book *The Spiritual Significance of*

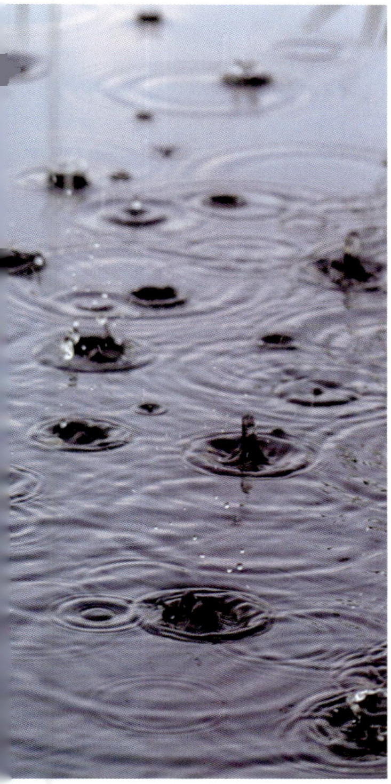

Flowers by the spiritual leader The Mother of Sri Aurobindo Ashram, the author has painstakingly elaborated on 898 flowers giving a deep spiritual meaning to each flower she has absorbed herself in. The book itself is an extremely immersive experience, where different ways of interpreting and understanding flowers which we usually look at through the concept of visual beauty alone. This book is indeed a special and iconic study shedding a special focus on deeper insights into the mere physical form of flowers. According to the Mother, 'each variety of flower has its own special quality and meaning. By establishing an inner contact with the flower, this meaning can be known. Flowers speak to us when we know how to listen to them,' the Mother said. 'It is a subtle and fragrant language.'[2]

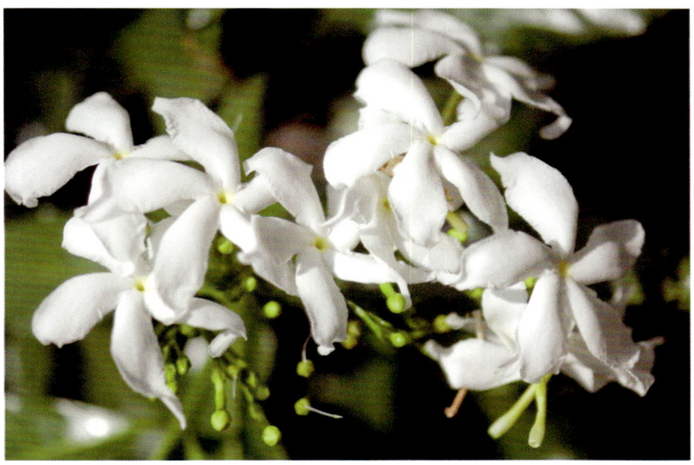

Clockwise from left:
Pink water lily

A bharatanatyam dancer poses, offering flowers to the deity

Kalhara flower—white water lily

White jasmine buds symbolise coolness and peace

COLOUR CODED

Brenda E.F. Beck, who has collected folktales across India, describes the importance of the choice of coloured flowers in South Indian temple ritual:

> Colour in Indian ritual derives primarily from underlying assumptions about the nature of heat. In essence, heat is associated with life and fertility. The energy which can both activate and nullify life is a kind of heat. The heat when taken alone, however, can be highly dangerous. It must therefore be focused and controlled in order to become a source of power which humans and super-humans can utilise. In ritual, as in mythology, heat must be encompassed or surrounded by cooling things. Red substances symbolise 'heated' states and white substances 'cooled' ones. In keeping with these larger notions the red used in South Indian ritual normally occurs against a background of white.[3]

This colour-cum-flower symbolism is used in divination too: in this context the white flower indicates an abating process, whereas the red flower indicates the progress and development of a state. In the case of the *pushpanjali shloka*, we see that the kalhara flower is employed to extinguish evil or eruptive influences. Hereafter, a new procreative process is indicated by the use of the red hallaka flower.

OTHER FAITHS AND BELIEFS

The use of flowers in Hinduism greatly influenced Buddhism and Islam. However, the Buddha refused the use of garlands for himself and for monks because cutting a bloom destroys a living thing. There is an ambivalence towards flowers in Buddhism because of their association with luxury and because they are potential temptations or distractions. In Jainism, flowers in garlands should be tied together, rather than sewn together. The best flowers to offer are those that have not been picked but have fallen from the tree or plant.

In many parts of Asia, the production and use of flowers was more elaborate than elsewhere because of an intensive agriculture that supported courts and temples. Flowers were offered in religious rites instead of slaughtering animals because it was a non-violent symbolic worship of the gods.

THE LAST JOURNEY

Upon the death of an individual, the family will call a funeral home to prepare the body as well as a priest to perform the last rites. In India, a funeral is generally performed before the sun goes down on the day of passing. Last rites are never performed in a temple. A priest recites mantras, final rituals are performed which includes a *havan,* eulogies are narrated and family and friends are given an opportunity to offer their last respect to the deceased with flowers and petals. Largely people express their respect for the dead and sympathy for the bereaved by sending flowers. This longstanding custom has served people as a means through which they can express their feelings.

Colours, shapes, sizes and the range of fragrances of flowers add colours and aromas to the rituals of life practised in India. From birth to death, be it ritual or decoration, perfumes or adornment, prayer or atonement, there is a flower for every occasion.

Our rivers, particularly those associated with temples and the divine, are flooded with the remains of prayer and ritual. Often these flowers would rot and pollute the river, but

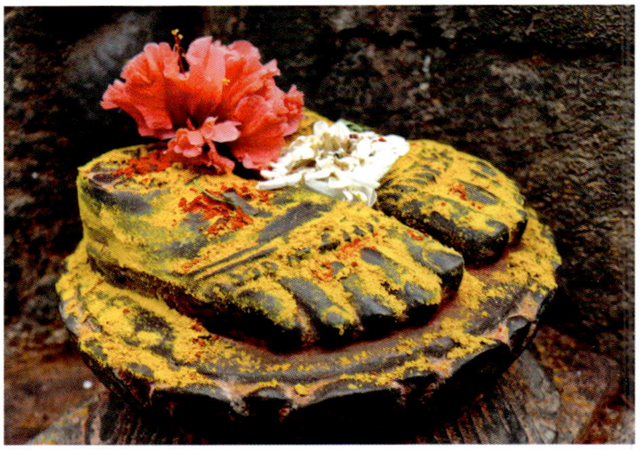

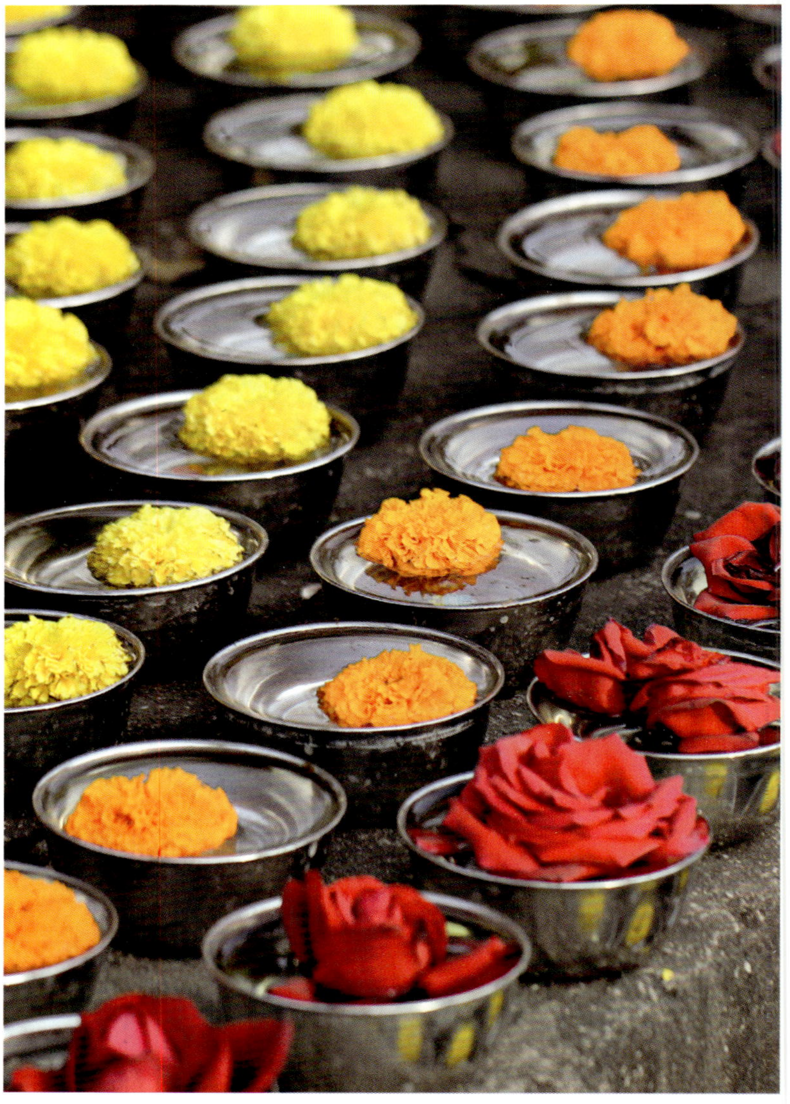

Facing page above: Hibiscus flower on Guru's Feet at Chamundi Hill, Mysore

Facing page below: Rows of containers with floral offerings at an Indian temple

Below: Floral tributes are placed on the funeral pyre of a deceased as part of the last rites in Hinduism

Bottom: The founders of HelpusGreen and their 'Flowercyclers' are making a positive difference

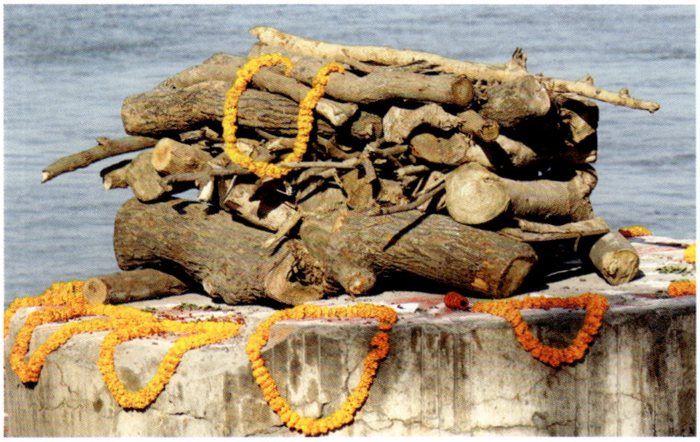

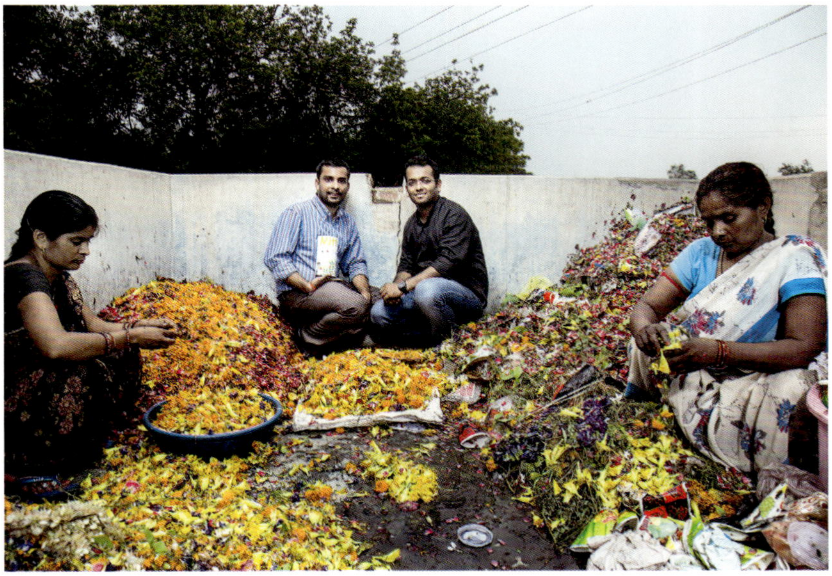

in contemporary India, organisations like HelpUsGreen have found alternate ways of extracting the very best from these by-products of rituals and worship, by transforming floral waste into incense cones and incense sticks, which can further be used in prayer and devotion.

ENDNOTES

1. Joseph Campbell, *The Power of Myth* (New York: Anchor Books, 1991), p. 206.
2. The Mother, *The Spiritual Significance of Flowers, Part I and II* (Pondicherry: Sri Aurobindo Ashram Publications Department, 2000). p. 36.
3. Brenda E.F. Beck. Peter J. Claus, Praphulladatta Goswami, and Jawaharlal Handoo, ed., *Folktales of India* (New Delhi: Motilal Banarsidass, 1989), p. 42.

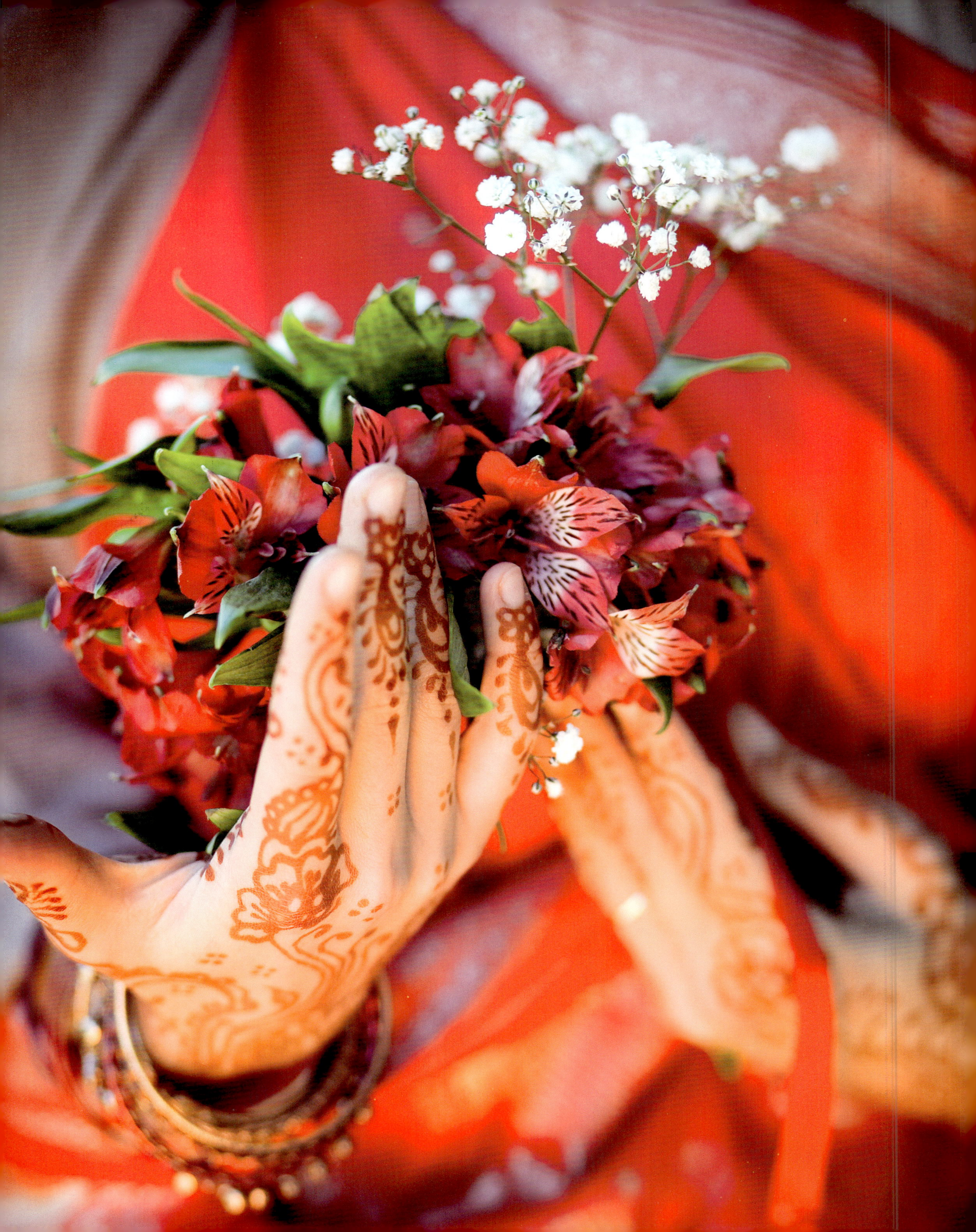

The Body Adorned

Little flower – but if I could understand what you are,
root and all, and all in all,
I should know what God and man is.
—Alfred, Lord Tennyson

Indian women and flowers are inseparable. In the Indian ethos, the human body is the temple and flowers add nature's divinity to it. Just as flowers are offered to the deity in ritualistic worship, so are they used as adornment to make a human appear divine. At the same time, one of the most sensual adornments of the human body can be with flowers. It uplifts the wearer's spirit and inner beauty, as well as heightens physical appeal.

The *Thiruvilayadal Puranam* (*The Sacred Sports of Shiva*) is a 16th century collection of stories about the activities of Lord Shiva when he appears on earth in different guises, in order to test his devotees. Written by Paranjothi Munivar, a Tamil saint, some of the stories mention the fragrance emanating from Parvati's hair. The stories indicate that it is an age-old custom for women to decorate their hair with flowers. In the ancient southern tradition, hair is said to be an erotic part of a woman, and decorating the tresses with flowers is a ritualistic act. Kalidasa in his *Meghaduta,* mentions that during the monsoon season, the women of Alakapuri embellish their hair with *kadamba* flowers, wear *sirisha* flowers in their ears and carry pink lotuses in their hands. In *Abhigyanashakuntalam,* Shakuntala's constant companion during her lovelorn state is the lotus.

Facing page: A bride, her hands decorated with floral mehendi designs, holds a flower bouquet

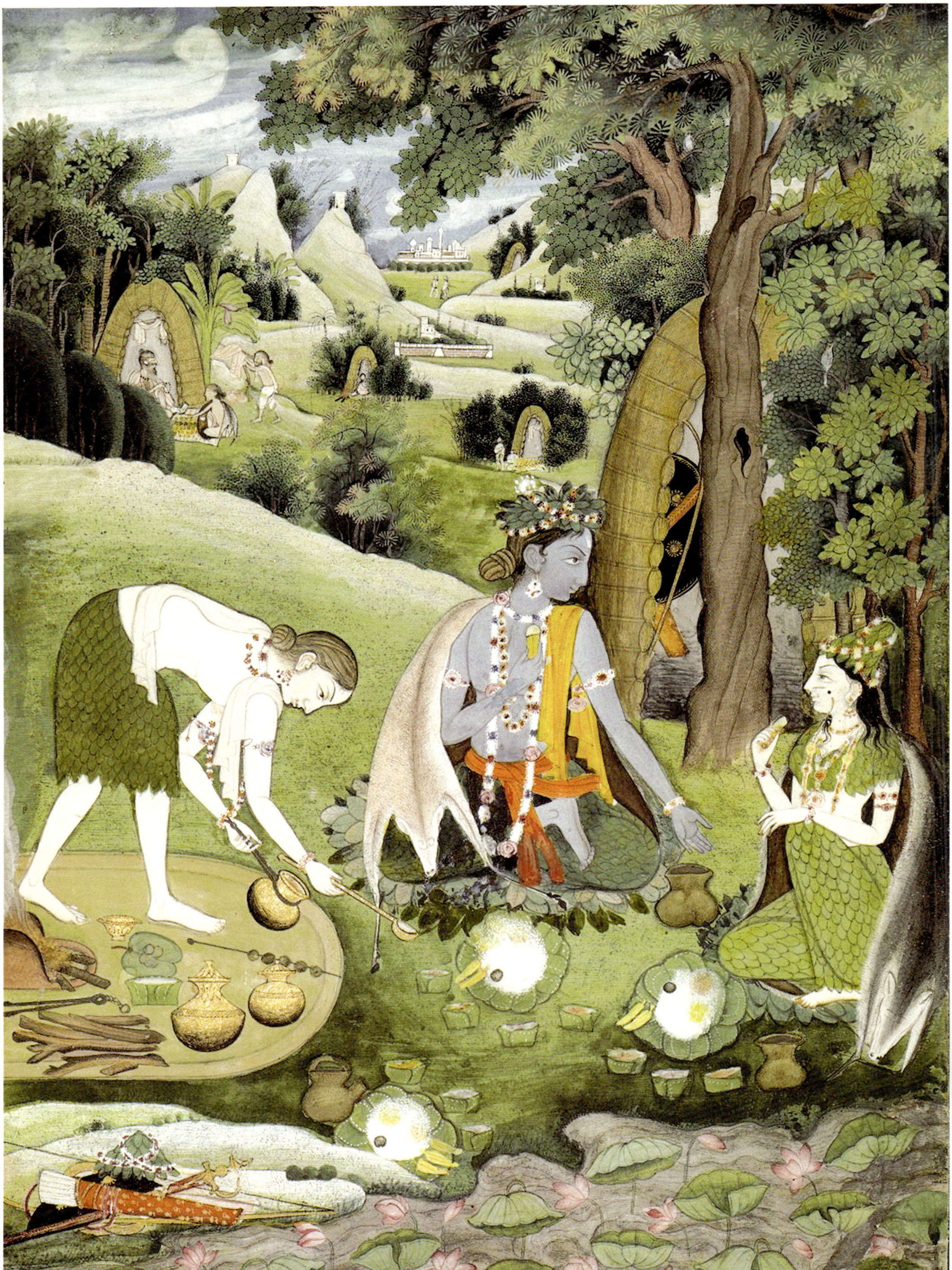

Facing page: Rama, Lakshmana and Sita in the wilderness, adorned with floral ornaments, gouache with gold on paper, c. 1815-1825, coll. Museum of Fine Arts, Houston, USA

Below: Shakuntala patralekhan (Shakuntala writes a letter to Dushyanta), Raja Ravi Varma School, oil on canvas, 1881

Suffering from the pangs of separation due to the absence of her royal lover, she wanders in a garden fragrant with the sweet scent of lotus flowers. She wears bracelets of lotus stalks and writes a letter on lotus leaves.

SHRINGARA -A SACRED DUTY

The floral garland is possibly one of the oldest methods of using flowers for decoration. In our epics *Mahabharata* and *Ramayana*, we read that the princess proceeds to the court for her *swyamvar* (to choose a husband of her liking from among the princes assembled) with a garland in her hand. Even today, the tradition of placing a garland round the neck on special functions, such as marriage and welcome ceremonies, is prevalent. In West Bengal, a floral crown is also used in ceremonies, especially during *annaprasana*, when the new-born takes his or her first grains of cooked rice at the age of around 6 months.

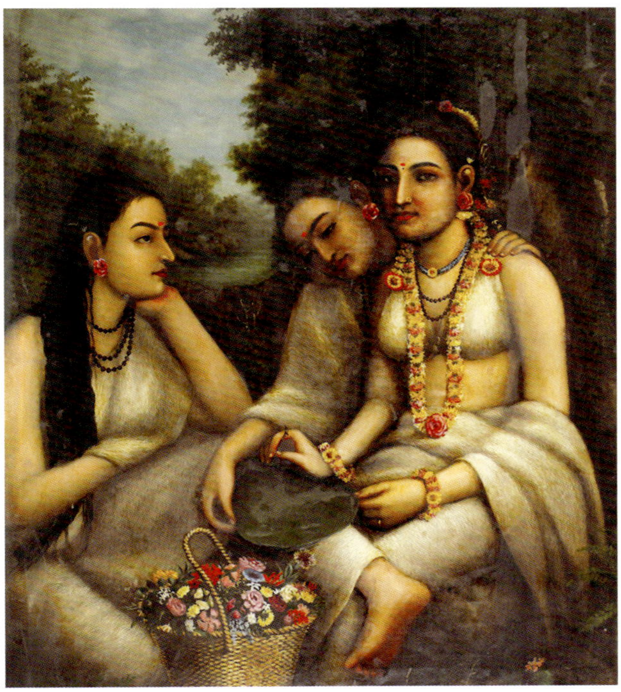

Another reference of using flowers as body adornment is found in the Dharmashatras, which exhorted Hindu women who were fertile to dress themselves with flowers, cosmetics and jewels. The purpose was to make themselves as alluring as the bright and fragrant flowers that attracted bees. Thus, adornment (or *shringara*) becomes a sacred duty, rather than just a feminine whim. It is mentioned in the *Ramayana*, when Sita is getting ready to follow her husband into exile in the forest and she begins removing her royal finery. She knows that as the wife of a hermit in the forest, she will have to wear clothes made from bark. However, the women of the royal household stop her and tell her:

> It is unbecoming for a married woman to remove the jewels that adorn her body. Even if circumstances have forced your husband to become a hermit, you must continue to wear flowers, jewellery, cosmetics and bright clothes.[1]

When Sita follows Rama into the forest, she is given enough clothes and jewels to sustain her during the 14 years of exile. Once she is in the forest, she meets Anasuya, wife of the sage Atri, who gives her similar advice. Anasuya also gives Sita magical clothes that were never tainted by use.

The *Natya Shastra*, an ancient treatise on Indian dramaturgy, prescribes various parameters for adornment of the dancer. It lays out the different kinds of ornaments for every part of the body, stating that the flower garland is an ornament for the entire form. It classifies the garlands into five types which are, encircling (*vestita*), spread-up (*vitata*), grouped (*samghaya*), tied-round (*granthima*), and hung down (*pralambita*).[2] Garlands are also among a number of other ornaments that are meant to be judiciously used as part of the costume to give rise to the proper emotions during a performance.

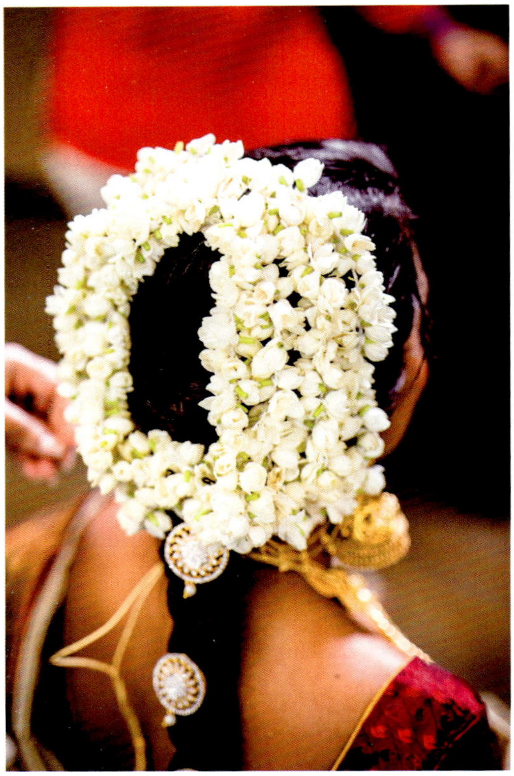

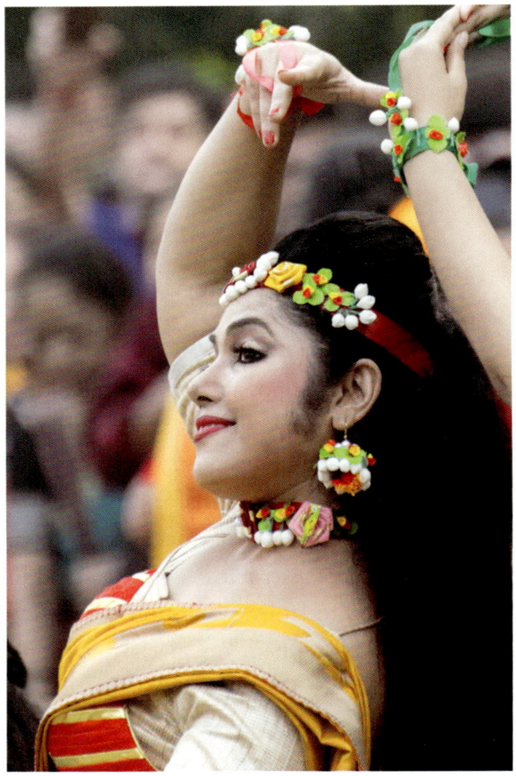

Far left: Indian bride adorned with flowers on her wedding day

Left: A young woman wearing floral jewellery, dances at a spring festival in Kolkata

Below: Ancient Greek statue of a woman wearing a flower tiara

Even adornment and embellishment in traditional Indian culture is not bereft of myths. Hair adornment with flowers carries a special significance in the Indian tradition. A woman's braid, according to ancient mythology, symbolises the three sacred rivers of India—Ganga, Yamuna and Saraswati. It also carries the significance of the holy trinity of Brahma, Vishnu and Shiva. There is yet another theory that one strand represents the bride's father's house, the second stands for her in-laws' house and the third represents the bride herself who unites both these families. Ancient Hindu mythological stories have many references to *malati* in the braids of women, and overhanging plant holders under which lovers met. This flower is favourite of the gods (especially Vishnu) and is used as decoration in the hair of goddesses as well as maidens.

ETERNAL CHARM

Flowers have been worn for personal adornment throughout recorded history. Their purpose has been for both beautification and ceremony. Earliest uses in ancient Greece varied from a single flower worn in the hair as a personal touch, to crowns of flowers and foliage for the head of both men and women as a symbol of achievement. Sometimes the wearing of flowers was in the belief that the fragrance would help ward off disease. Today the emphasis is more on personal beauty, as the flowers enhance the appearance of the wearer and the apparel, with the bonus of elevating the inner beauty of the spirit.

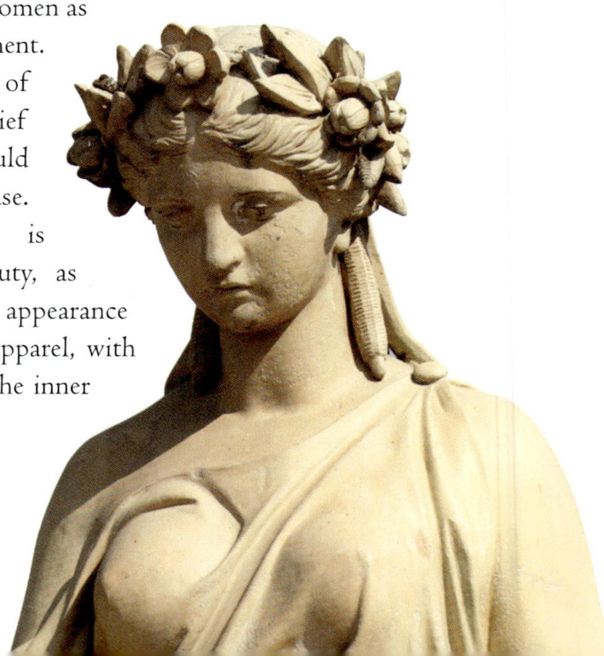

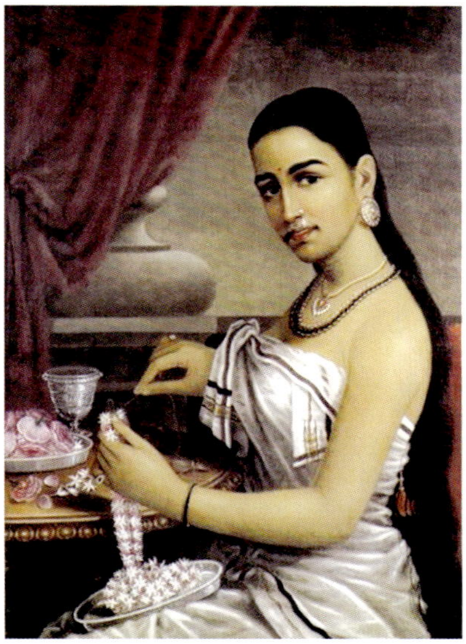

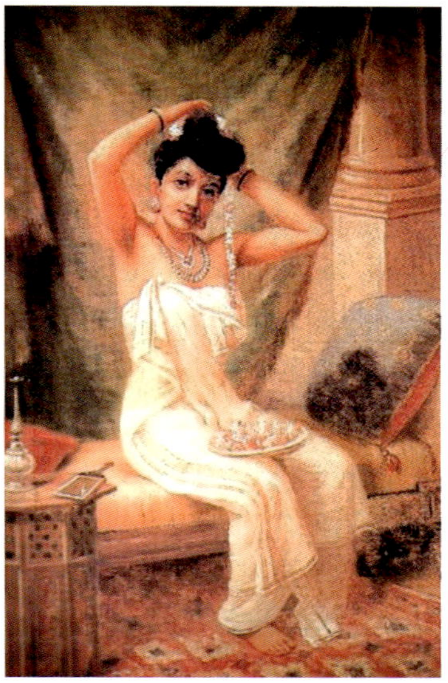

Above: Lady with a flower garland, Raja Ravi Varma, oil on canvas, c. late 1800s

Above right: Nair woman adorning her hair with a garland of jasmine, Raja Ravi Varma, oil on canvas, 1873

When a woman wears flowers in her hair, it is believed that it will bring happiness to her household. The sweet-scented flowers are more preferred for making a garland. The most sought-after flowers are jasmine and tuberose (*rajnigandha*), though other flowers such as marigold, chrysanthemum and the rose are also used. Garlands often have a pendant or locket which is also made of flowers of contrasting colour, and is generally given a heart shape. Besides garlands, the other important floral ornaments are bangles and floral crowns, which can also be made from jasmine or tuberoses.

Flower garlands are also used as decoration by Indian dancers, especially women. A floral crown may also be worn by a female character such as a queen or a princess in traditional dance-dramas. At first a crown-shaped structure of cardboard is made, over which the flowers are sewn or glued. Fine wire strings may be used to make the structure and the flowers may be sewn inside the wires. In the countryside, the frame is made of a large leathery leaf, for instance jackfruit or *kadamba*, on which the flowers are sewn with the help of a cotton thread and needle. Earrings and *bajubandhs* (upper arm bands), made of flowers, are also used in ceremonial dances and other functions. It is quite common to see photographs or paintings of the epic heroes and heroines such as Rama and Sita wearing floral ornaments such as garlands, bangles and *bajubandhs*.

Though contemporary women may not put *sirisha* flowers in their ears, the tradition of adorning one's hair with flowers still continues. Here again, the fragrant flower gets preference over other flowers in hair adornment, with a few exceptions, such as *kanakambaram* and *vajradanti*. The practice may be to put one or more flowers singly in the plait or to make a mini garland (*gajra*) and dangle this from the hairdo or to wrap around the hairdo. The chain may be made out of a single fragrant species of flowers such as jasmine, tuberose, or *champaka* or a mixture of flowers and decorative foliage. The *gajra* is purely ornamental and meant to decorate a hairstyle; it is not used to hold a bun in place.

THE LANGUAGE OF FLOWERS

Every culture in India has attached a different meaning to the flowers worn by women. In India, a girl is supposed to wear flowers in her hair because it brings happiness to the family and prosperity to all the members. It is the symbol that Lakshmi is residing in the house and that wealth will never leave the house. Wearing flowers, therefore, becomes not only a way to beautify oneself, but also a mark of the well-being of the family.

JASMINE Known as the queen of flowers because of its breathtaking fragrance, it is also named as God's own flower as no festival is deemed complete without using jasmine flowers.

ROSE The rose is a symbol of love and passion. It is also a mark of mourning for loved ones. So, a girl wearing a rose in her hair may symbolise her passion in life or the memory of a lost one.

CHRYSANTHEMUMS These colourful flowers signify happiness. A girl wearing chrysanthemums in her hair is said to bring happiness to her family.

HIBISCUS A symbol of strength since it is used in the worship of Kali and other forms of Shakti.

VENI is widely used in South India to decorate the long plait of hair (*veni*) during special ceremonies, such as marriage or at the time of a Bharatanatyam dance recital. This special flower arrangement is about 90 cm long (depending upon the length of the plait) and 7-10 cm broad. The braid-like structure may be made over a frame of cardboard or some leathery and tough leaves, over which different flowers are arranged decoratively so as to completely cover the background frame. Several decorative materials, such as tinsel thread, may be used to make the *veni* more gorgeous.

RITES OF PASSAGE

In the southern state of Tamil Nadu, when a girl reaches puberty, she is dressed in a half-sari and adorned with a *mallipoo jade*. This is a floral adornment comprising of jasmine buds affixed to a length of cardboard. It is meant to run along the back of the head down to the end

Facing page above and below: A child undergoes the Hindu religious ceremony called *mundan*, where his head is shaved and receives the blessings of his mother

Below left: South Indian woman adorned with a *veni*

Below: Heaps of colorful jasmine and rose flower garlands on display in a street in Velankanni, Tamil Nadu

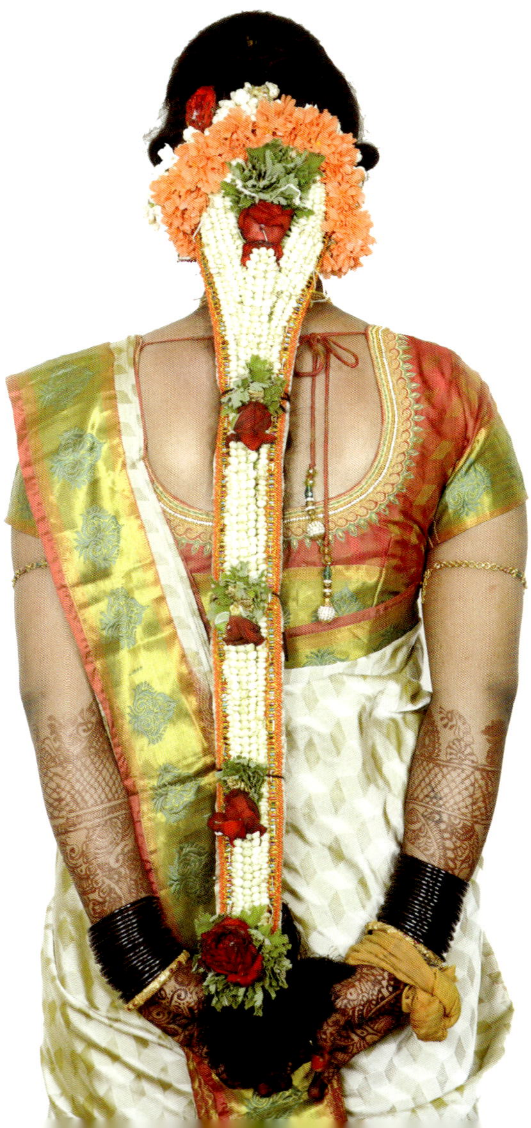

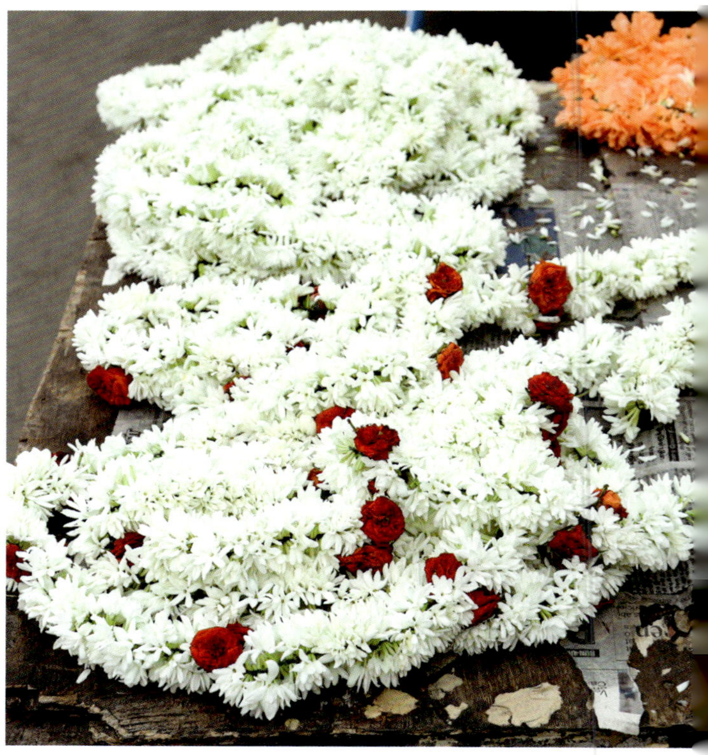

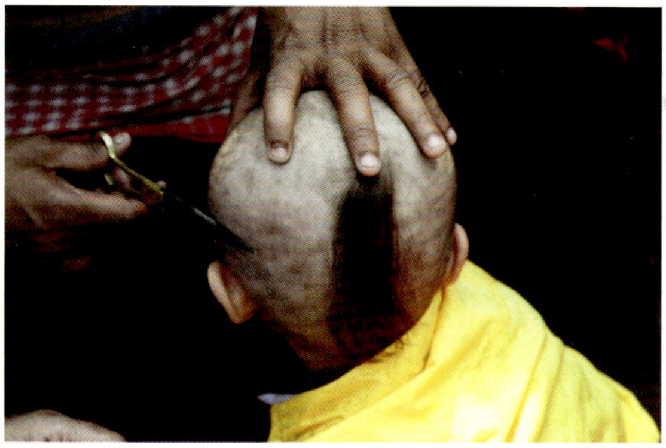

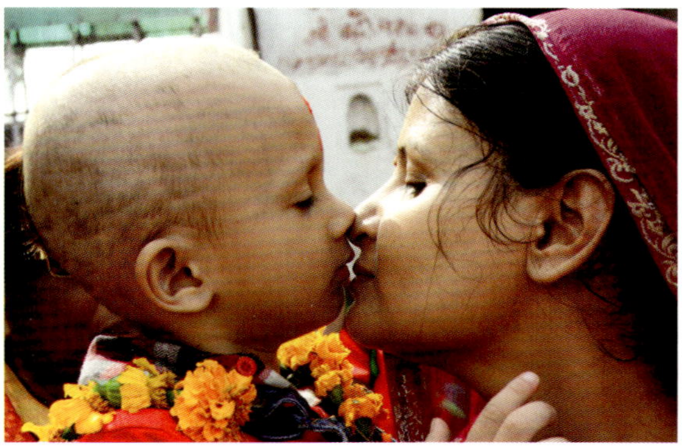

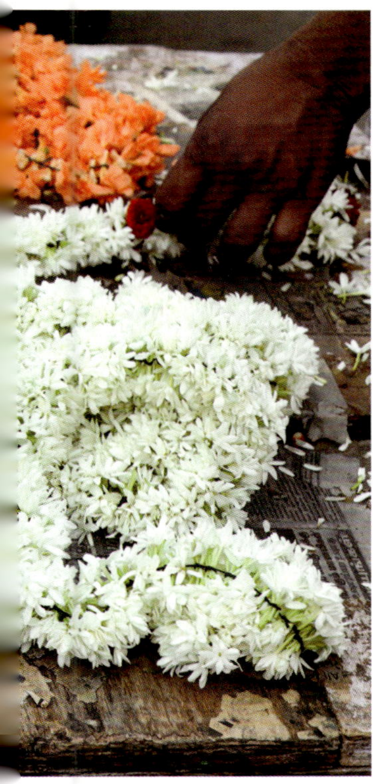

of her plait of hair. While modernisation has seen this practice fall out of favour, the jasmine flower, once part of so many women's daily hair-dressing ritual, is no longer a very popular adornment. However, outside the modern metropolis of Chennai, the jasmine flower still finds favour, especially among the flower sellers who sit outside temples with their fragrant wares. Tamil Nadu is still the largest producer of this flower in the country.

Simantoyannayanam is a ceremony performed to celebrate woman's pregnancy few months before the delivery date of the baby. It is performed for the protection and health of the mother and the unborn child. Literally, *Simantoyannayanam* means 'parting of the hair.' A priest is invited, *mantras* are chanted, and a fire ceremony called *havan* is performed as the husband parts the hair of his wife and places vermillion in the parting of her hair. The flowers play an important role as they are compared with the newborn's softness, purity and beauty. People gift them to the expectant mother as their blessings.

Mundan or the first hair cut of the child is another important function. Different communities perform the ceremony at different ages, but the ideal time is when the soft spot in the skull of the child vanishes, usually around 14 months. As in all *samskaras*, family, friends and a priest get together in the home or in a temple. Mantras are recited, a sacred fire is evoked and the priest with the help of the father or maternal uncle cuts small locks of hair from the four sides of the child's head. After this ritual cutting, a barber shaves off the child's hair completely. It is believed that the first hair from the time of birth is impure and shaving it off is purification. However, the real point of the *samskara* is that it is a sacrifice. The parents sacrifice the beauty of their child by shaving his or her head, and in exchange they ask God for long life and prosperity for their child. The first haircut often occurs at a temple or during a festival when the hair is offered to a deity. Flowers can be given on these occasions as a gift also.

Upanayana is a crucial event in the life of the orthodox, upper-caste Hindu male. It is an initiation ceremony which takes place between the ages of six and 18 to mark the transition to awareness and adult religious responsibilities. At the ceremony itself, the family priest invests the boy with a sacred thread to be worn always over the left shoulder, and the parents instruct him in pronouncing the *Gayatri mantra*. The initiation ceremony is seen as a new birth; those groups entitled to wear the sacred thread are called the twice-born. Flowers play an intrinsic role in all these rituals.

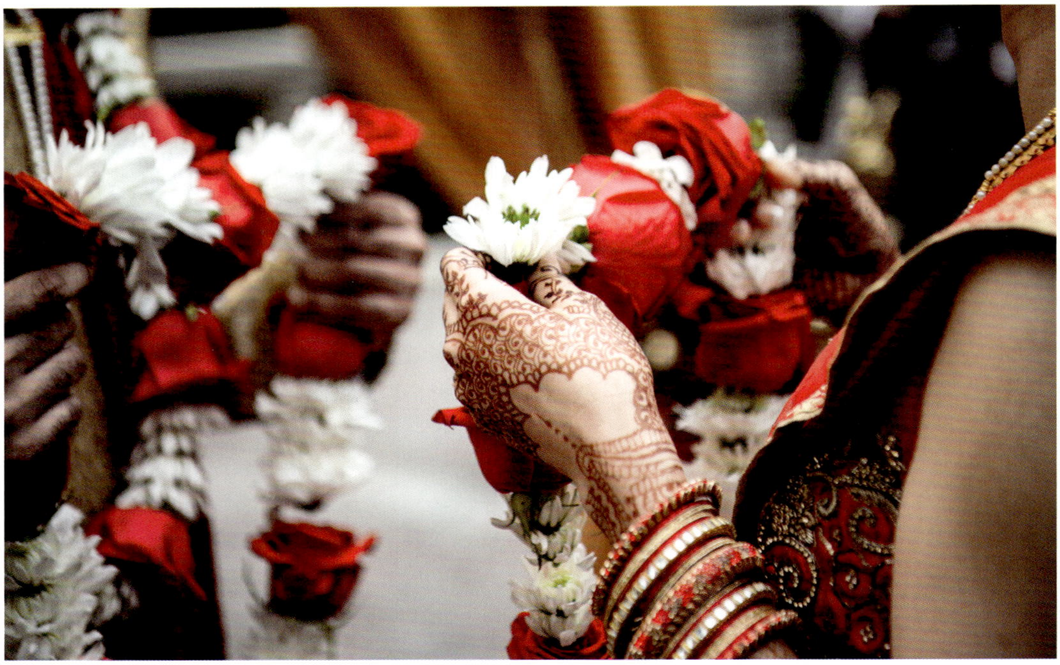

Left: An Indian couple at their *Jaimala* ceremony. The bride's hands are adorned with floral mehendi designs

THE UNBREAKABLE BOND

Flowers play an important role as décor items at any Indian celebrations, however their presence is most appreciated or used more at weddings. Weddings are grand, last as long as a week and red, white, and gold flowers are the blooms of choice to celebrate the union of two families. Bouquets and adornments are customary for the wedding party. As in virtually all cultures, the marriage ceremony or *vivaha* involves a bride and groom coming together in the presence of family and friends and then taking an oath of dedication to each other. In Sanskrit, the word *vivaha* comes from the root '*vah*' which means to carry. The oath of dedication that the bride and groom make to each other is a contract that 'carries' them along for the remainder of their lives. The role of flowers is crucial in all the functions.

Let us begin from the *baraat*, which is the procession of the family, relatives and friends of the groom who accompany him to the bride's home for the official wedding ceremony. The groom makes his way to the bride's home on a horse or car richly decorated with flowers. This has become an art form in India, and the floral decoration depends on the taste and spending capacity of the parents. The groom is given a welcome by the bride's family with flower garlands and rose petals are thrown upon the procession by the bride's side consisting of the sisters, cousins and friends.

Most Hindu weddings involve an exchange of garlands between the bride and groom. The ritual of *jaimala* also called *var mala* is when the bride welcomes and greets her groom by putting a flower garland around his neck, and he puts a garland round her neck in reciprocation. This is a big moment in the Hindu marriage ceremony because it is the first ritual that signifies that the bride accepts the groom as her husband. The *var mala* is usually a garland made of fresh flowers. But in the present times it has taken on a colourful and exotic look, and is sometimes made of artificial flowers. While the tradition of *var mala* is practiced in all the parts of India,

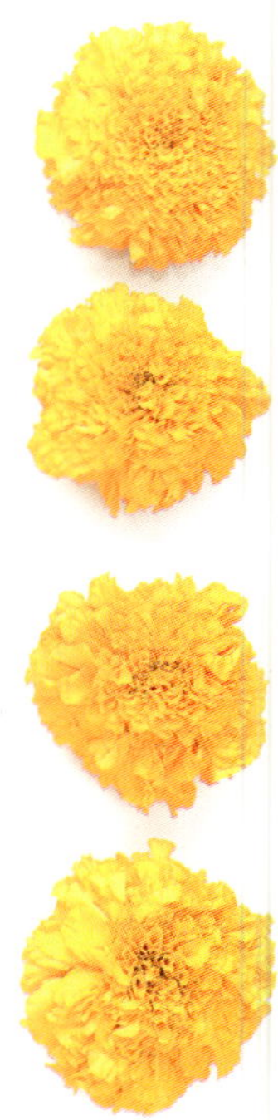

Below: The bride and groom exchange circular objects like rings, necklaces or garlands

Bottom: *Mangalsutra* worn by married Hindu women, arranged with a traditional sari and turmeric, vermilion and jasmine flowers

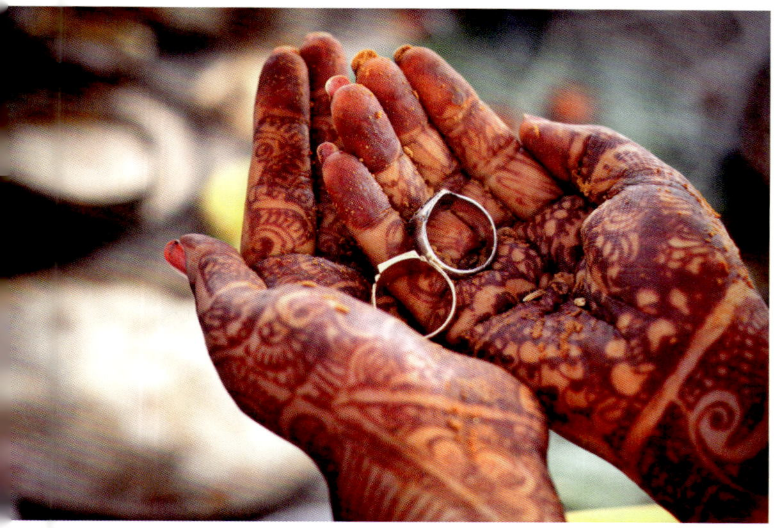

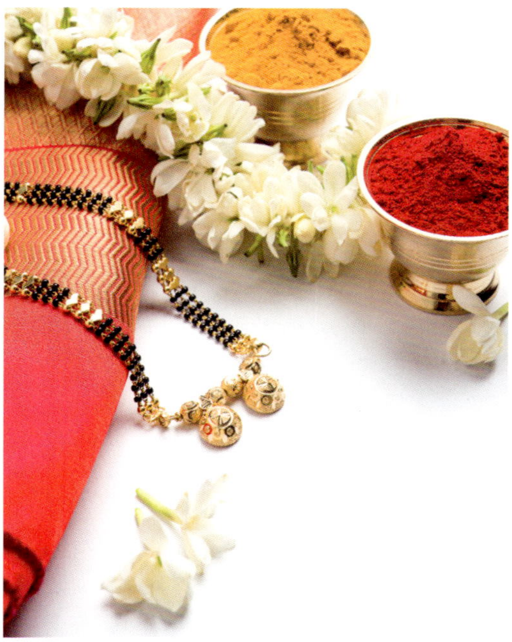

the garland used varies from region to region. The garland used in the southern parts of India is extremely heavy and long, as compared to the ones used in North India. Red roses are the most obvious choice for wedding garlands, though one can even go for orchids, carnations and so on.

The *jaimala* is followed by a bestowal of the bride by her father to the groom (*kanya daan*), the lighting of a sacred fire (*havan*), circumambulation of the fire (*parikramanam*), and the taking of seven vows (*sapta padi*). In general, southern marriages have the addition of a sacred necklace given to the bride by the groom (*mangalsutra*), and northern weddings have the addition of a sacred mark of vermillion applied by the groom to the bride in the parting of her hair (*sindhura*). There are, of course, huge regional variations between wedding customs and rituals, but the one common factor is the exchange of floral garlands between the bride and the groom.

Most wedding ceremonies around the world have a tradition where the bride and groom exchange a circular object like rings, necklaces or garlands. The circle represents that there is no beginning and no end to the union that is being created.

Suhaag Raat refers to the couple's first night together (after the bride has left for the groom's house) and they make their first erotic encounter of conjugal bliss memorable with a lingering fragrance. It is customary to decorate the nuptial bed of newly married couples with jasmine, so that the night of love making is associated with the aroma of the crushed blossoms. The bedroom is decorated with garlands, while roses or rose petals are laid across the bed. Usually strings of tuberoses or jasmine are used as bed curtains. These fragrant flowers add that essence of the sensual, which can best describe the moments of the couple's first time together. Apart from adding a sensuous

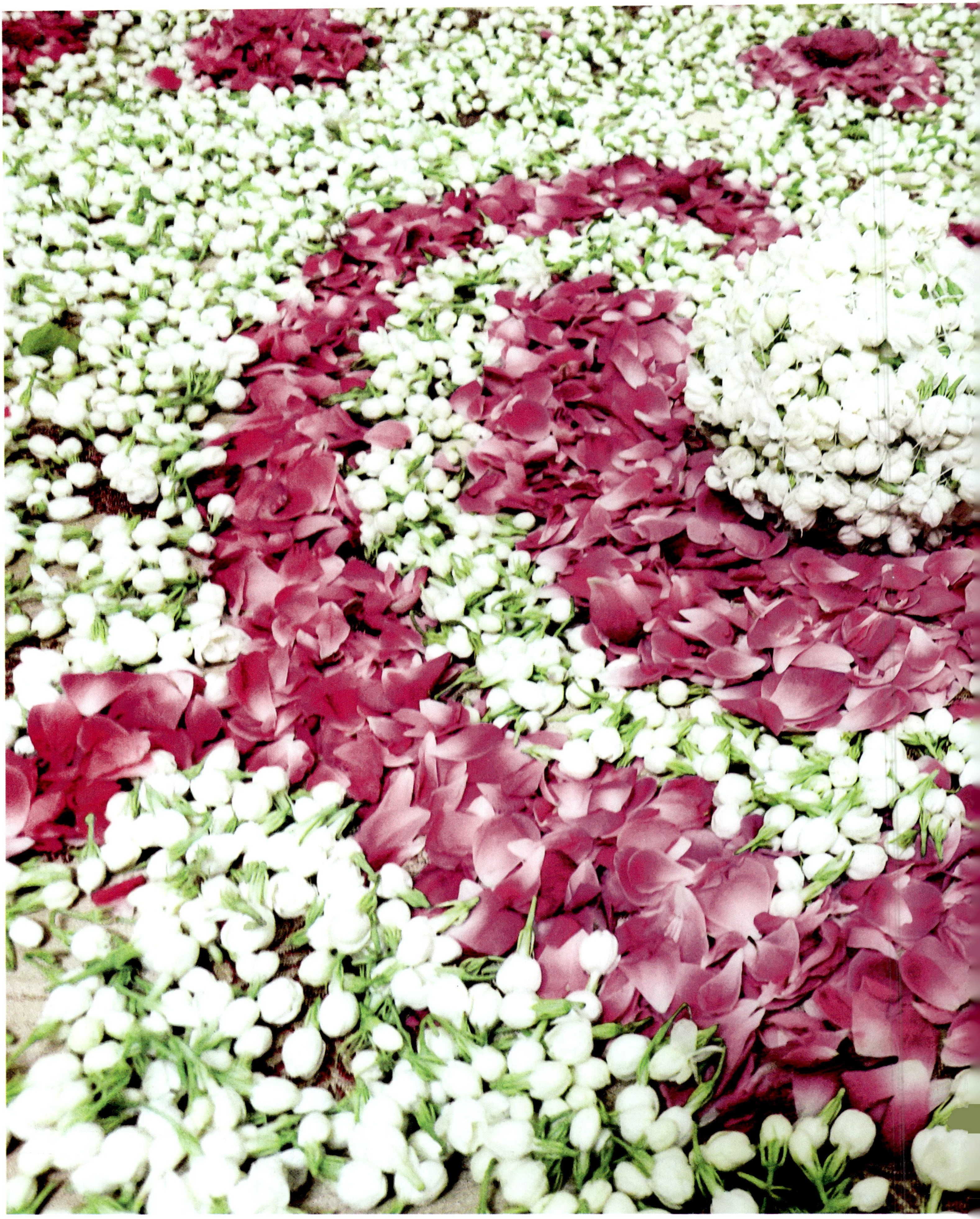

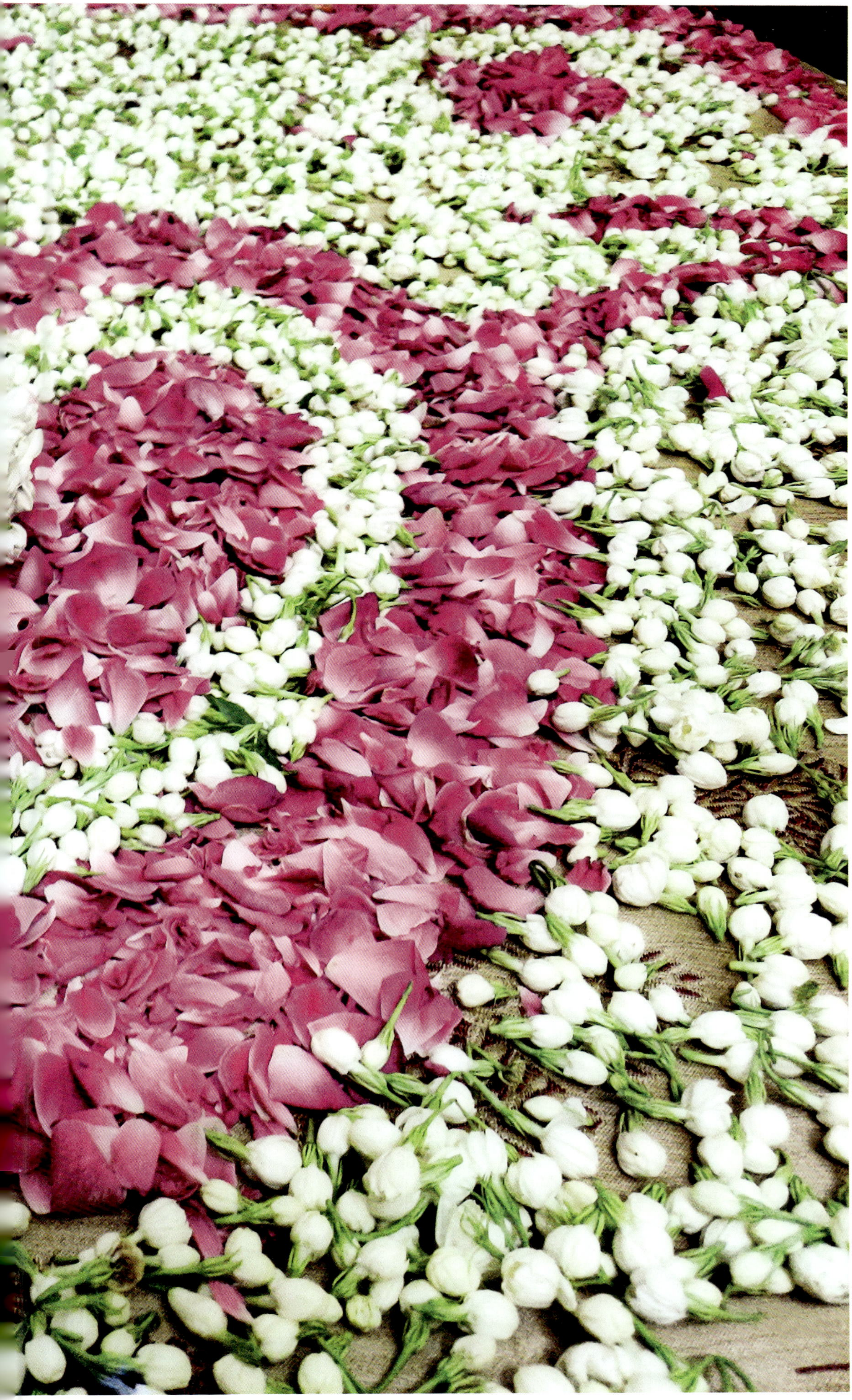

A bed decorated with jasmine and rose petals for a newly married couple's first night together

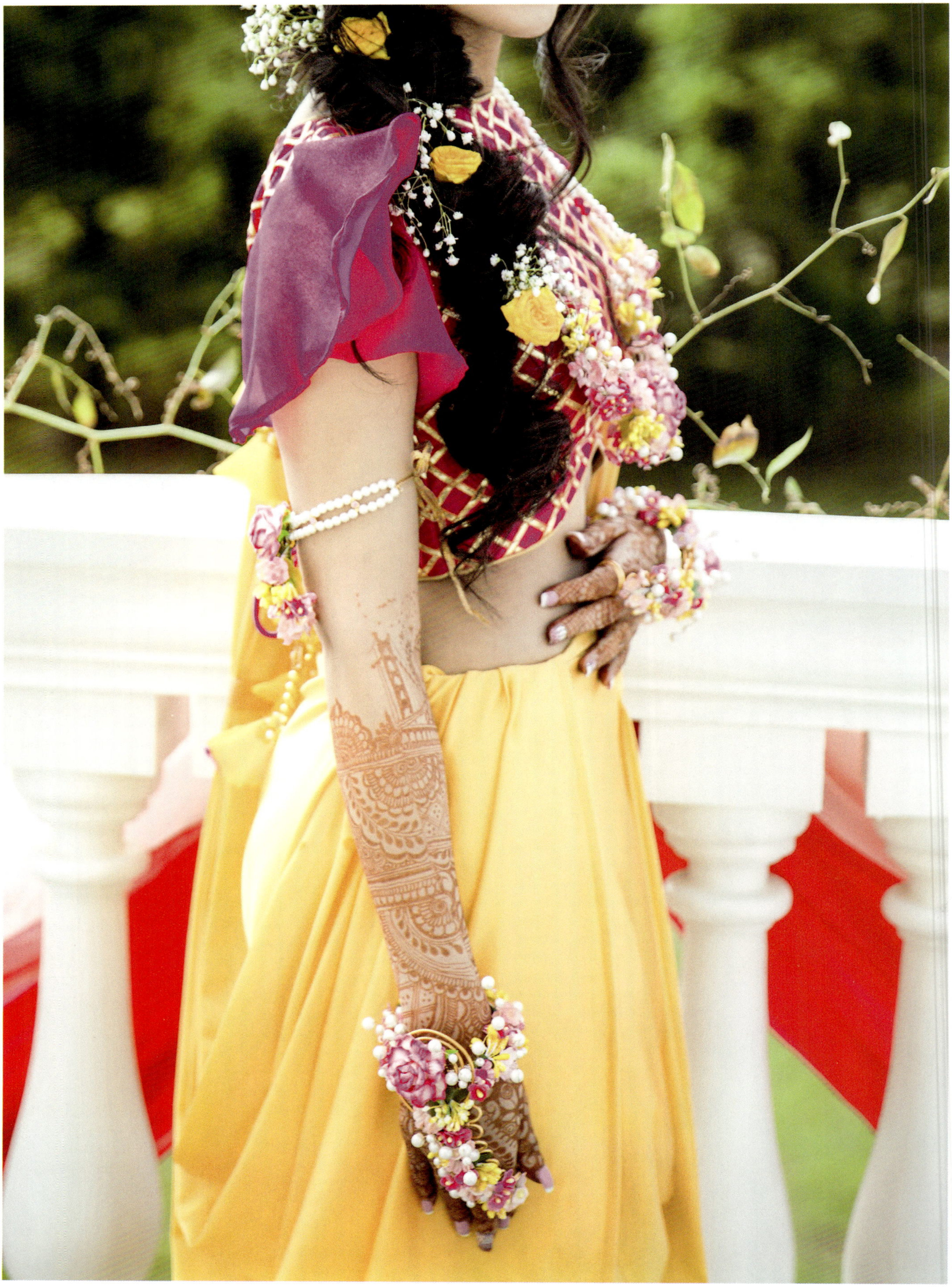

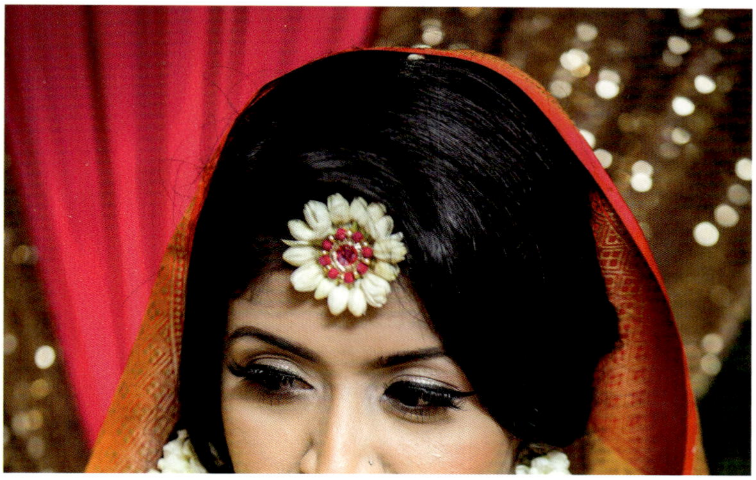

Above: An Indian bride wearing a flower *teeka*

Facing page: An Indian bride wearing *Phoolon ka Gehna*, flower jewellery

Below: Jasmine *gajras*, garlands for the hair

beauty to the occasion, the custom signifies a comparison between a virgin girl and a bud, because of the naturally beautiful physical appearance, softness and warmth. By the time the girl settles herself into the new family with the love of her husband and his family, she emerges into a woman, just as a bud blossoms into a flower.

Floral designs are much in vogue for contemporary weddings and include:

Heart ring boxes – for engagement ceremonies

Wedding garlands – made with a modern twist for the couple

Hair accessories – look amazing and standout in the crowd – matching floral bracelets also available (popular for proms, hens night)

Mehndi floral jewellery – for all the girls in the party house… this includes floral earrings, centre head piece (tika) and bracelets

Mehndi garlands – for the bride to be, her friends and cousins.

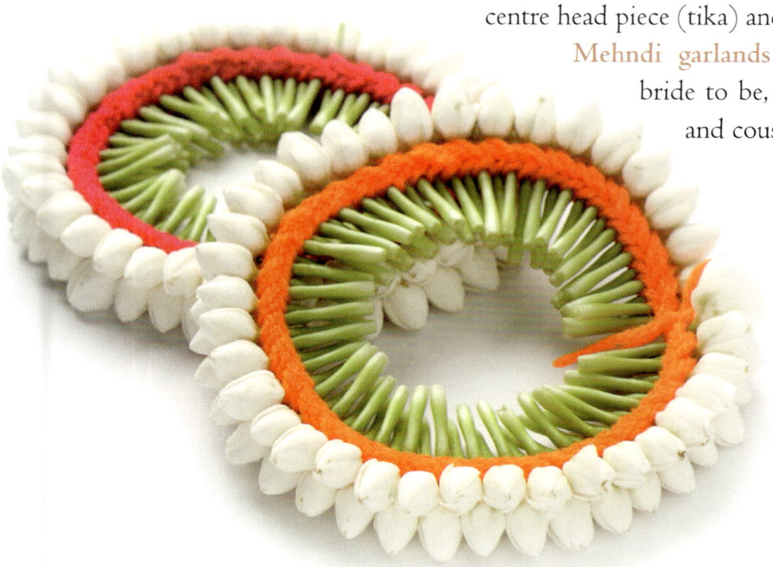

PHOOLON KA GEHNA

The cultural diversity of the 29 Indian states, each with their individual customs also extends to the floral traditions of each state. Kashmiri Pandit weddings have unique body adornment rituals where flowers are extensively and beautifully used. In the ceremony, known as the *Phoolon Ka Gehna*, the bride is decorated with jewellery made from flowers. Around two days before the wedding, the groom's family sends flower jewellery and tinsel to the bride. The girl wears this jewellery as a symbol of her first *shringar*. *Phoolon Ka Gehna* was an indicator of what jewellery the groom's side would offer to the bride. Young ladies from the groom's side would carry floral jewellery and a new sari for the bride on a silver plate. They would go to the bride's home in a ceremonial procession. A band playing traditional *shehnai* music would accompany them. Fresh red roses and jasmine flowers are used for preparing floral jewellery—necklaces, earrings, the *teeka*, ornaments for the bride's hair, anklets, *bazubands* (arm bands), *tagadis* (waist bands), and a special thumb ring called *arsi*. The latter has a small mirror embedded in it to help the bride view herself in it. Toys and dolls were also presented to the bride in this function.

The bride's family serve tea and sweets to the guests from the groom's side. In certain cases where the groom's side resides in the same city, the *Phoolon Ka Gehna* ceremony would take place one day before the *baraat*. In this case, the bride's side serves dinner to the ladies carrying *Phoolon Ka Gehna*. During the season, many flowering annuals or perennials such as chrysanthemums, especially the small-flowered ones, carnation, snapdragon, sweet pea, dianthus (pink), etc., are used for hair decoration.

MALE ADORNMENT

Traditional Indian society can boast of truly metrosexual men. The men wore gems and jewels and even adorned themselves with flowers. The descriptions of the gods and goddesses are sterling examples of male sartorial elegance. As in the *Adhya-tma-Ra-ma-yan.a: Concise English Version* by Chandan Lal Dhody, Rama is thus robed, 'She beheld Rama, the Most Exalted Raghu, in his Vishnu form, dressed in silken yellow robes...Like the lotus petals were His eyes. Resplendant with *Shrivatsa* was His chest.'[3]

Even the *nagraka,* the man about town character in the ancient love manual, the *Kama Sutra,* is described as wearing perfumed garments, jewels and is seen as a connoisseur of taste and refinement.

Over a period of time when the male sense of attire became minimal and masculinity was associated with an austere elegance, the modern, Indian male started emulating their colonial masters by wearing a three-piece English suit. Silk dhotis, brocade turbans, accompanying bejewelled *sarpenches* (turban ornaments), heavy rings and long necklaces of emeralds, sapphires, diamonds, were replaced by the bow tie. Pandit Nehru brought in the trend of wearing a flower bud in the button hole.

Buttonholes are a fascinating item in terms of floral decoration. They are worn by men in their coat collars on the chest, where a provision is made by crisscrossing two threads to hold the buttonhole. The first prime minister of India, Pandit Jawarharlal Nehru, popularised the wearing of a rose in the traditional attire of the achkan or long coat worn over fitted pyjamas, also known as the churidar. The most suitable flowers for a buttonhole are possibly roses and orchids, but other flowers, such as carnations, camellias, and daisies may also be used. A buttonhole is prepared by first taking a small portion (about 5–10 cm long and 4–7 cm broad) of a leaf of *Asparagus plumosus,* the most commonly used leaves for this purpose, and then fixing the desired flower or flower bud in position with the help of threads. Often a small coconut broomstick is attached at the base of the buttonhole to facilitate insertion in the appropriate place. *Zari* threads may be used for brightening up the look. To camouflage the broomstick, aluminium foil may be wrapped around it.

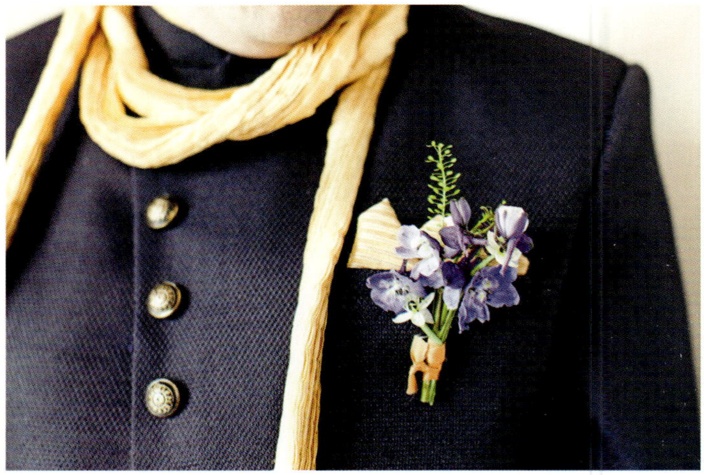

Above: A buttonhole adds a festive touch to a groom's attire

Facing page: Indian jewellery designs inspired by floral motifs

THE LANGUAGE OF JEWELLERY

Throughout the evolution of Indian jewellery designs, we see substantial inspiration taken from Mother Nature. From flowers and buds to leaves and vines, designs were based on the objects of natural beauty that caused humans to wonder. The motif of the flower was the most inspirational. Usha R. Balakrishnan writes in her seminal book *Alamkara,*

> The delicate elegance of jasmine buds, the sublime beauty of the lotus and the perfection of the narcissus metamorphosed into jewels

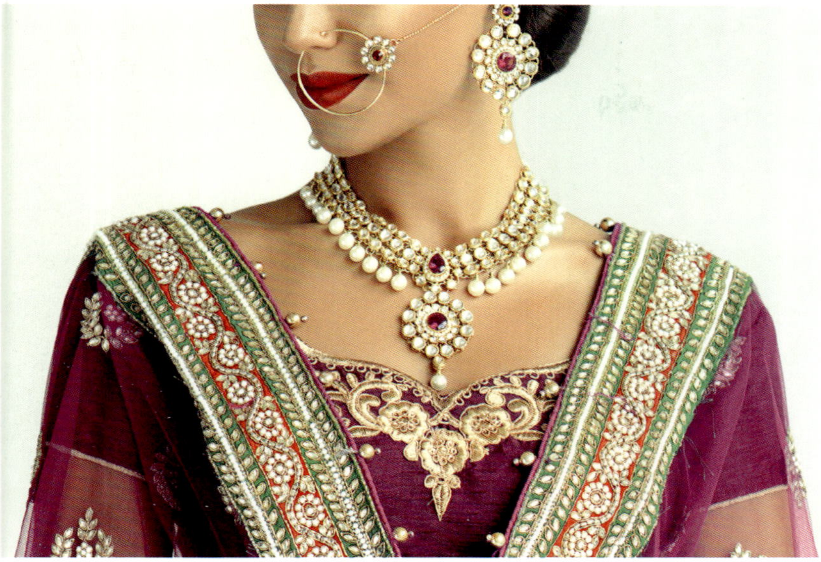

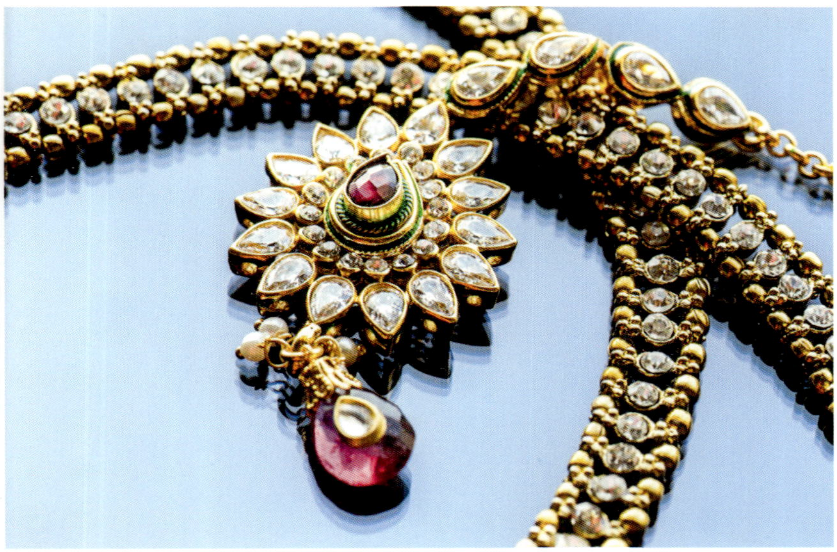

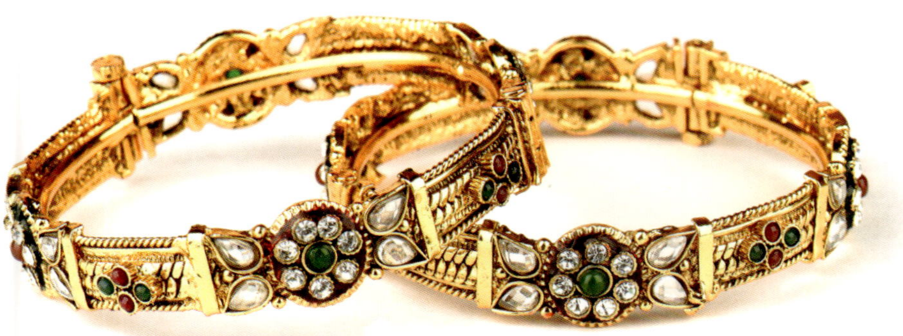

> and designs. Lotus flowers outlined in gold and set with diamonds decorate the surface of armbands while the narcissus occurs in the enamel decoration on the reverse of so many jewels...Necklaces in the form of jasmine and Magnolia champaca buds ensured that the delicate beauty and heady fragrance of these flowers passed to the wearer...Entire flower blossoms were transformed into earrings and forehead ornaments.[4]

Natural imagery translated into jewellery, highlighted metaphorical meanings and symbolism. The imagination became as fertile as the natural gems that inspired these pieces of art. The root of this idea lay in the belief that flowers spoke of fecundity and seeds denoted regeneration and rebirth or new possibilities. In the language of jewels, jasmine was meant to signal a pureness while the lotus was a symbol for perfection in an imperfect universe.

ENDNOTES

1. Devdutt Pattanaik, *The Goddess in India: The Five Faces of the Eternal Feminine* (New Delhi: Simon & Schuster India., 2000), p. 46.
2. Manmohan Ghosh, *Natyashastra Ascribed to Bharatamuni Volume 1* (Calcutta: Asiatic Society of Bengal, 1951), p. 411.
3. Chandan Lal Dhody, *The Adhya-tma Ra-ma-yan.a: Concise English Version* (New Delhi: M.D. Publications Pvt. Ltd., 1995), p. 18.
4. Usha R. Balakrishnan, *Alamkara: The Beauty of Ornament* (New Delhi: National Museum, 2014), p. 22.

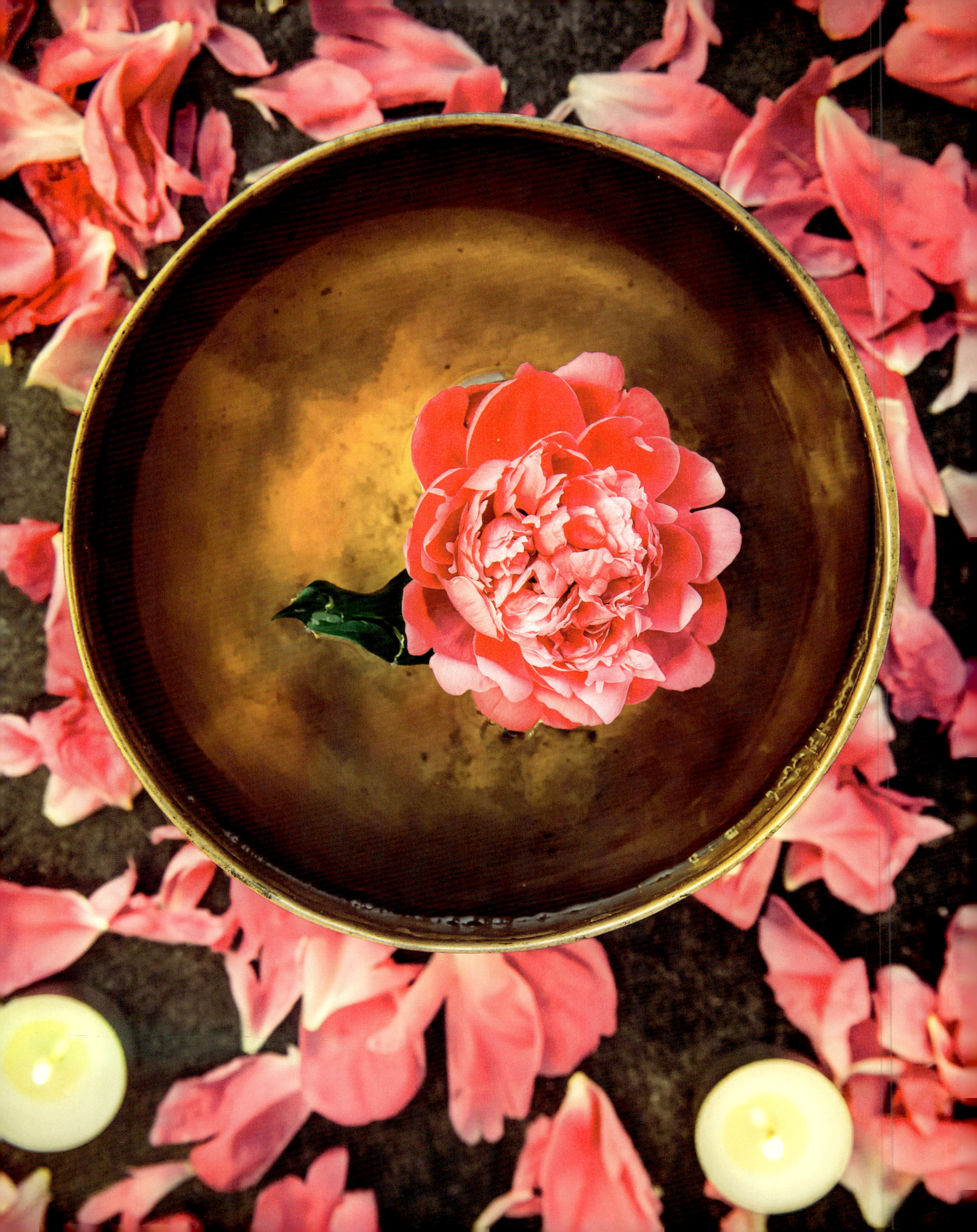

Enhancing Spaces

Art is like a border of flowers
along the course of civilization.
—Lincoln Steffens

Human nature is such that it does not remain content by only growing flowers in a garden, but also by bringing the beauty and aroma of flowers into their living spaces in the form of cut flowers. The staunch nature lover may advocate that flowers look best only on plants. While this is true to a great extent, the practice of using flowers for decoration is an age-old tradition which is followed today and may remain so for ages to come. Beauty is the realm of play and appearance. It is the unification of the spiritual and the sensuous. The spiritual aspect comes from nature, whereas the sensuous is the significant form and embellishment given by the artist to the selected objects from nature. A creeper of *raat ki raani* trailing outside your bedroom window or the honeysuckle creeper in the garden can make even the hottest summer months refreshingly fragrant and apparently cool.

Facing page: A pink peony flower floating in water in a Tibetan singing bowl, with candles and petals in the background, sets the mood for meditation

The height of luxury is decorating the rooms in one's personal space with fresh flowers. Fresh roses and lilies, garlands of jasmine and marigold can lighten the spirits of the most serious of men and women—such is the magical power of flowers. Rarely is there an individual who does not inhale more deeply when encountering the smell of fragrant blossoms.

India being so geographically diverse, with great variations in bio-diversity, different cultures in different regions produce singular nuances in the appreciation and aesthetics of flowers. A similar genus of flower has separate manifestations in different states because of differing traditions. Flowers might be used in private altars, in gardens or as room decorations. It used to be that every city, especially the smaller towns, had at least one flower *haat* or open-air market. However, the market is changing, as is demand, and the *haats* are disappearing. All cities and towns once had numerous flower *haats*, but sadly these have been reduced to a shadow of their former selves—the victims of urban expansion and changing tastes.

INVITATION TO THE GODS

Floral rangoli is an art form which is a world infused with vibrant hues and lovely fragrances. Rangoli, one of the most well-known forms of Indian folk art, brings flowers into the interior. In the South and South-west of India, especially Kerala, where it is called *pookalam*, flowers are used to create this form of floor art. Consisting of bright colours, the designs are meant to be sacred welcoming areas for the Hindu deities or the legendary King Mahabali who is believed to visit Kerala during the festival of Onam. The ancient symbols which form the pattern of the rangoli have been passed down through the millions of ages, which keeps the tradition alive and popular

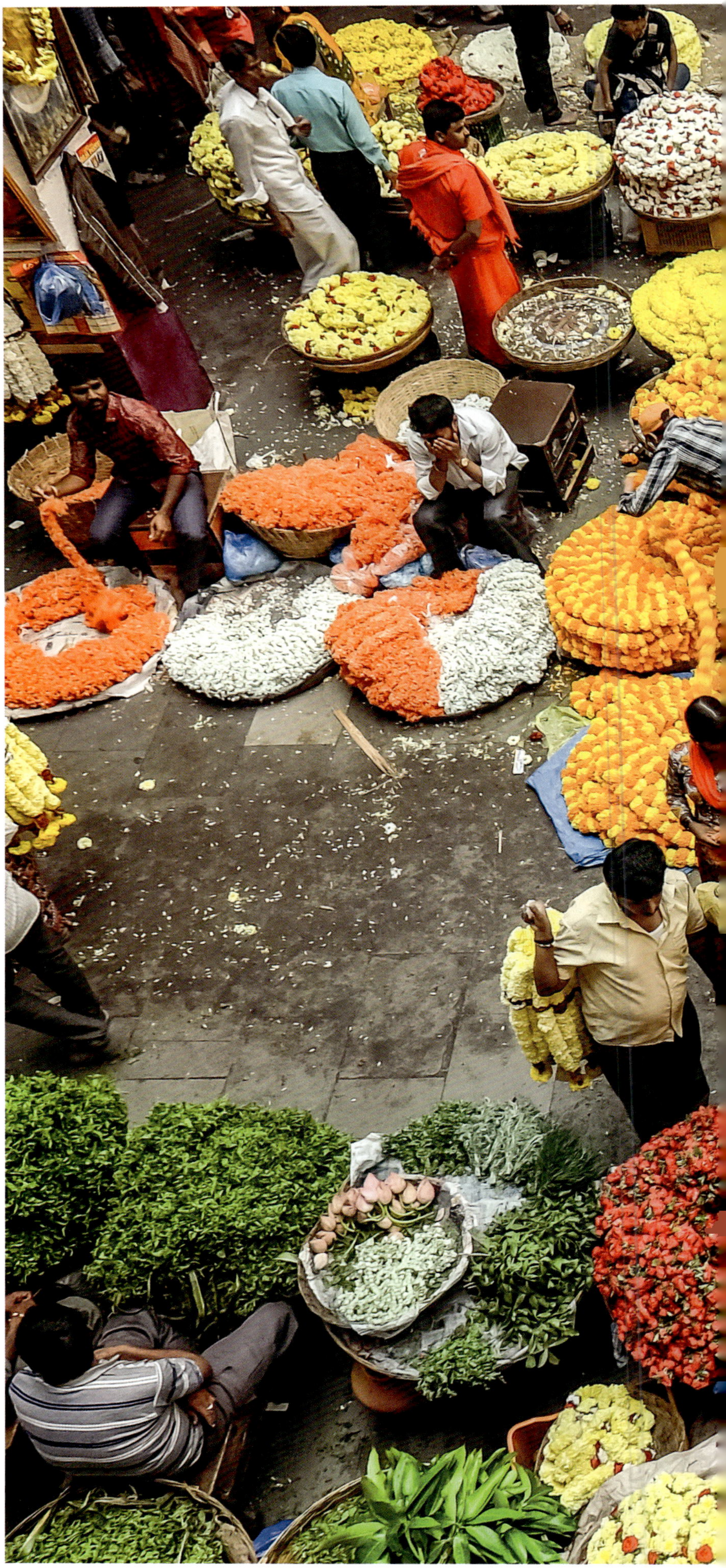

Flower sellers at the colourful KR Market in Bengaluru, Karnataka

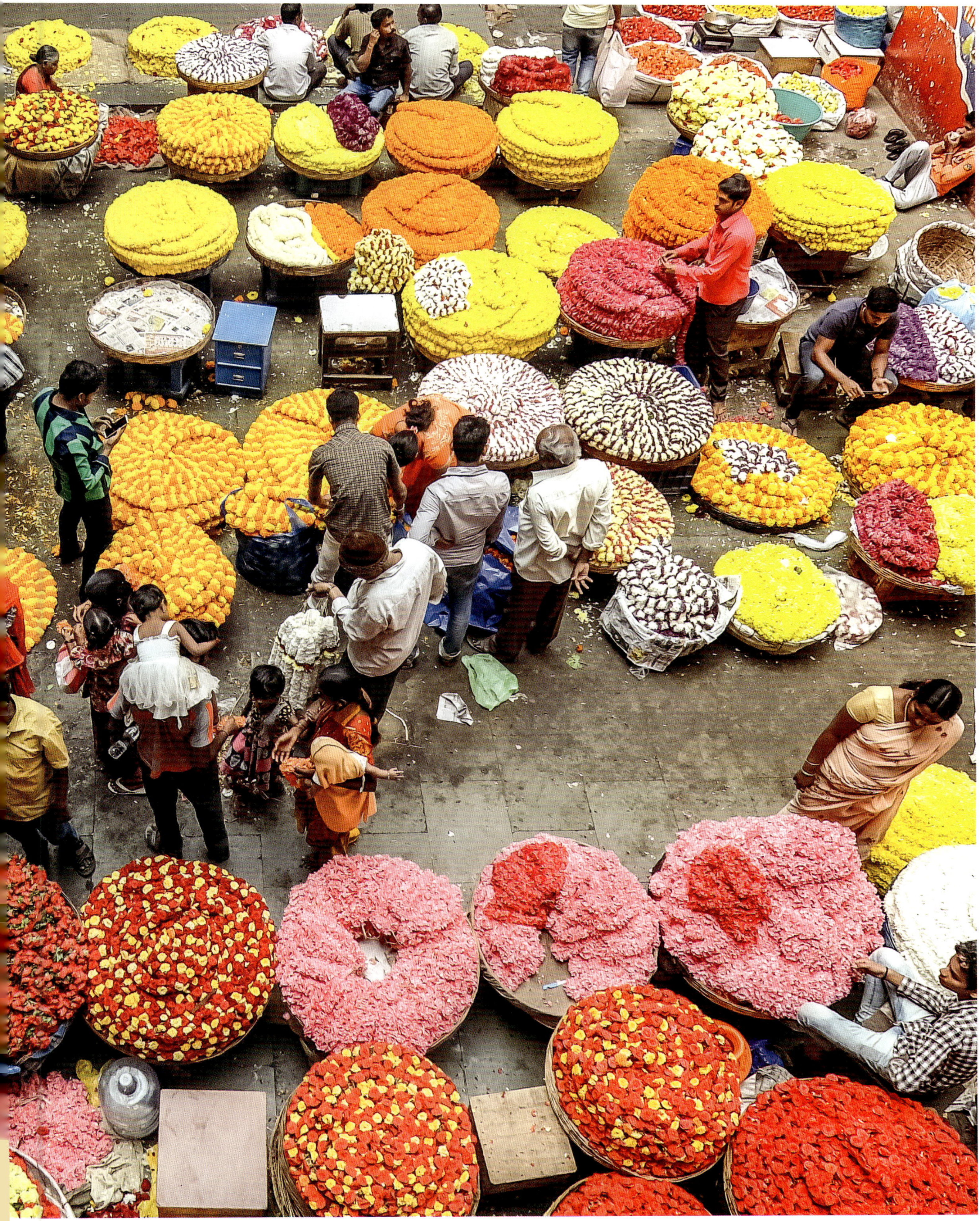

with the contemporary generation. The purpose of rangoli is not just the decoration of the house, but it is also believed to bring good luck to the home. It is traditionally done by women. Common festivals or family occasions inspire the creation of a rangoli. This practice is performed during Indian auspicious occasions, marriage celebrations and many more events.

Rangoli designs are special geometric shapes embellished with deity impressions and flowers according to the specific celebrations or festival. They can also be very elaborate designs made by a group of persons. The base material is typically dry or wet rice or dry flour, to which *sindoor* (vermilion), *haldi* (turmeric) and other natural colours can be added. The chemical colours are a modern-day variation of this traditional set of colourful material.

Specifically with Hindu festivals rangoli is used as a symbolic expression of religious and cultural beliefs. Women may make rangolis at the entrance to every room of the house. Symbols of a lotus flower and its leaves, mango, fish, different kinds of birds: parrot, swan, peacock, as well as human figures are found in almost all Indian rangolis. Rangoli created with icons such as the swastika or the goddess Lakshmi's footstep, are considered indicators of prosperity. The designs are symbolic and can include geometrical patterns, with lines, dots, squares, circles, triangles, creepers, leaves, trees, flowers, animals and anthropomorphic figures. These motifs often are modified to fit in with the local images and rhythms.

Rangoli has different names in different regions of India; for example, *purna* in Uttar Pradesh, *mandana* in Rajasthan, *aripan* in Bihar, *alpana* in Bengal, *rangavallie* in Karnataka, *kolam* in Tamil Nadu, *pookalam* in Kerala, *muggu* in Andhra Pradesh, *arpan* in Kumaon, and *saathiya* in Gujarat.

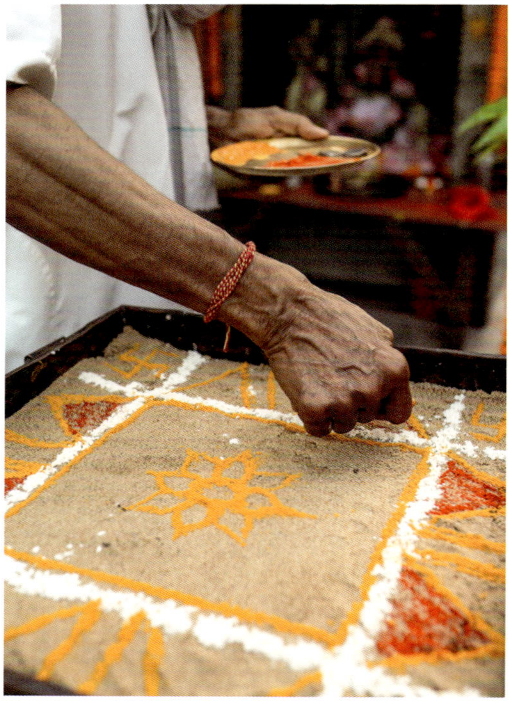

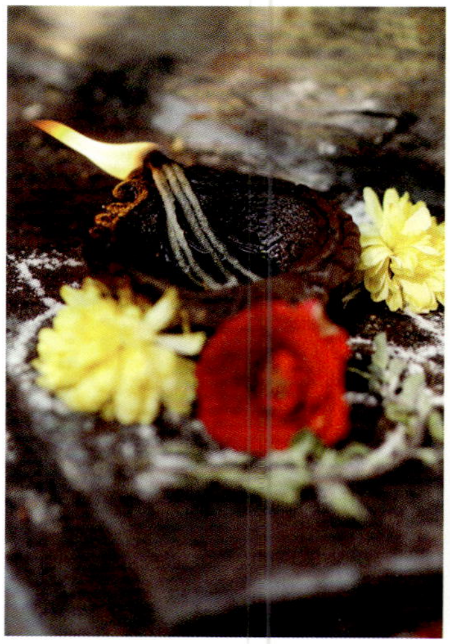

Above left: A Hindu priest creating a rangoli with colours from natural pigments like *haldi* (turmeric) and *sindoor* (vermilion)

Above: A lit earthen oil-lamp surrounded by flowers, placed on a rangoli is a traditional offering to Hindu gods in a temple

In ancient days rangoli was done in small patterns, but now entire floor areas of rooms and hotel foyers are covered in intricate and detailed designs. Rangoli can be improvised into diverse visual art forms. For instance, it can be used to make elaborate images that look as if they were painted. The one important point that all rangoli artists follow implicitly is to see that the entire pattern is created by an unbroken line, with no gaps to be left anywhere, so there is no opening for evil spirits to enter. This is a strong belief in the Indian culture. This is important, as rangolis are usually drawn as an auspicious expression of hospitality to invite gods and goddesses and also guests to bless one's home.

The Sri Chakra rangoli is an example of a sacred form of the art. It is a geometrical figure, which has many intertwined triangles, and is considered the symbolic representation of the Goddess. In many parts of southern India, the Sri Chakra kolam is drawn elaborately, a lamp is lit, and the sacred

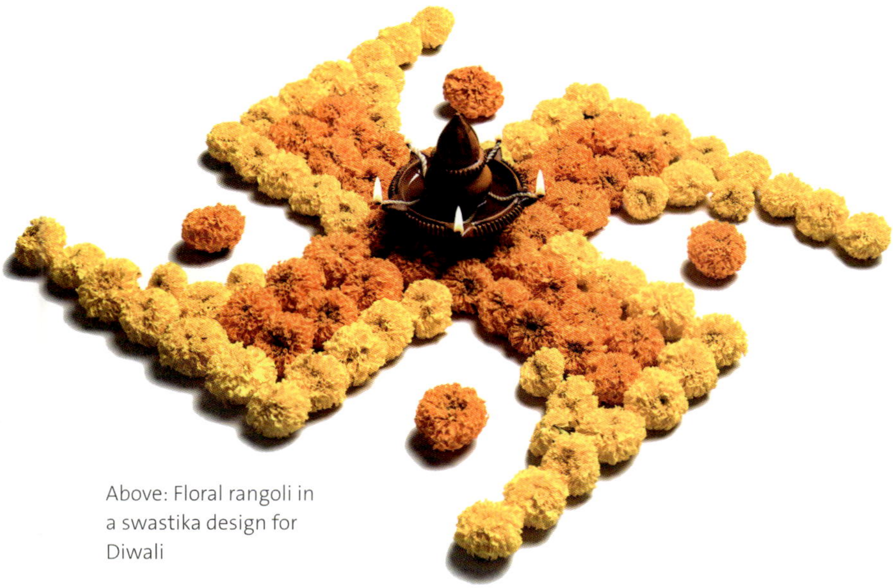

Above: Floral rangoli in a swastika design for Diwali

Below: A Malayali woman creates a *pookalam* for Onam in Kerala

Below right: A hibiscus flower provides a finishing touch to a kolam at a house's entrance in South India

Lalitha Sahasranamam (the 1,000 names of the Goddess) is chanted. Another well-known sacred rangoli is the *Hridaya Kamalam*, or Lotus Heart. Many people argue that the spirit of Hinduism is represented by rangolis. Hindus are encouraged to feed all forms of life before feeding themselves; true to this belief, rangolis are drawn using rice flour to feed even the tiny ant before the members of the household start feasting.

Rangoli is made extensively during the Festival of Lights, Diwali. During this festival, people take time off for cleaning and decoration of their houses, in order to please Lakshmi, the goddess of wealth. Homes are lit up with *diya*s or lamps and every nook and corner of the house is decorated.

People decorate their *puja ghar* or *mandir*, their backyard, verandahs, rooms, kitchen, walls and entrance with beautifully designed Diwali torans and rangolis. Some of the popular themes of Diwali rangoli are the signs of Om, swastik, leaves of the *Ashoka* tree, mangal kalash, lighted diyas, shree, lotus and other flowers.

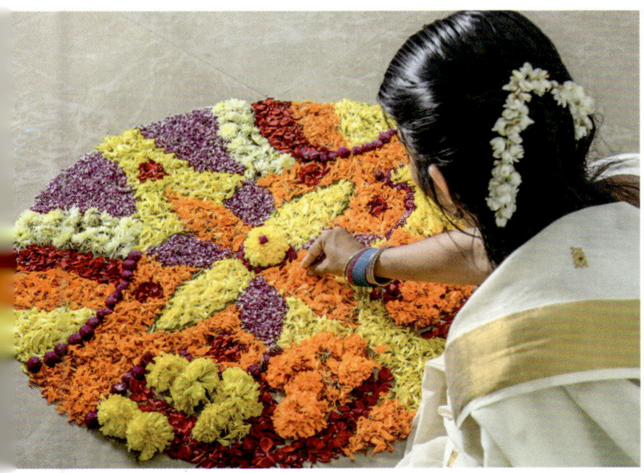

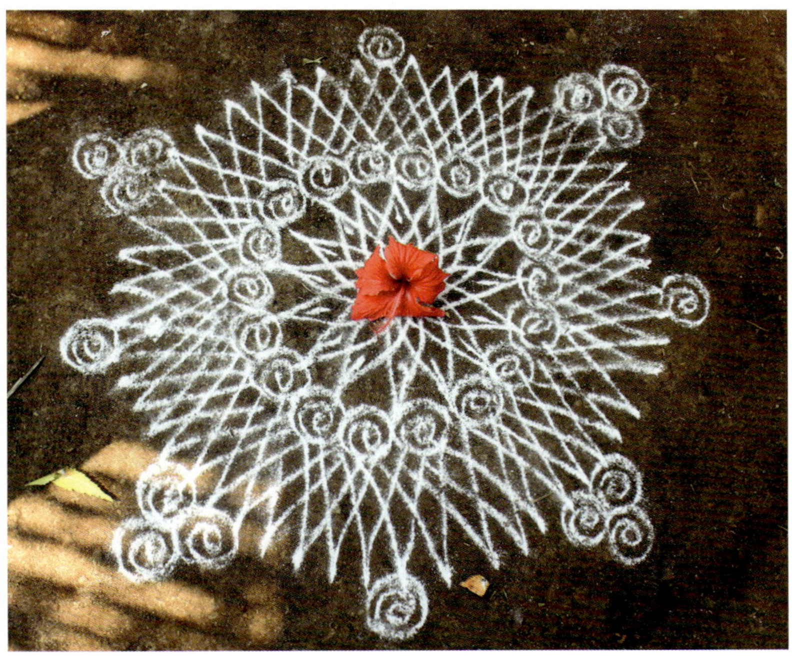

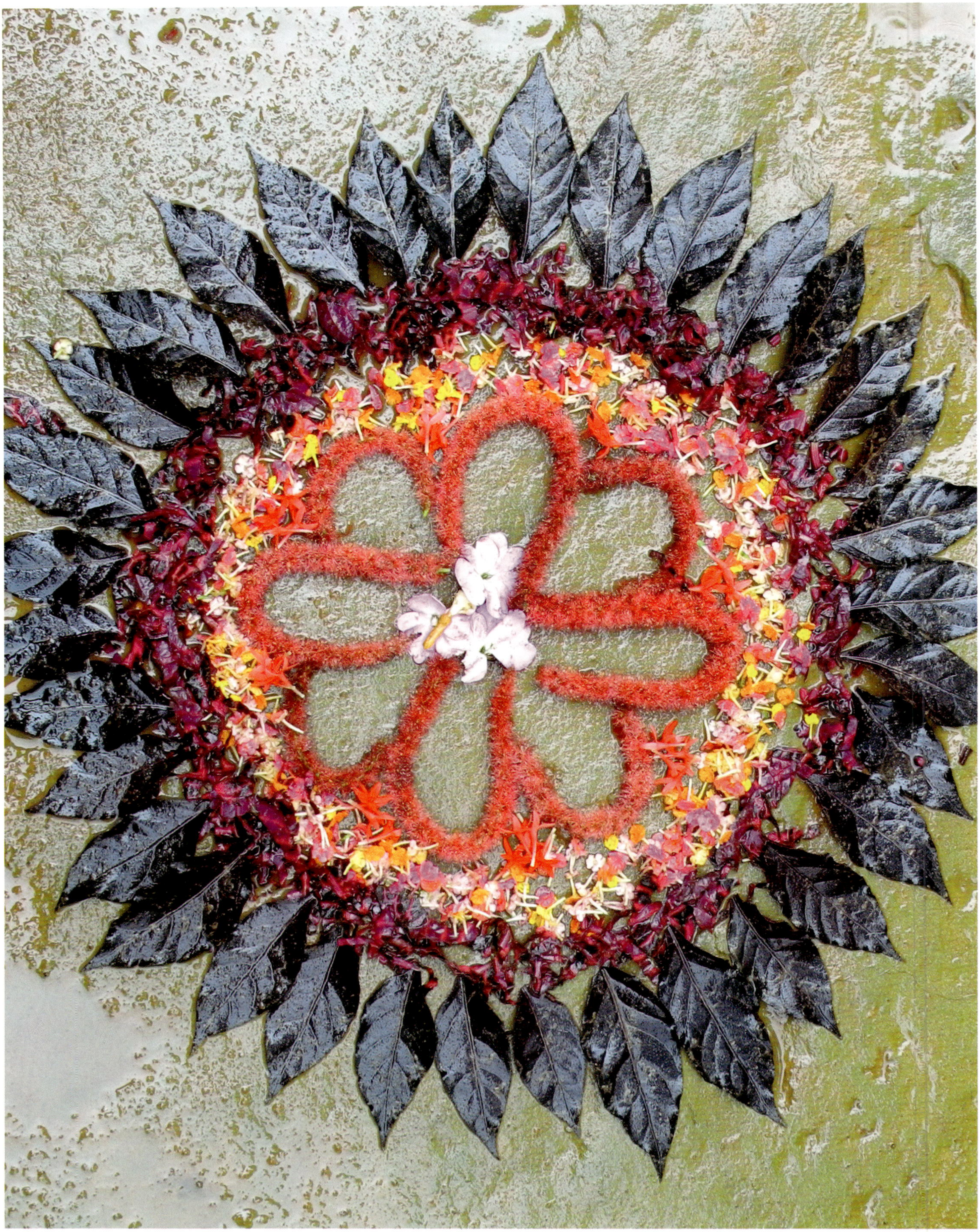

Facing page: A *pookalam* or floral decoration, for Onam in Kerala

Below: Colourful flowers of *Ixora coccinea*, also called jungle flame

SOUTHERN TRADITIONS

In south India, *kolam* is generally drawn at the entrance of a home or any other institutional building. Dried rice flour or other types of white powder is used for drawing such *kolams*. In today's times colour is introduced with the use of different flower petals. Although there are several types of traditional *kolam* patterns, a lot more can be done depending on the creative mind of the person who draws it. In the states like Tamil Nadu, Andhra Pradesh and Karnataka, the *kolam* is drawn upon the ground or floor daily. Nowadays there are also *kolam* competitions on festive occasions.

During the popular festival of Onam in the month of September, every household in Kerala decorates the home with *pookalam*. *Pookalam* is an intricate and colourful preparation with flowers laid on the floor. It is also known as *aththa-poo* since it is considered auspicious to prepare it as a welcome for Mahabali, the legendary king who visits Kerala during Onam.

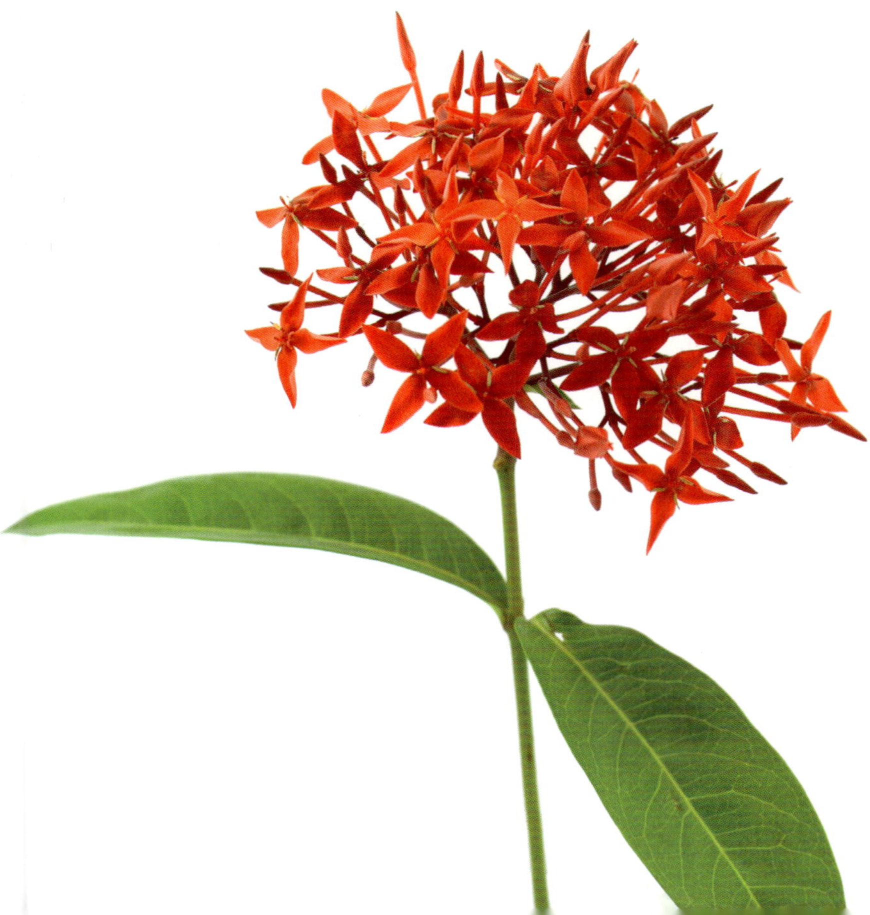

The aromatic floral rangolis of Onam can linger for days on end. During the festival, flowers are laid down for each of the ten days of the celebration, the design growing larger and more complex every day. A specific flower is dedicated to each of the ten days of Onam. The flowers include *thumba* (*Leucas aspera*), *kakka poovu* (Gromphera), *thechipoovu* (*Ixora coccinea*), *mukkutti*, *chemparathy* (Shoe flower), *aripoo* or *konginipoo* (Lantana) and *Hanuman kireedom* (Red pagoda plant). The *thumba* flowers are given most importance in Pookalam as they are small in size and glitter in the soft rays of the sun. Lord Shiva's favourite flower is *thumba poo*, and interestingly King Mahabali was a devout worshipper of the deity Shiva.

THE SCIENCE OF DESIGN

Vastu Shastra is the traditional Indian science of design for temples and also includes Vaastu Vidya, which is literally the knowledge of dwellings. Vastu Shastra also includes guidelines on the layout of houses, towns and cities as well. In pre-modern India, the traditional doctrine or shastra of construction and architecture consisted of precepts born out of indigenous wisdom and the traditional Hindu view about the cosmos. This philosophy believed that the laws of nature have a direct effect on human dwellings. In fact, traditional Indian designs are based on the integration of architecture with nature, comprising geometric patterns, symmetry and directional alignments.

Vaastu Shastra is becoming popular once again and many alternate minded people are looking towards the principles of vaastu to bring peace and harmony to the interiors of their home environment. Whenever an impressive building, beautiful in every way from the material aspect to its aesthetics, does

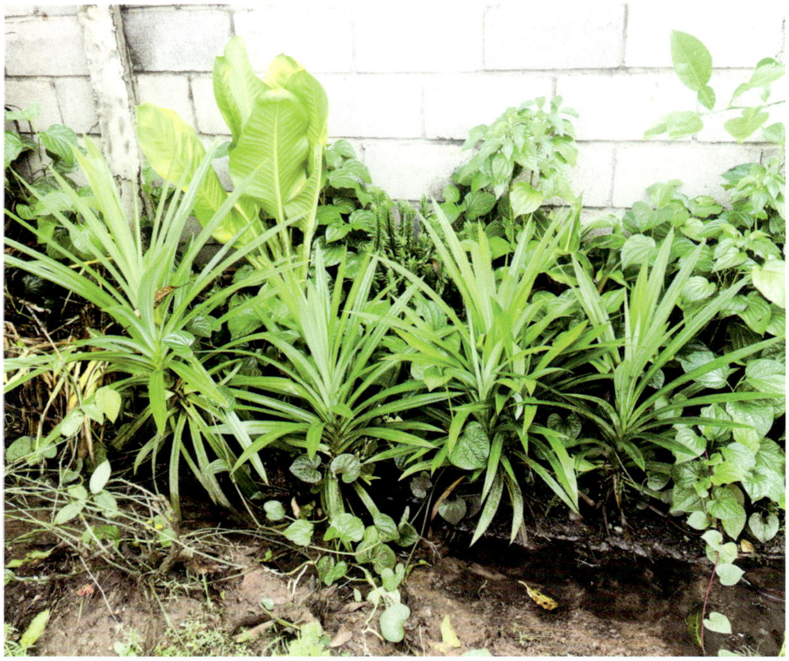

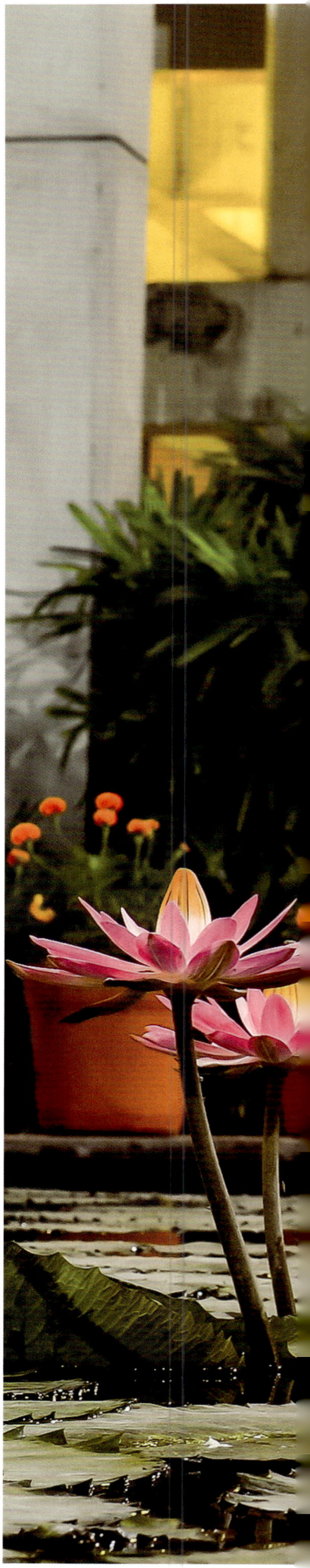

Left: Pandan (screw pine) is considered a good plant to grow in a home

Right: A water body with flowers is a refreshing feature at India Habitat Centre, New Delhi

Below: neem flowers have therapeutic value

not bring happiness to the inhabitants of the building, it is often remarked, 'Probably the vaastu of the house is not right.'

Naturally, vaastu also includes the use of fresh flowers that add beauty and aroma to the house. Flowers also enrich the ambience due to their association with certain emotions and philosophies, playing their own role in the larger picture of Vaastu Sastra. Jasmine, for example, is associated with spirituality and is good for a house. Vaastu pandits advise that flowers like the *champa*, *mogra* and *parijat* can be planted in any corner of the house except in the southwest, northwest and east. Neem flowers are favoured for they have therapeutic values.

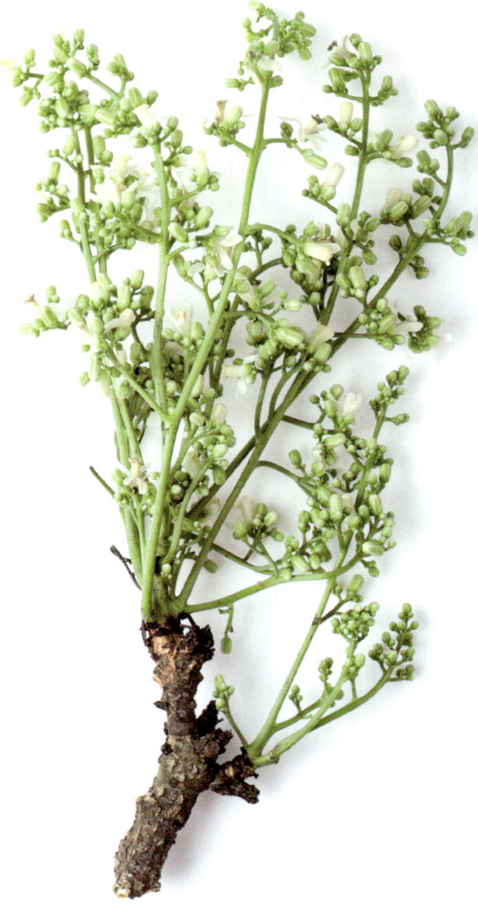

The pundits advise individuals not to keep dried or dying flowers since they are not good for the home environment. Trees such as khadir (acacia), sindhu, pandan (screw pine), mango, arjan, and tinduka (Indian persimmon) are considered to be good. And most importantly in vaastu, it is auspicious to have a pond with lotus flowers in the north-east part of a house.

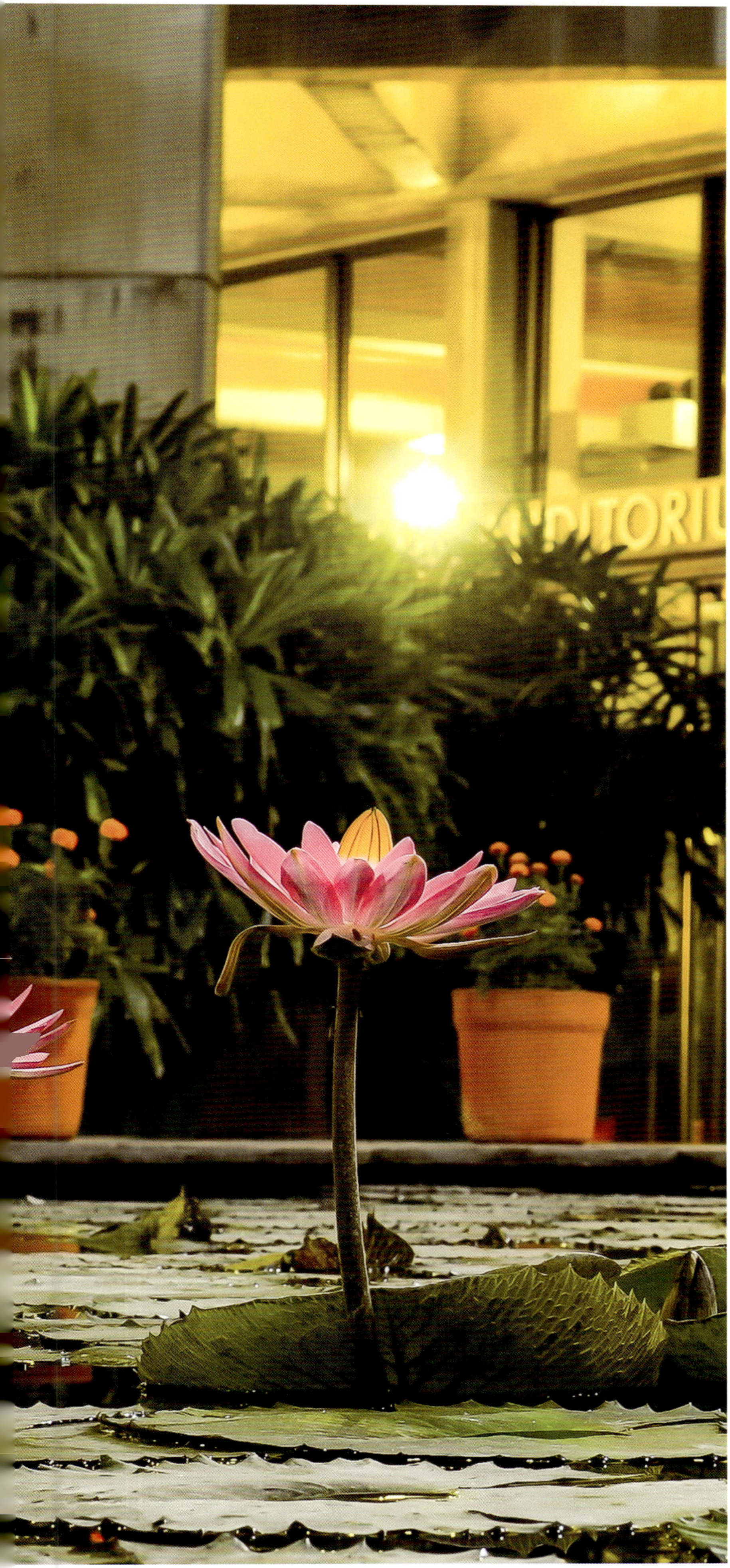

WIND AND WATER

Feng shui, the ancient art and science developed in China, is similar to vaastu in that it is a complex body of knowledge on how to balance the energies of a given space to guarantee health and good fortune to anyone inhabiting it. It is slowly gaining popularity in metropolitan India.

Feng, which means wind, and *shui*, which means water, is based on a Taoist understanding of nature, especially the concept that the earth is alive and filled with *chi*, or energy. In Chinese culture, wind and water are associated with good health.

In feng shui, flowers are used because of the universal feelings they evoke in everyone—a sense of grace and beauty, as well as a delicate sensuality. In addition, flowers also bring good luck, numerous blessings and positive energy to the home. For these reasons, live flowers are preferred because they impart a strong healing energy. However, feng shui also uses images of flowers or high quality silk flowers in its practice.

The colours and numbers of flowers are important in feng shui, as is the symbolism of specific flowers. Among the most popular flower symbols used in feng shui applications is the quintessential lotus. It symbolises the ultimate perfection because its purity is not soiled by the mud from which it springs. In fact, in Chinese medicine, every part of the lotus is used, making the flower a very powerful feng shui symbol for a healthy and harmonious home.

The peony is also a very popular feng shui symbol. This sensual flower has long been used as a cure for love and romance, especially the pink variety. The peony is considered a metaphor for female beauty. Other flowers used in cures to promote love and a harmonious

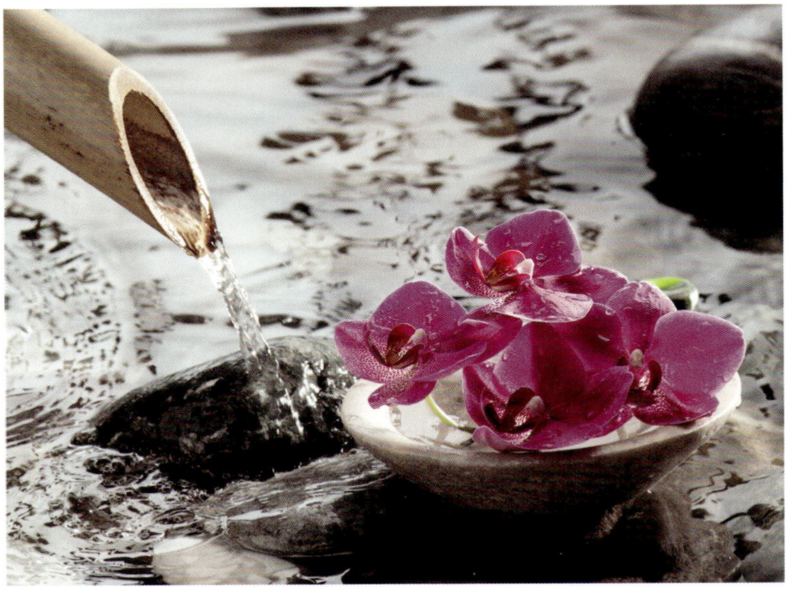

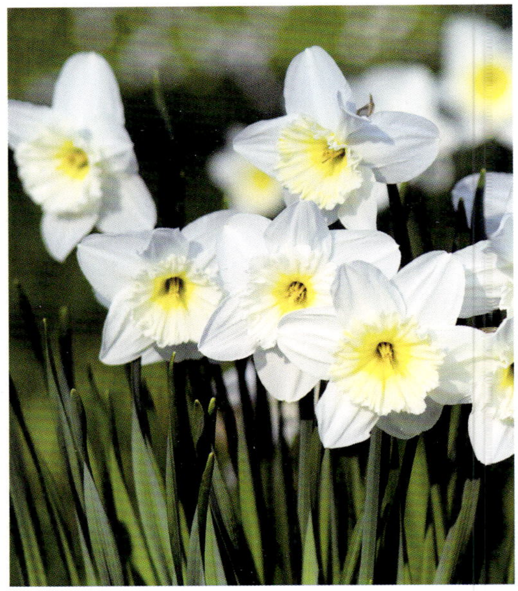

marriage are the blossoms from trees such as the cherry, apple, dogwood and peach. They are believed to bring in the energy of new beginnings, innocence and freshness.

The orchid, meanwhile, is the classical feng shui symbol for fertility. It brings the energy of natural symmetry and has become a symbol of the quest for perfection in any area of your life. Thus the orchid has become so popular worldwide—it symbolises perfection, abundance, beauty, purity and spiritual growth.

In Feng Shui, for improvement of the career, talents and abilities of a client, a feng shui master will advise the client to place the narcissus in their home. This flower is frequently used to help one get the proper rewards for their hard work. The white narcissus is used more often than the yellow variety in career cures. Once the client's career goals are achieved, then the chrysanthemum, which is a symbol of a life of ease and balance, can be used. Interestingly, this flower is thought to have a strong energy that attracts good luck to anyone who places it in their home.

Above left: Orchids symbolise perfection and beauty

Above: Narcissus flowers are believed to bring good luck to one's career

Below: The peony is considered a metaphor for feminine beauty

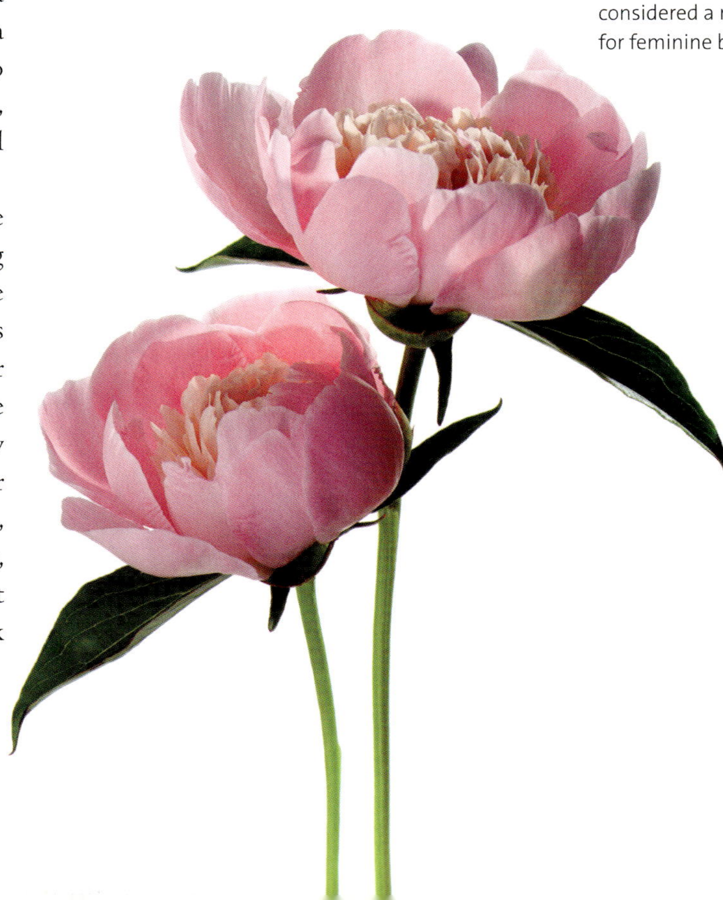

NEW FRONTIERS

Till a few decades ago, floriculture was just a pastime of the rich and a hobby of flower lovers, but now it has opened a new vista in agri-business i.e., commercial floriculture. The trend is growing simply because people don't have enough time and space to grow flowers, particularly in urban areas and metropolitan cities. With the increase in buying capacity, flower lovers have now started ordering large quantities of exotic blossoms to beautify their homes. Bouquets are also a way of acknowledging someone you love, as we all know that flowers play an essential role in celebrations as well as everyday life—be it at weddings, birthdays, anniversaries, funerals, and festivals. Flowers, arranged in formal structures, have become great vehicles of communication and even as messengers of love.

Flowers once inhabited very specific spaces in India—in temples or private puja rooms; as garlands for deities; in women's hair; for marriages and as floral rangolis. With the western influence, flowers are now entering many new spaces. We are also seeing an increase in formal arrangements like posies, bouquets and ikebana. There was a time in the 1970s and 1980s in India where almost every homemaker prided herself on her prowess as a flower arranger. In fact, in some cities there were flower-arrangement themed kitty parties where the participating woman competed with each other based on their personal creativity in flower arrangements. This tradition continues to flourish. Chandigarh has its Rose Festival, while Delhi has its annual YWCA Chrysanthemum Show, Rose Society of India Rose Show and every spring the Mughal Gardens at Rashtrapati Bhavan are opened to the public, with their renowned rose garden and bonsai collection. Although the contemporary world is becoming more pragmatic, particularly within its built and lived environment, flowers, because of their ability to refresh our spirits with their fragrance and aesthetic appeal, have not lost their relevance to our daily lives.

Below: A beautiful arrangement of protea and pink rose flowers

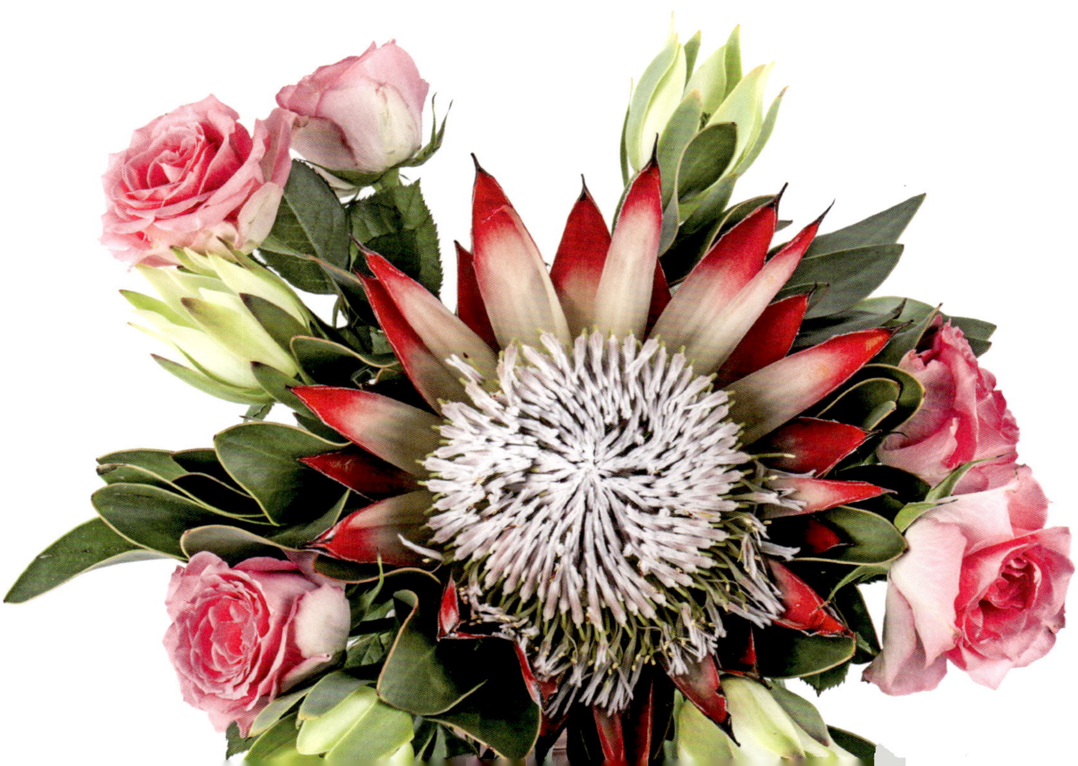

OFFICIAL FLOWERS OF INDIAN STATES AND UNION TERRITORIES

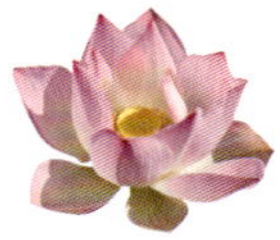

ANDHRA PRADESH
Common name Lotus
Nelumbo nucifera

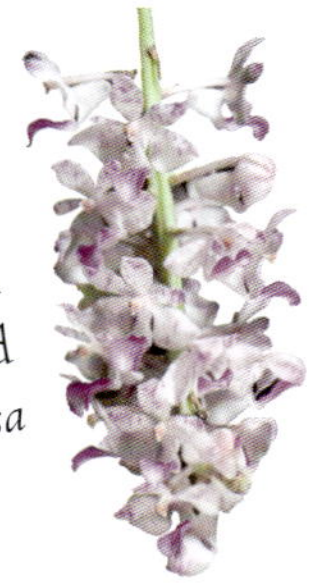

ARUNACHAL PRADESH
Common name Foxtail orchid
Scientific name *Rhynchostylis retusa*

ASSAM
Common name Foxtail orchid
Scientific name *Rhynchostylis retusa*

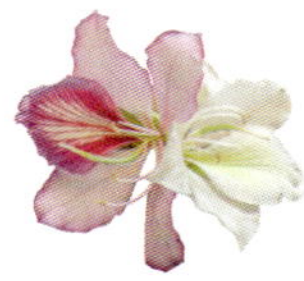

BIHAR
Common name Orchid tree (Kachnar)
Scientific name *Phanera variegata*

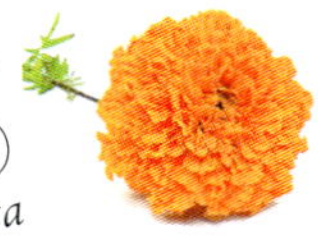

GUJARAT
Common name Marigold (Galgota)
Scientific name *Tagetes erecta*

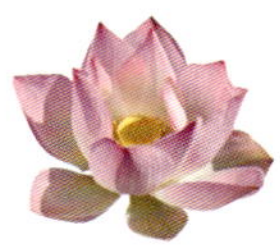

HARYANA
Common name Lotus
Scientific name *Nelumbo nucifera*

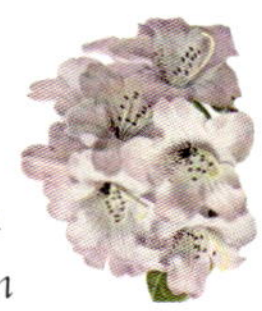

HIMACHAL PRADESH
Common name Pink Rhododendron
Scientific name *Rhododendron campanulatum*

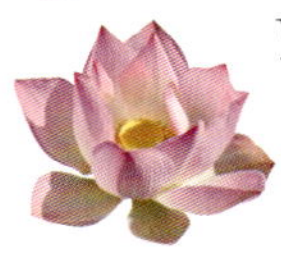

JAMMU AND KASHMIR
Common name Lotus
Scientific name *Nelumbo nucifera*

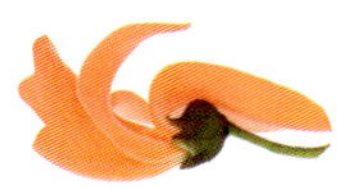

JHARKHAND
Common name Flame of the Forest (Palash) flower
Scientific name *Butea monosperma*

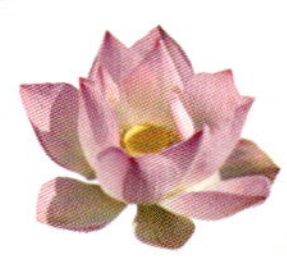

KARNATAKA
Common name Lotus
Scientific name *Nelumbo nucifera*

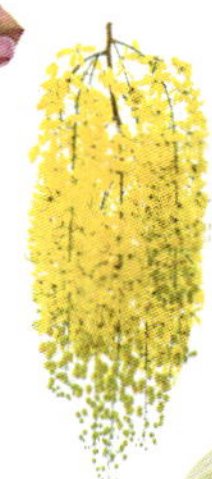

KERALA
Common name Golden Shower Tree's flower (konna poo)
Scientific name *Cassia fistula*

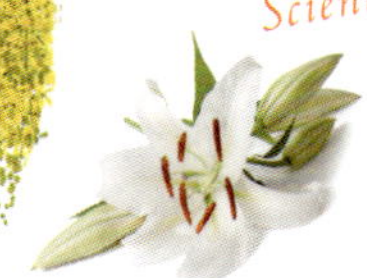

MADHYA PRADESH
Common name Madonna Lily
Scientific name *Lilium candidum*

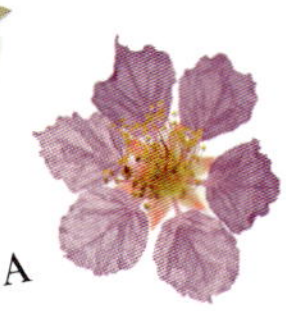

MAHARASHTRA
Common name Pride of India flower
Scientific name *Lagerstroemia speciosa*

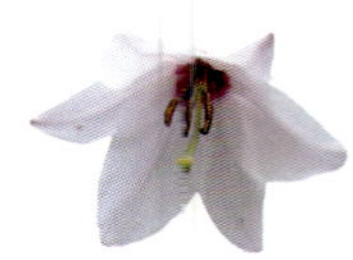

MANIPUR
Common name Siroi Lily
Scientific name *Lilium mackliniae*

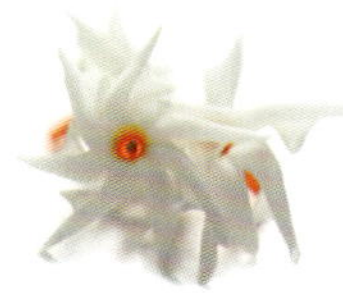

WEST BENGAL
Common name Night-flowering Jasmine (Parijat)
Scientific name *Nyctanthes arbor-tristis*

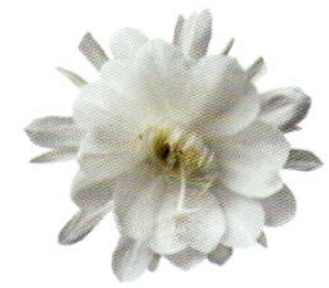

UTTARAKHAND
Common name Brahma Kamal
Scientific name *Saussurea obvallata*

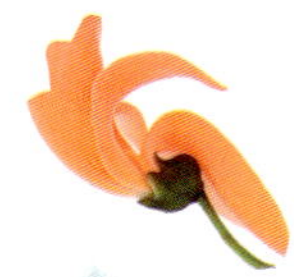

UTTAR PRADESH
Common name Flame of the Forest (Palash) flower
Scientific name *Butea monosperma*

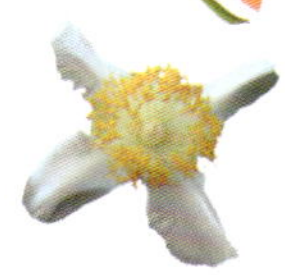

TRIPURA
Common name Indian rose chestnut flower (Nag Kesar)
Scientific name *Mesua ferrea*

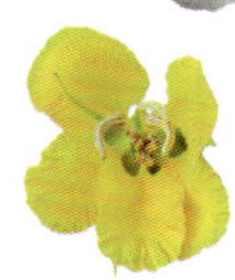

TELANGANA
Common name Ranawara flower
Scientific name *Senna auriculata*

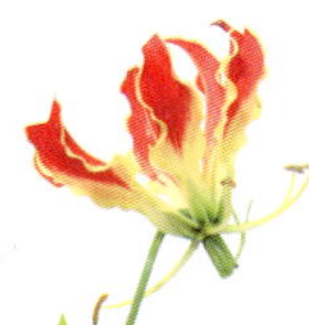

TAMIL NADU
Common name Glory lily
Scientific name *Gloriosa superba*

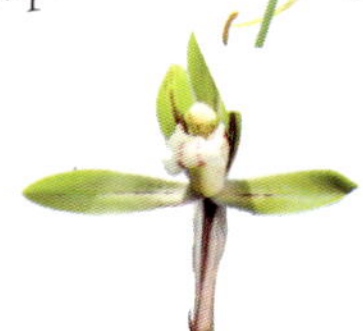

SIKKIM
Common name Dendrobium Noble
Scientific name *Cymbidium goeringii*

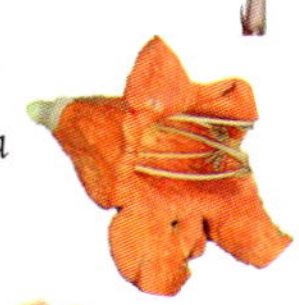

RAJASTHAN
Common name Rohida flower
Scientific name *Tecomella undulata*

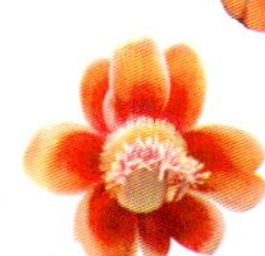

PUDUCHERRY
Common name Cannonball tree's flower
Scientific name *Couroupita guianensis*

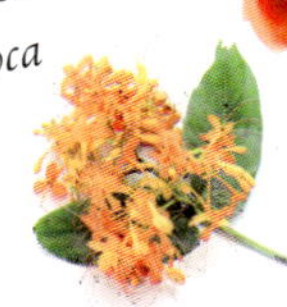

ODISHA
Common name Ashoka flower
Scientific name *Saraca asoca*

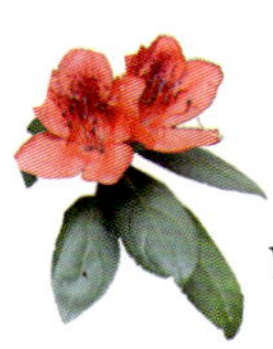

NAGALAND
Common name Rhododendron
Scientific name *Rhododendron arboreum*

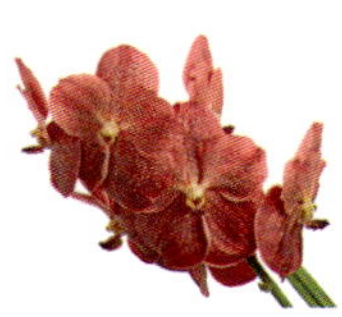

MIZORAM
Common name Red Vanda
Scientific name *Renanthera imschootiana*

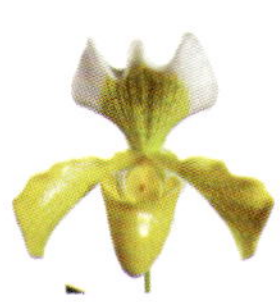

MEGHALAYA
Common name Lady's Slipper orchid
Scientific name *Paphiopedilum insigne*

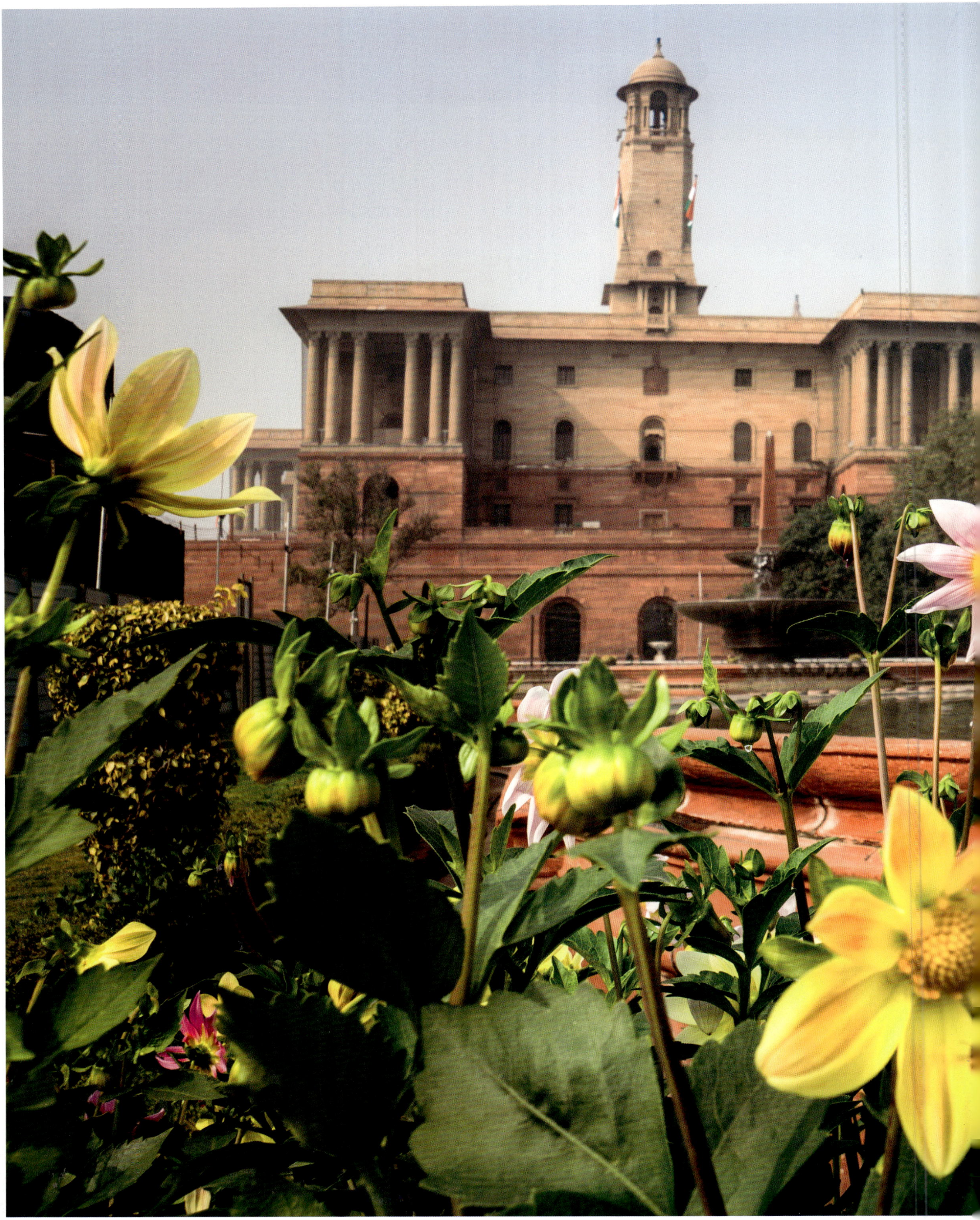

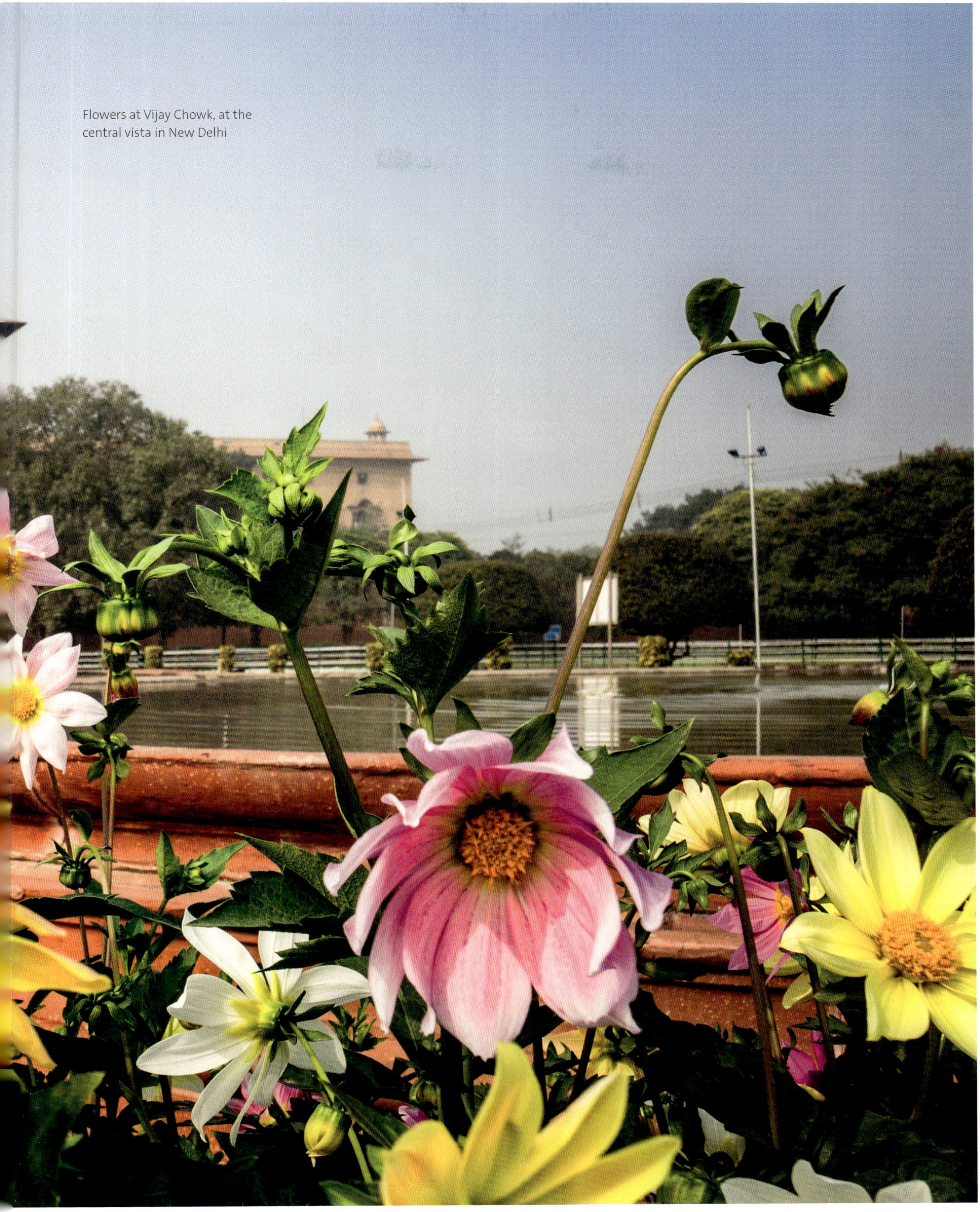

Flowers at Vijay Chowk, at the central vista in New Delhi

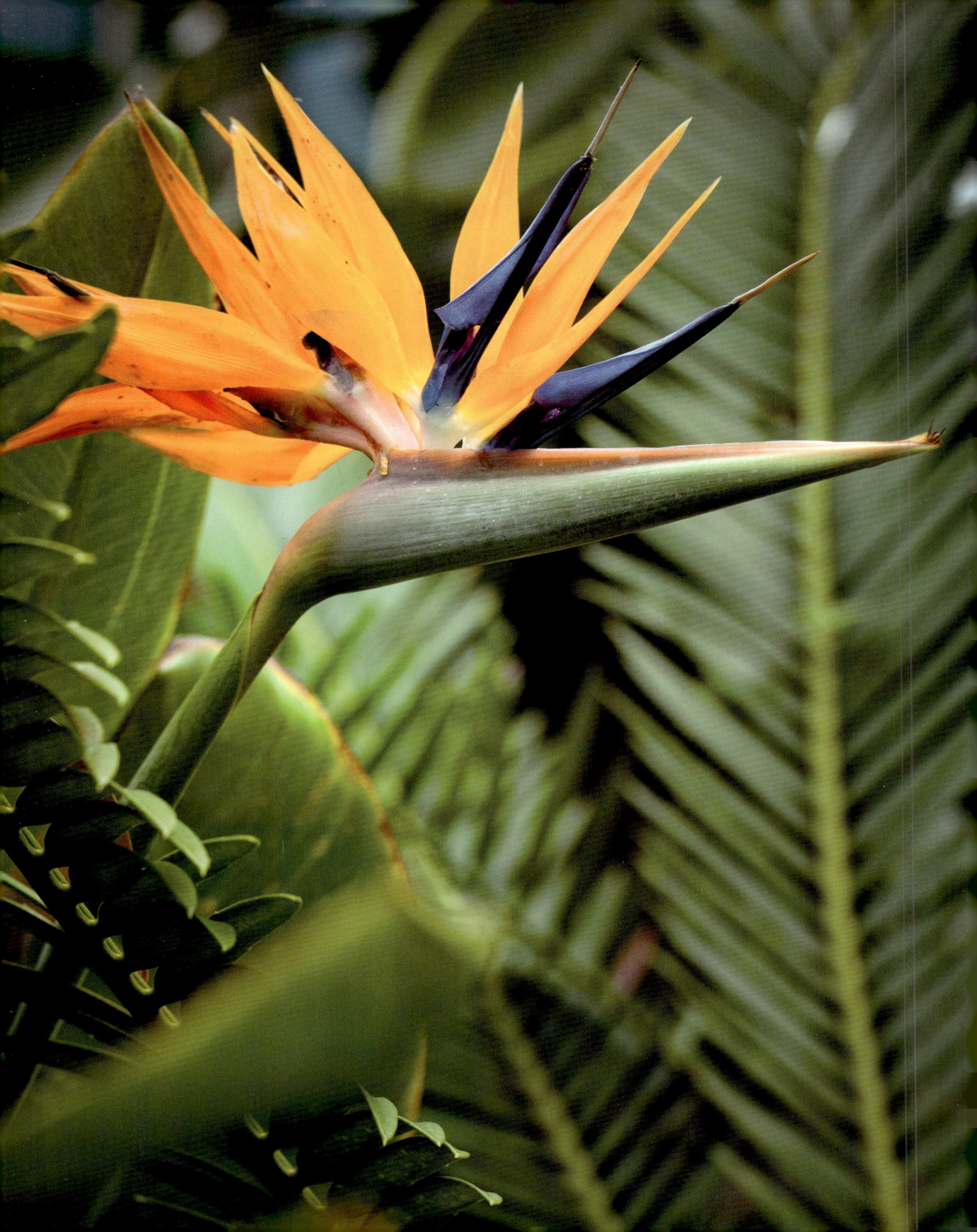

Gardens of Paradise

Beauty surrounds us but usually we need to be
walking in a garden to know it.
—Rumi

The classical Sanskrit texts, the epics and the ancient poetry of the subcontinent are replete with examples of poets who extol the wonders of the natural world. While forests, groves, orchards, bowers and jungles are all prominently featured, the garden is what is most fascinating. It is a space that is simultaneously the domain of man and nature, a place where the two coexist in perfect harmony.

Indian literature is dotted liberally with references to various gardens and groves, the ideas of which have seeped into cultural thought. As is demonstrated in the creative tradition of the subcontinent, lush green gardens are a theme and trope throughout our epics and myths. The exquisite murals of Ajanta are a record of life from the 2nd century BCE to about 480 or 650 CE and depict many flower-filled gardens. The paintings on the walls of the Mattanchery Palace in Kerala have some of the main protagonists painted in the stylised gardens as conceived by the artists. Gardens have always been an inspiration for miniaturists as well, as seen in the illuminated Jain and Buddhist manuscripts where they feature prominently.

Facing page: Bird of Paradise flower

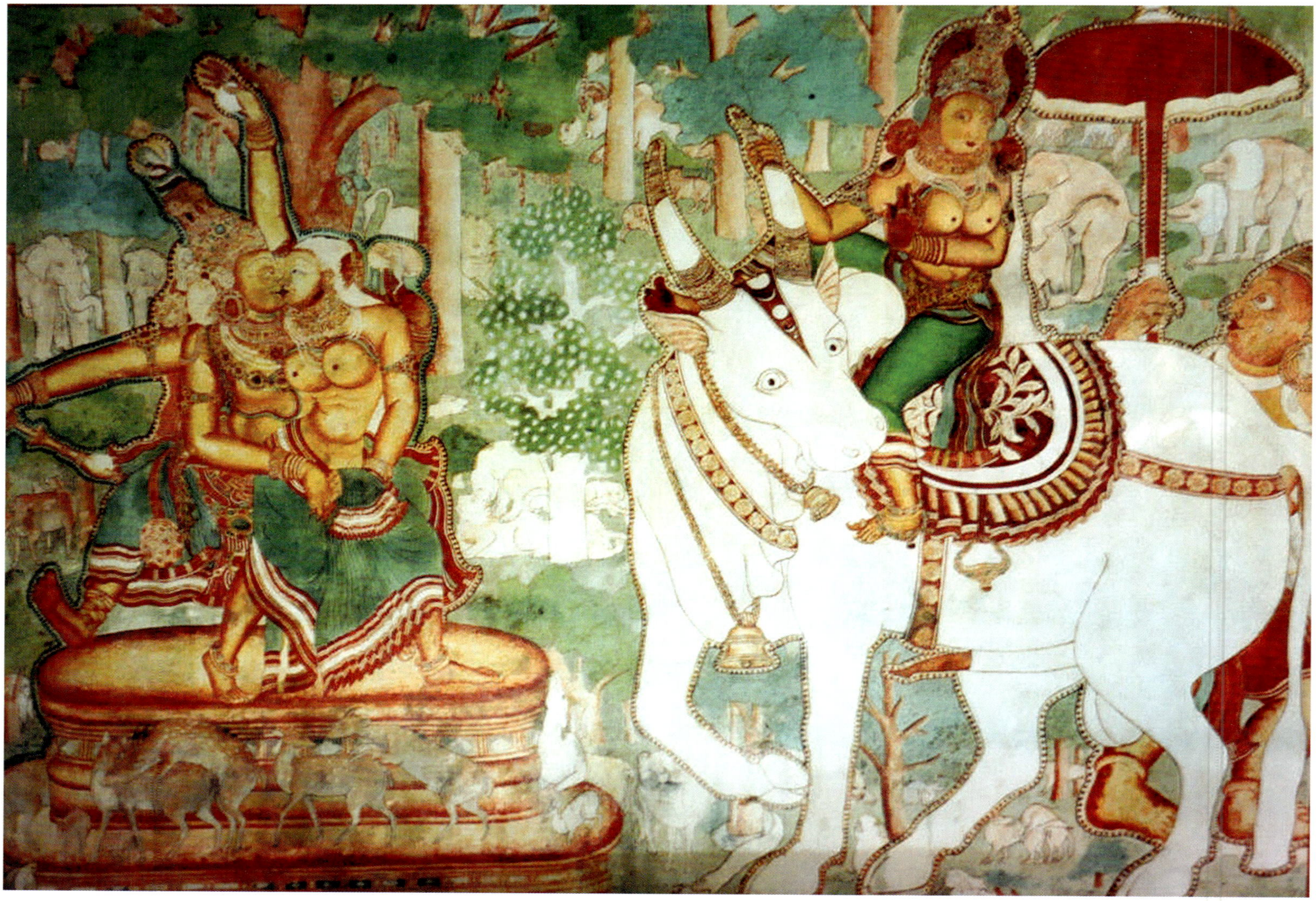

'Shiva and Mohini embrace in a garden, while Parvati looks away in distaste', mural at the Mattancherry Palace, Kochi, Kerala

As Chintaman Vinayak Vaidya puts it,

> There are descriptions of gardens all over the country in both the epics…The *Champa* garden to the north of the capital of the Angas and the *Priyaka* tree garden at Ujjain have…been spoken of. It was also a custom, which has unfortunately fallen into disuse now, for the young girls of a town to go to these gardens for airing and play in the evening. It was also considered of great merit to plant groves of trees usually mango trees in the vicinity of towns and cities which could be used as pleasure resorts for men and women.[1]

CELESTIAL SPACES

Gardens, mythical and mundane, are extremely various:

> Edenic, Arcadian, pastoral, paradise, natural, wild, formal, plantation, vegetable, fruit, flower, botanical, physic, orchard, arboretum, allotment, walled, country, city, park—each of these terms is familiar in relation to gardens; each gives rise to other issues, such as: myth, cultural specificity, historical context, nation, topography, class, race, religion, gender.[2]

Right: 'Mohini', Raja Ravi Varma, oleographic print, c. late 1800s

Below: 'Arjuna Subhadra', Raja Ravi Varma, oleographic print, c. late 1800s

A cursory glance through the stories of the world leads one to realise that heavenly gardens of paradise are a common motif throughout the mythological stories in most ancient cultures. Famously Adam and Eve resided in the Garden of Eden before their fall from grace. There is the ancient Greek and Roman Garden of Hesperides. Indra, the King of the Gods in the Indian pantheon, presided over Swarga or Heaven, within which was the heavenly city of Amravati, featuring the celestial gardens of Nandana. In this garden is said to be located Lord Indra's palace, Vaijayanta, which is surrounded by splendid fruit trees of every kind.

The Indian literary and artistic tradition is replete with garden imagery. The important text, *Bhagavata Purana,* which was a constant source of inspiration for traditional Indian artists, has numerous interpretations of the gardens.

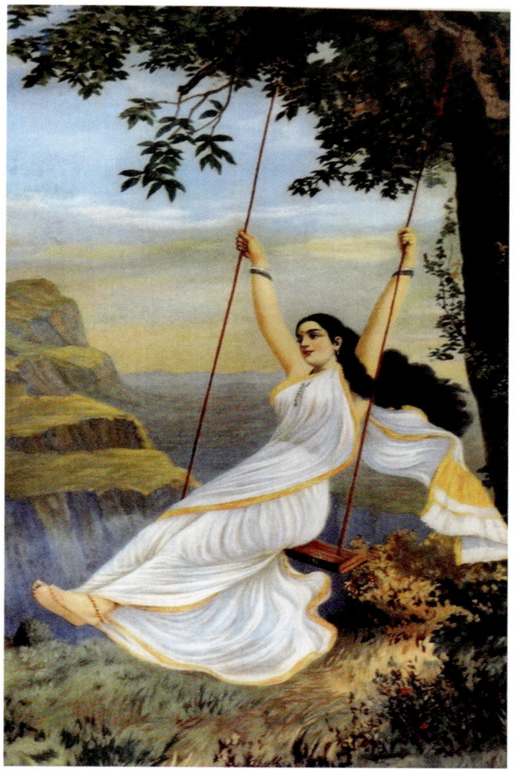

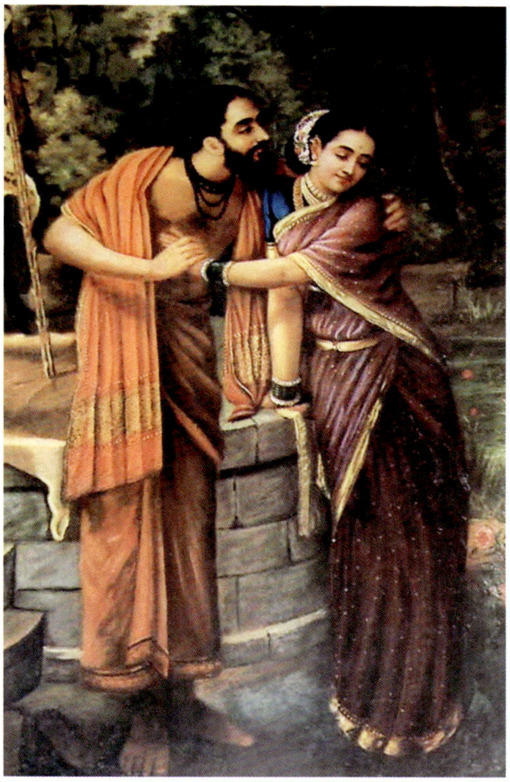

Gardens were also fertile ground for lovers' meetings and subsequent trysts. From Dushayanta and Shakuntala who met while the latter was watering plants in a sacred grove, to Rama and Sita who first crossed paths in the *Pushpa Vatika,* and finally to Radha and Krishna, who would play love games in the Vrindavan gardens, these were verdant locations where love blossomed. The gardens of Vrindavan were where the *madhavilata* were said to flourish in great profusion; the tri-coloured flowers filled the gardens with a heavenly smell, quite literally evoking the *shringara rasa* or erotic essence.

The *Ramayana* and *Mahabharata* have lush descriptions of heavenly gardens which have a deep paradisiacal imagery to them. The Tulsidas *Ramayana,* includes a scene in the *pushpa vatika,* where Ram encounters Sita for the very first time among the creepers and charming trees. In yet another tale at the conclusion of the Lankan war in the *Ramayana,* Kubera, Ravana's brother, eventually left Lanka to move to Kailasa where he built Alaka, the secret city. In this city was established the splendid pleasure garden known as the Chaitra, which was said to be as marvellous as, if not more than, Indra's Nandana in the latter's heavenly abode of Amravati. Thus, there is also a deep connect between the sacred, the celestial and gardens.

PLEASURE SPOTS

Classical Sanskrit poetry is rife with references of the *dolagriha*, the pleasure gardens with swings, where *rati krida* or love play is often set, with amorous couples enjoying the pleasure of love making. *Svapnavasavadatta* by Bhasa, which can be traced to the 1st, 2nd or 3rd century by varying accounts, even speaks of the erotically charged symbolism of the birds and the bees that populate the gardens in Indian art.

Kalidasa, the king of *shringara rasa* or erotic mood, sets many of his writings in a garden. Much of Act I of his iconic work, *Abhigyanashakuntalam*, is set in a bower where Shakuntala is watering plants as Dushyanta surreptitiously admires her beauty. The dialogue is seamlessly woven with natural imagery to great effect. Complaining to her friends that she has been pricked in order to tarry a while next to the King Dushyanta, Shakuntala says: 'Anasuya, my foot is cut by the point of a fresh Kusa blade and my bark-dress is caught in a Kurabaka twig. Wait for me while I loosen it.'[3]

In this manner, the trees and flowers almost become another character in the story, providing as they do, a reason for the lovers to stay awhile in each other's company.

Kalidasa also favoured the *dolagriha*, an imagery that is most associated with Radha and Krishna, the rope coiled around with creeping flowers, hanging from a sturdy tree bough against a backdrop of lush green. In Kalidasa's work *Malvikagnimitram*, the king abandons his plans to swing with his queen Iravati, so he can spy on the object of his affection, Malvika, hiding himself in the thickets of the garden. It transpires that Malvika is in the garden to encourage the *Ashoka* tree to blossom, something that is said to happen when a lovely and beautifully attired maiden touches the tree with her foot. Kalidasa's earliest epic poem, the *Ritusamhara*, delves into the intricacies of the six Indian seasons, where much action is in the natural environs, and the gardens are integral to the landscape, connecting and contrasting them against an erotic love story, brimming with the *shringara rasa*.

Another of Kalidasa's protagonists, King Agnivarna, copulated with his female attendant among garden creepers, in the work *Raghuvamsa*.

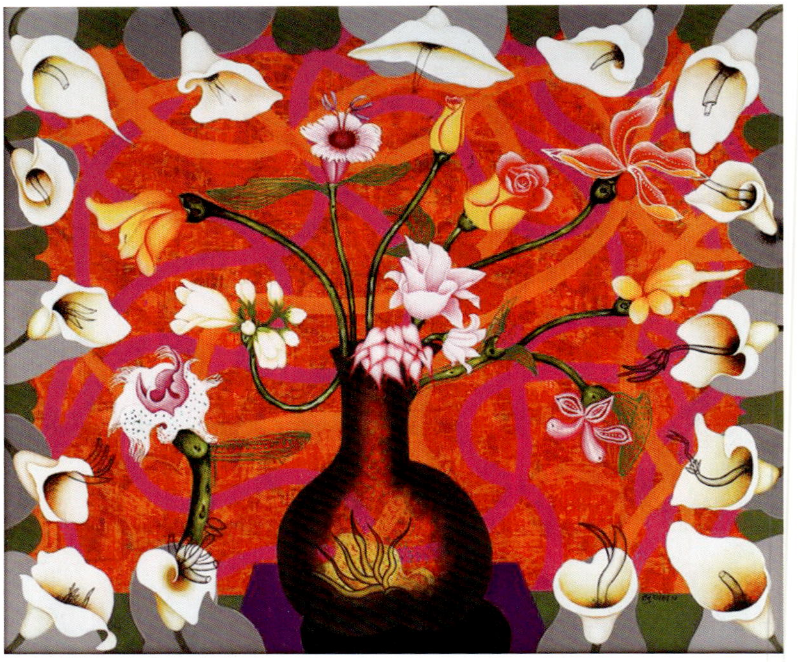

'Flowers from heaven 28', Manu Parekh

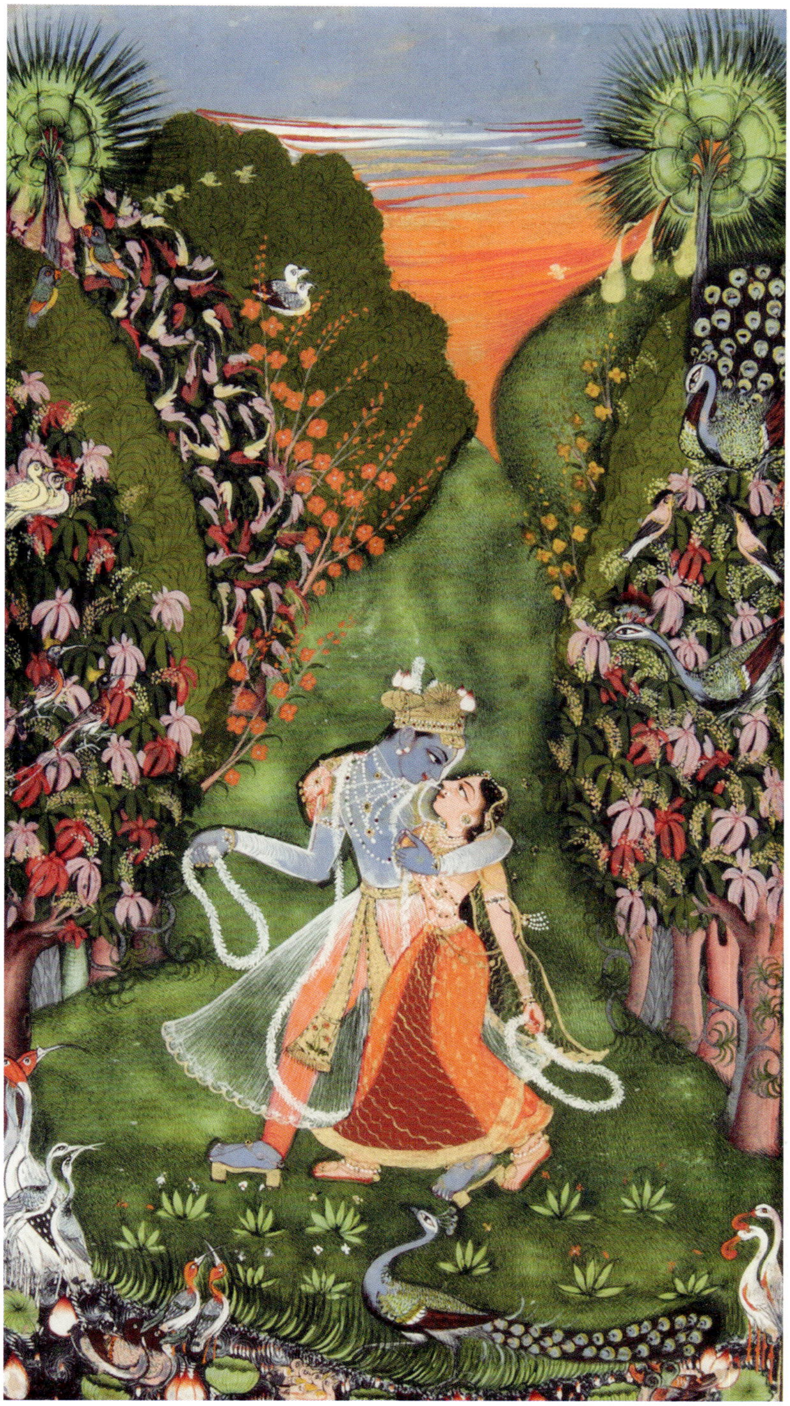

'Radha and Krishna walk in a flowering grove', The Kota Master; Ink, opaque watercolour, and gold on paper, Rajasthan, early 18th century

EVERGREEN VRINDAVAN

The lyrical poetry of Jayadeva sets the scene for the divine lovers, Radha and Krishna in Vrindavan.

> In the Vrindavan garden flowerful,
> Sore to those who are love-forsaken,
> With bowers jovial with roving bumble-bees
> And lively with the call of cuckoos,
> Caressed by the merry wind cool and redolent
> Waving from the verdurous Malaya mountain
> That sets the slender lavanga vines,
> Spring thrill'd anon to toss and swing
> Around which bees vibrantly wind,
> Multi-tinted butterflies flutter,
> And tiny song-birds twitter.[4]

The holy city of Vrindavan, translates into '*tulsi* forest' and is named for a sacred *tulsi* grove. The name Vrindavan and all the verdant connotations that are associated with it immediately bring to mind a young, virile Krishna frolicking with *gopis* and making intense, passionate love to his beloved, Radha; their love story being the basis for many devotional poems in the Bhakti genre.

As Ruth Vanita writes,

> In Radha-Krishna poetry in Sanskrit and Brajbhasha, the rural field or grove is primarily meant to grow food, though lovers also tryst there. But the city garden of *rekhti* and nonmystical *rekhta* is only for pleasure—it typifies that most urban of indulgences, the excursion *(sair)* that has no utilitarian purpose. Gardens and orchards were generally on the outskirts of medieval cities and were thus liminal spaces between urban and rural.[5]

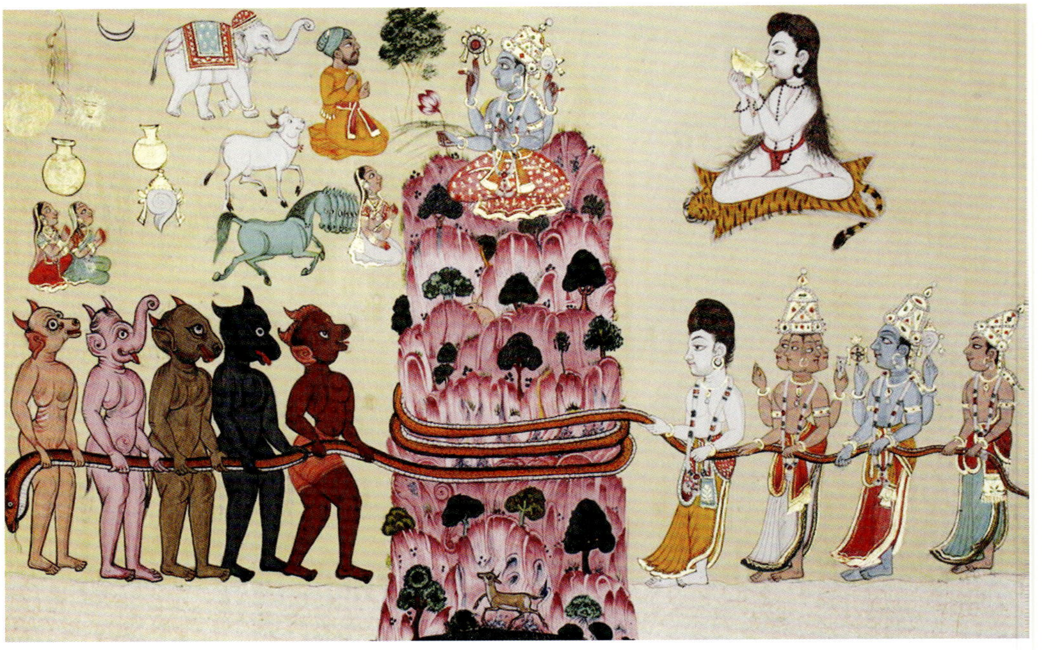

'*Samudra manthana*—the Churning of the Ocean', Maharashtra, opaque watercolour with gold on wasli, c. 1800; note the Kalpavriksha, sacred wishing tree, emerging from the churning process

On close observation it becomes exceedingly clear that the flora and fauna of India has much to do with the idea of beauty itself and the *nayika* is most often compared to various flowers, breasts like the *bilva* fruit, eyes like lotus petals, teeth like the jasmine flower, a vulva like the lotus bud. The abundant natural imagery was inherently connected to sensuality and especially to highlight her feminine beauty and her sexuality. In the reading of pre-modern and early medieval texts it is observed that the ideas of love, beauty, fertility and flora and fauna are intricately and inextricably linked.

KALPAVRIKSHA —THE TREE OF LIFE

One of the important residents of the garden is the tree. There are various kinds of trees from flowering trees to trees that bear fruit and even those that grant wishes. While this book concentrates on flowers, it is also important to understand how trees are also part of the kingdom of flora.

The mythic icon of the Kalpavriksha goes back to the legend of the *Samudra manthan* (churning of the ocean).

Many stories tell of the treasures that were churned from the depths of the ocean of milk. This *samundra manthan* yielded Kamadhenu, the divine cow; Lakshmi, the goddess of wealth; *amrit* (heavenly nectar) and Kalpavriksha, the sacred wishing tree. This tree was believed to be able to grant an individual everything they wanted, the *devdasi* and poetess Muddupalani even comparing Lord Krishna's generosity with the Kalpavriksha as he rained his largesse on everyone. The Kalpavriksha is also called the World Tree. There are accounts that narrate that Indra, the chief of the gods, took the tree to his abode Amaravati. Amaravati is said to be a paradise because of the divine presence of Kamadhenu (also called Surabhi), the divine cow, as well as of the wish granting Kalpavriksha. It is supposed to be Krishna, however, who, at the bidding of his wife Satyabhama, went head to head with Indra to bring this celestial tree to his abode on earth, a well-known representation of this episode

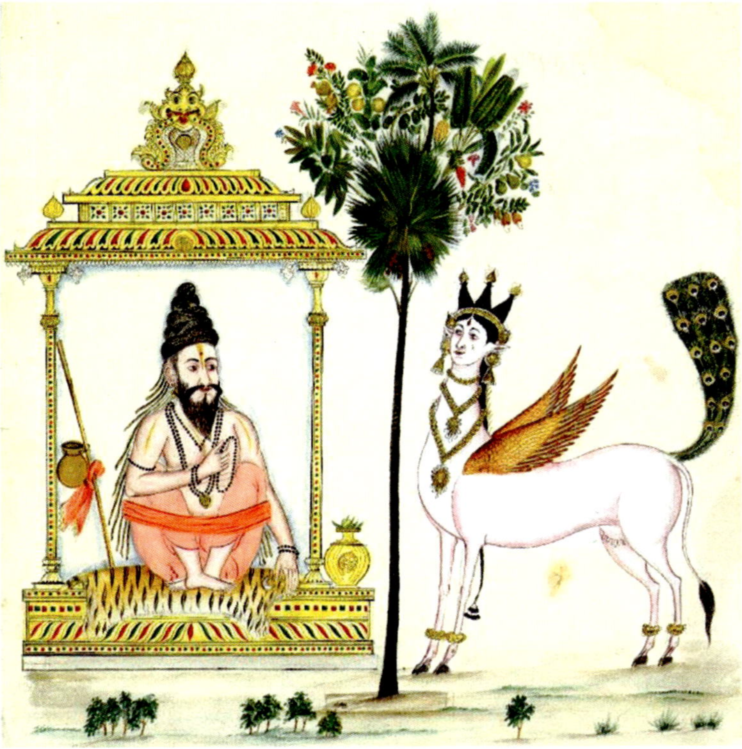

Painting of a rishi, Parijata tree and Kamadhenu, paint on paper, c. 1820

being at the Angkor Wat temple complex in Cambodia, a part of the Pan-India culture that touched many other countries.

The idea that trees are hallowed is deeply rooted in Indian thought, making it of prime importance in the customs and rituals of the subcontinent. It is believed that sacred trees root themselves deep into the earth and thus absorb the creative juices into themselves. This sustenance enables them to reach for the sky, the abode of the gods. In one myth, the earth assumes the form of a mountain floating on water and it is this mountain that remains the cosmic centre, fixing the earth in its place.

In another account, Indra, the Chief of the Gods, used a cosmic pillar or tree to hold the sky up. Thus, there was able to be life on earth, when the earth and sky were separated. The tree establishes an order and keeps things in place, thus forming a stage for all of life to play out on.

Early versions of the Tree of Life feature the date palm, the fig tree, the pine and cedar. In India, the Tree of Life symbolises fecundity but also connects death and life, the reason for which can be traced back to the belief that trees house spirits and that the living must honour them. On the subcontinent it is the flowering *Pipal* or Indian fig tree that is most commonly believed to house spirits.

Other trees identified with the Kalpavriksha are the sandalwood tree, the parijat tree and even the coconut palm (as the latter seems to fulfil so many of our necessities from food to drink to fuel, fibre for use and more), which are all flower-giving trees.

There are more tales of this ancient, magical wish-fulfilling tree variously called the Kalpataru or Kalpa-Druma or Kalpavriksha. It is a mythic icon and there are plentiful visual manifestations of it to be found in textiles, carpets, sculptures and the visual arts. It is a constant within Buddhist and Jain and, of course, Hindu iconography.

It is often depicted with precious treasures at its roots. Among the most marvellous examples is the one at Besnagar, possibly a crucial part of the ornamentation of a pillar adjacent to the Lakshmi or Kubera temples (both deities being patrons of wealth), which depicts the tree with gold, a conch and lotus at its roots.

One of the oldest representations of such a wish-fulfilling plant was perhaps the *kalpavalli*[6] or wish-fulfilling creeper, a representation of which is found at the Bharhut railing. It is depicted sumptuously hung about with vestments worked with stunning patterns, precious jewels, heady wines and, of course, profusions of flowers. The description of this vine closely resembles the divine flora that Kalidasa spoke of at Alaka, where Lord Kubera, the God of Wealth resides. These plants attend to all the needs of the maidens residing there. A similar representation of the tree, which yields bounties of jewels, silks and treasures, can be found at Nachna in Madhya Pradesh.

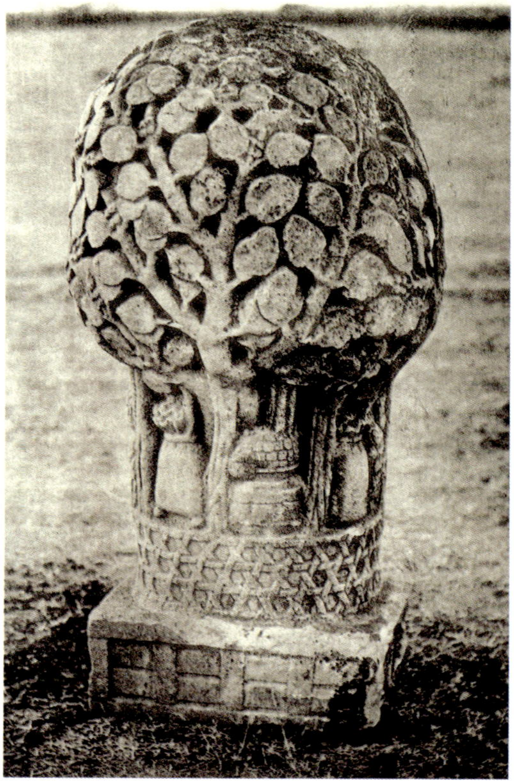

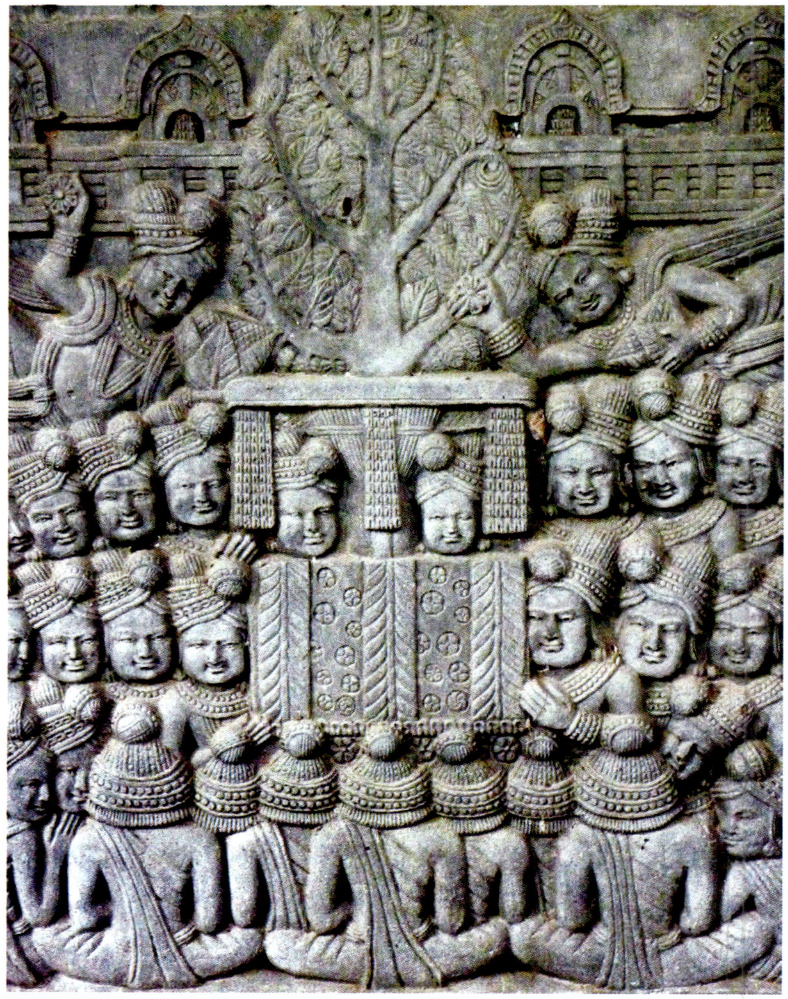

An important literary motif, as well, the Kalpavriksha has made an appearance in Kalidasa's play *Abhigyanashakuntalam*. Shakuntala was a humble forest-dwelling girl and when she was setting out to win back her love and husband, the king who had been cursed to forget her, her friends the woodland spirits (whom she had been kind to) offered her costly silks and wonderful jewels from the Kalpavriksha, so she could swathe herself in finery before she left for the royal palace. In a way that is similar to the celestial wish-granting plants at Alaka, a hand extends itself from the branches and offers the boons, whether food and drink or fine clothes.

According to Buddhist and Hindu mythos, the guardians of the Kalpavriksha are the Kinnara (male) and the Kinnari (female). These are a flying *devata* and *apsara*, half-avian, half-human, sometimes depicted alongside the tree; the divine beings, like the tree, indicating prosperity. A famed example of this can be found in Java, Indonesia in an 8th century shrine at the Pawon temple, a Buddhist temple somewhat close to the Borobudur and Mendut temples (Indonesian culture having beautifully assimilated many Indian influences under the sub-continent's Pan-Indian influence). The Buddhist *jataka* tales also describe wonderful wish-fulfilling trees, in which dwell helpful otherworldly beings called *vanadevatas*/tree goddesses that come to the aid of passing travellers.

The Kalpavriksha is also an important motif in Jain cosmology and is seen often in the Jain miniature paintings, as also the surface

Above: Copy of Barhut relief depicting the Kalpavriksha, Wat Umong, Buddhist temple, Chiang Mai, Thailand, 1297

Above left: Besnagar sculpture of Kalpavriksha, Bhopal, 1897

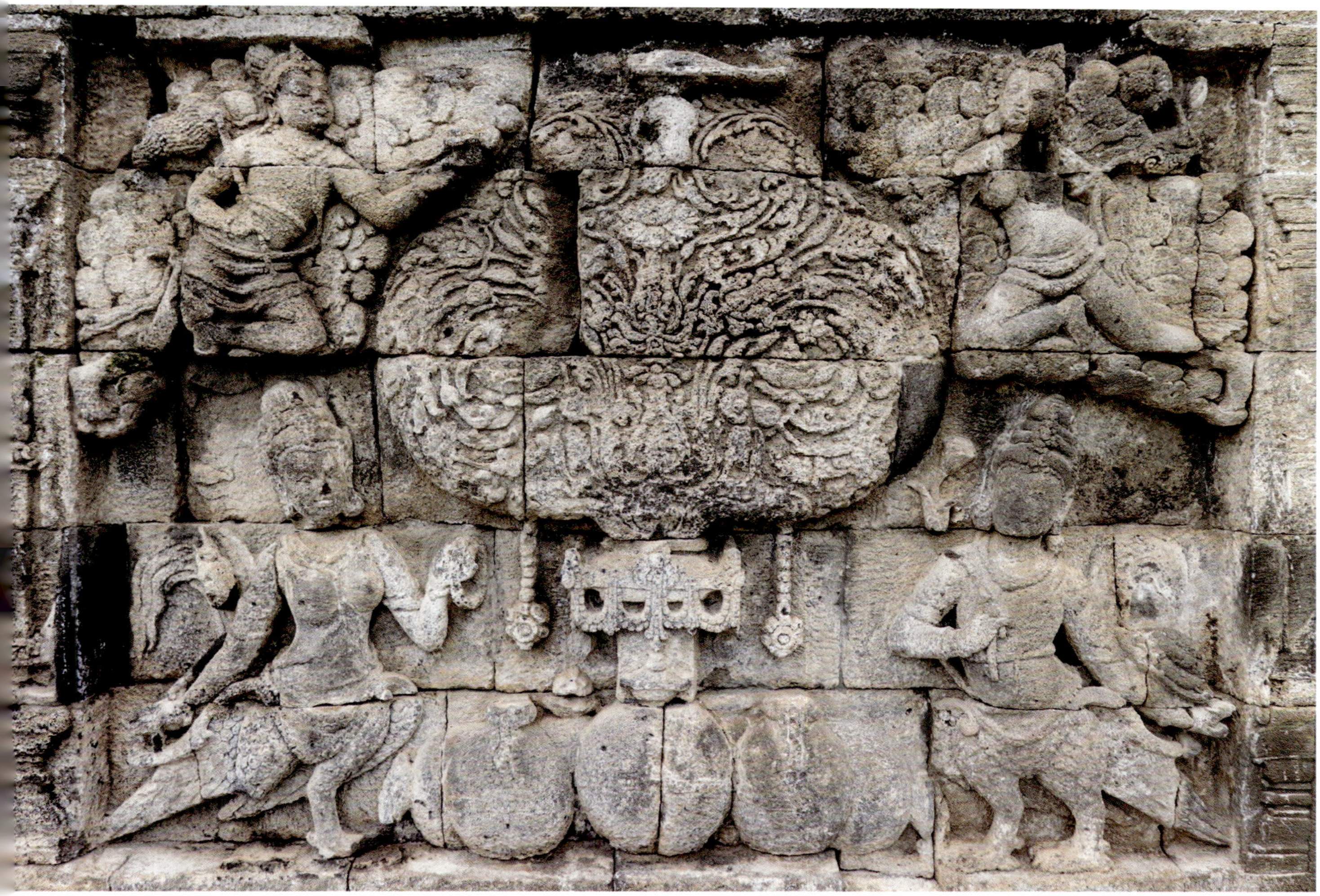

Relief of Kalpataru Tree, on the outer wall of Pawon Temple, Borobudur, Java, Indonesia

sculptural traditions of the numerous Indian temples. Rarely is the Kalpavriksha absent from traditional Indian painting and sculpture through the art historical narrative.

ISLAMIC GARDENS AND LANDSCAPES

Islamic gardens were not something the Mughals gifted to India. Long before their time there were signs of a Tughlaq garden that dates back to the 14th century. This garden even included pavilions and a system for water. A number of Islamic gardens from the 16th century were also attributed to the Lodi monarch, Sikandar Lodi. It cannot be said that the Mughal gardens were uninfluenced by those that came before them, given the Mughal thirst and curiosity to understand and absorb ideas from around them.[7]

Babur is generally given the honour of being the one to introduce the concept of the paradisiacal gardens to India.[8] The Timurids had used them as splendid camping grounds. At Dholpur, in Rajasthan, Babur built the famed Lotus Garden as he was so taken with the beauty of the area. This intimate and private garden was unique in its design, so named for a number of pools that illustrated the lifespan of a lotus from bud to flower and then to withering blossom.

Alongside the garden, built from the rocky outcrops, was a tank and lotus ponds that were linked with channels of water. The effect was one of unity built upon a platform of level

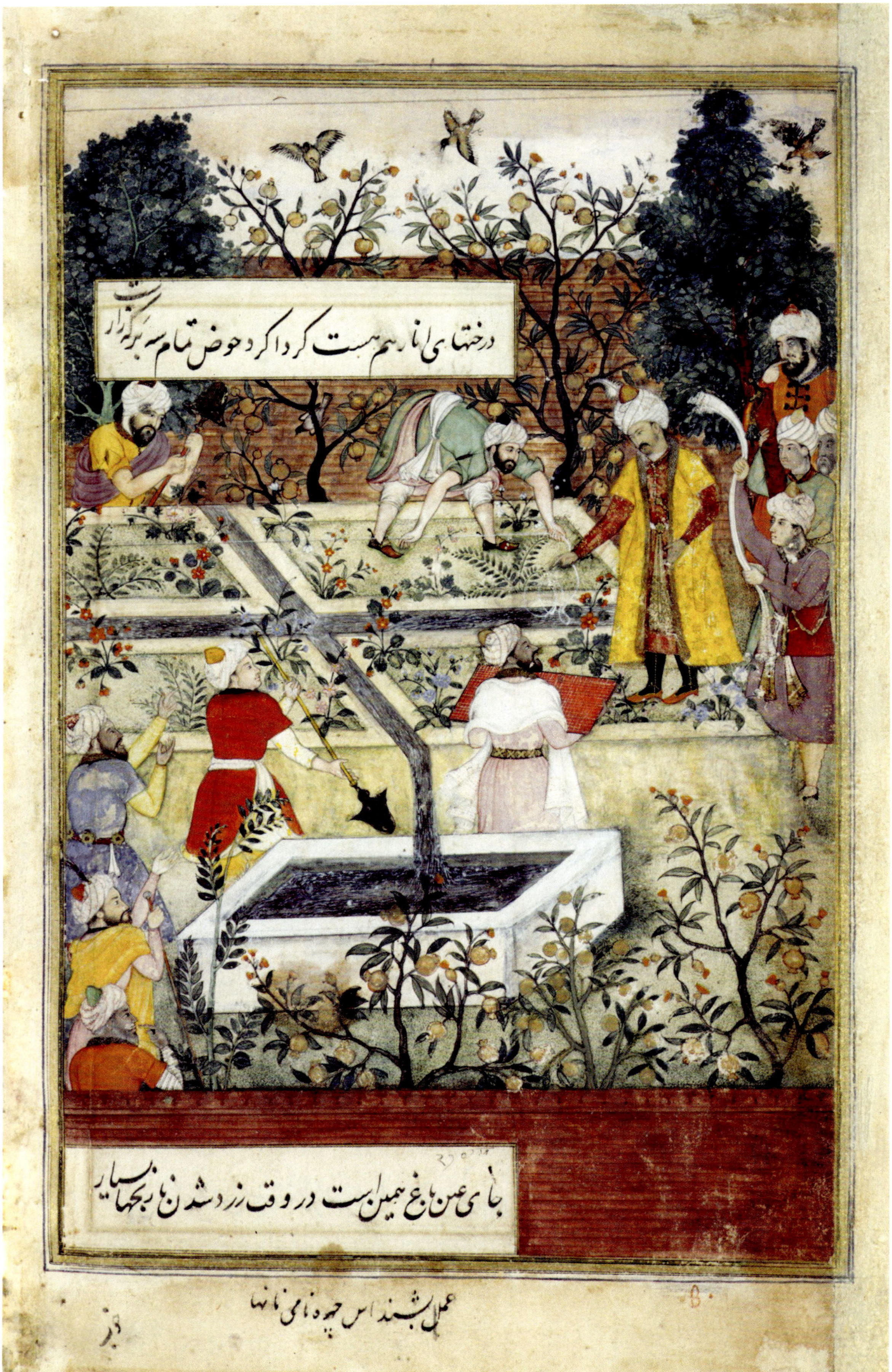

درختهای انار هم هست گرد اگرد حوض تمام سبزه زار
جای عجب باغ همین است در وقت زرد شدن بگهها بسیار
عمل بشنداس چهره نامی نانها

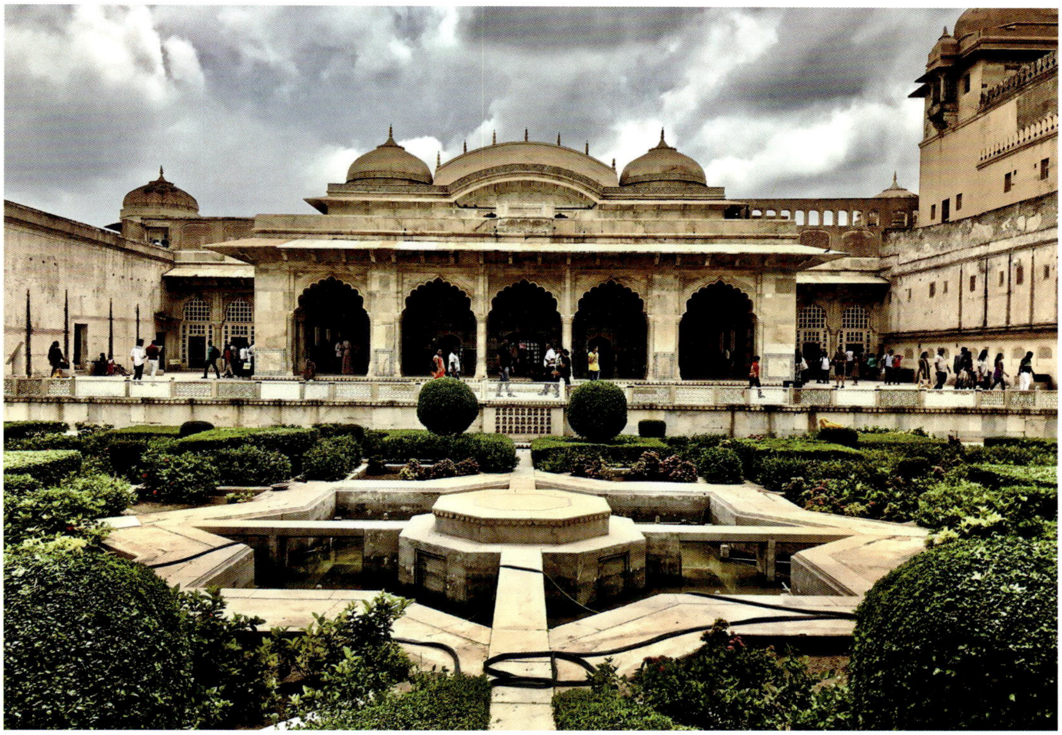

Above: Babur's Garden is a *charbagh* styled green space located in Dholpur, Rajasthan

Facing page: 'Babur supervising the laying out of the Garden of Fidelity', folio from a *Baburnama*, 16th century

stone. It also had a mosque and an aqueduct of dressed stone. The latter was the only one of its kind in Mughal architecture and was as beautiful as it was functional, an aesthetic solution for transporting water forward as the channels and pools mounted on the platform were nourished by the watercourse and the fountains then fed the surrounding beds. One of the first examples, or perhaps even the first example of the marriage of the Indian and Mughal design aesthetics, it was a holy alliance between the best of both worlds.

Babur was greatly enticed by the wonders of nature, which is where his love for gardens drew from. He spent a good deal of his time camped out in the outdoors, living as one with nature. His sense of space was intuitive and his aesthetic unusual. He had a gift for selecting the best sites on which to build his gardens. A large number of those that he built were spread across hillsides. The view of the valleys these gardens then offered was breath-taking. Through his life Babur built many versions of the Timurid gardens.

Babur ushered in the era of the *charbagh*, so called for their geometrical and symmetrical conception based on the terraced Kabul gardens which took on their own special characteristics in the Indian climate. Their division was as per a four-fold plan, being essentially quartered. Some other early examples of the *charbagh*s were Humayun's Tomb and Bagh-i-Wafa.

It was Akbar who followed in his grandfather's footsteps to establish the spiritual successor of the Lotus palace in red sandstone at Fatehpur Sikri; the spirit of the former palace seeped into the architecture of the newer one, even in the spatial organisation as well as the buildings divorced from each other and the platform with the watercourse. This did well to establish that of Babur's Mughal successors it was Akbar who would make more headway with regards to garden architecture than any of the other emperors.

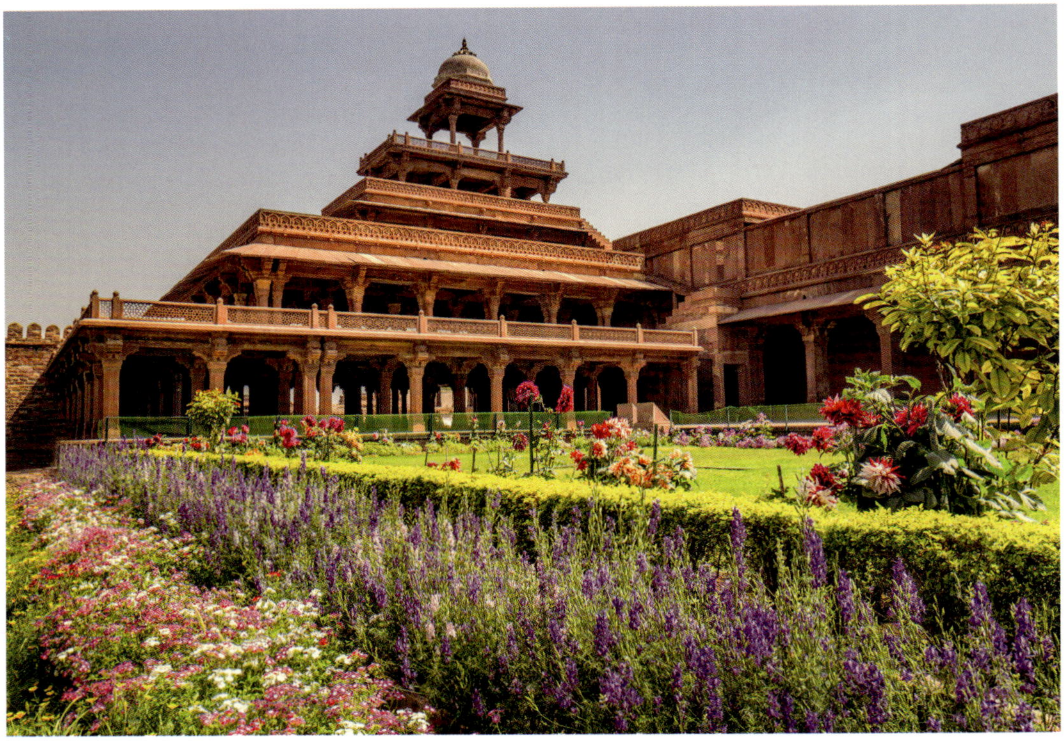

Flower garden at Fatehpur Sikri, near Agra, Uttar Pradesh

FLOWERS IN THE PHYSICAL LANDSCAPE

The gardens did not just provide food for thought for architects during the Mughal period. They went on to influence artists and craftspeople in equal measure, as their motifs were incorporated into textiles and art. Babur's fascination for the natural world led him to seek out and record the details of rare plants. He would then make sure that artists had immortalised these findings through their paintings. His great fondness for nature can be seen in his personal records which contain lyrical praises of autumn apple trees and spring flowers that could be found near the Hindu Kush range.

As we have established, his grandson Akbar the Great, was Babur's spiritual successor, given that the two developed many of the same avenues during their respective reigns. Their shared love for gardens, one might posit, is what led Akbar to handpick an image of Babur as he oversaw the building of an Irani garden as the illustration for the Persian translation of the *Baburnama*, which was carried out by a Mughal courtier, Abdul Rahim Khan-e-Khana, during Akbar's reign.

The first three Mughal emperors were legendary patrons of the arts and of architecture. Their intellect was vast and their view astonishingly liberal with regard to matters of philosophy, literature, poetry, music and the like. Their links with Iran were kept alive in their lifetime and this seeped into the literature, thought, language and even the art of the time.

When Humayun was dethroned by the Afghan soldier Sher Shah Suri, he took refuge in the Iranian court of Shah Tahmasp in approximately 1544. When he won his empire back, Emperor Humayun carried back with him the Hamza brothers and a host of Persian artists, thereby laying the solid foundation of the Mughal School of painting. This ushered in a lot of Persian floral imagery that went on to find its way into the carpets woven by artisans of the time.

The Iranian artists and craftsmen were warmly welcomed to the Mughal courts. Their work resulted in an amalgamation of styles that went on to be recognised as characteristically Mughal. This style incorporated elements of pietra dura, delicately weaving together the sensibilities of the Iranian style and of India before the Islamic era, resulting in a gentle yet vividly exultant aesthetic. Flowers were portrayed with the sensitivity based on the Iranian sensibilities of the 15th century that came together with India's inclination towards nature.

Akbar was also quite enamoured with the European aesthetic, making it his mission to promote it through his own craftsmen, even facilitating knowledge exchanges between his craftsmen and Portuguese settlers in Goa. He avidly studied the style himself through the art and textiles the European visitors to his court would bring. This gave impetus to Indian artists who were given free rein to absorb western concepts from the Europeans. The idea of perspective thus came to be borrowed and incorporated into Indian art.

Humayun's tomb, an example of early Mughal architecture, in Delhi

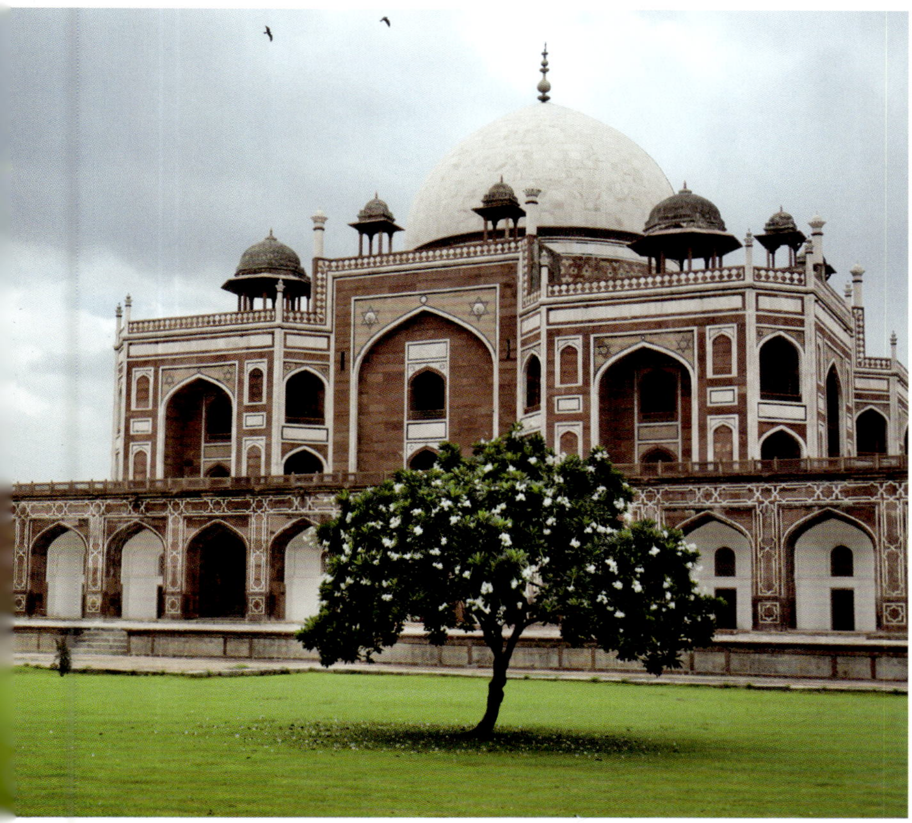

Dutch botanical diagrams and drawings were another surprising yet rich source of inspiration for the Indian artists as they came to be woven into the floral motifs in paintings, as well as in textiles and beyond. Like his sire, Akbar, and great grandsire, Babur, the Emperor Jahangir was another monarch who displayed a fierce passion for mother nature and her wonders, instructing and even travelling alongside his court artists so they could record the natural beauty, both flora and fauna, in all its glory. Under his reign blossomed Mansur, an artist whose masterful methods combined the ways of the Iranian artists with his own eye for nature. Notably, one of his works depicts a scarlet tulip. Similarly many Mughal paintings of the time depicted the subjects with a flower, often a rose, clasped in their hands.

Robert Skelton, Keeper of the Indian Department at the Victoria and Albert Museum from 1978 to 1988, was of the opinion that this artist had almost certainly developed the Mughal-style flowering plant motif. It was one that became inextricably linked with Mughal court arts during the last years of Jahangir's reign and the subsequent years as well. Skelton posits two possible reasons for why the motif of a flowering plant, at once formal yet fashioned to seem naturalistic, seemed to develop under Jahangir's rule. One was apparently the influence of Pierre Vallet's floral pattern book *Le Jardin du Roi Tres Chretien Henri Qautre*. The second influence

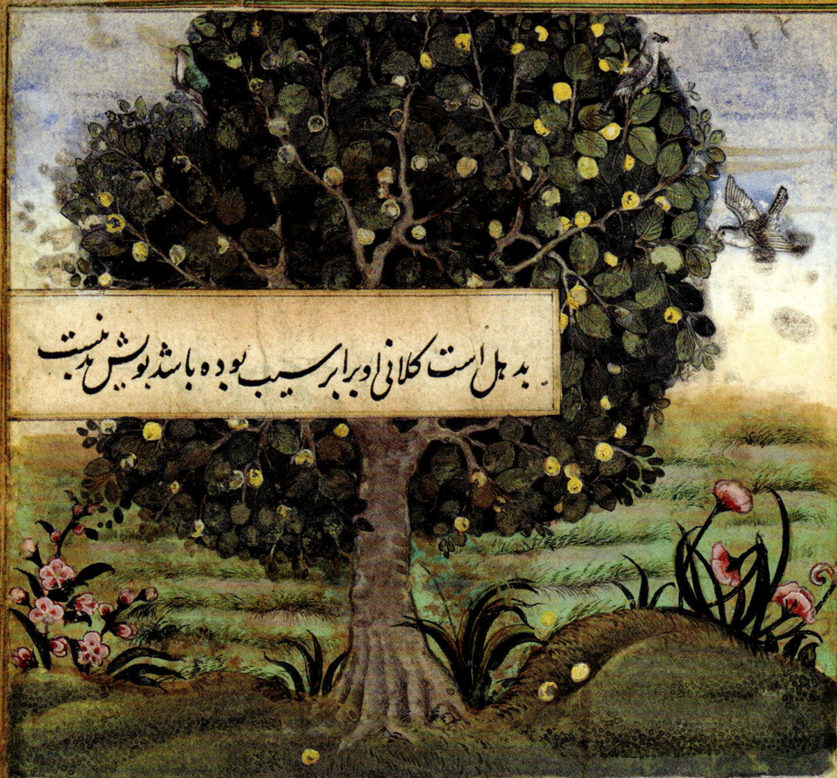

بد هل است کلانی او برابر سیب بوده باشد بویش بد نیست

غریب پست و بیمزه چیزیست یک دیگر بیر است در فارسی

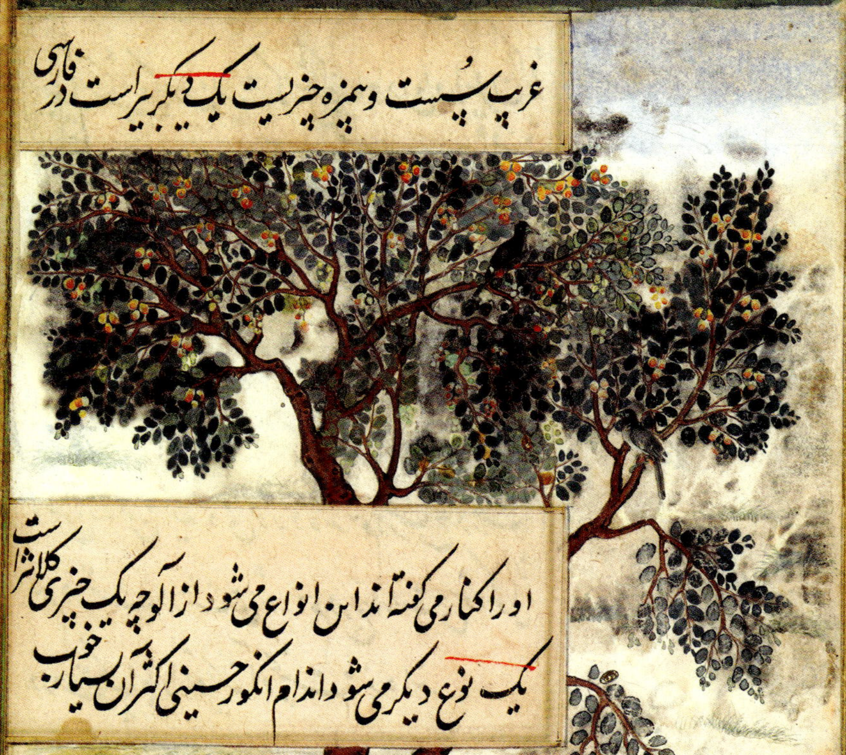

او را کنار می گفته اند این انواع می شود از آلوچه یک چیزی کلانتر است
یک نوع دیگر می شود اندام انگور حسینی اکثر آن بسیار خوب

came about during the Emperor's journey to Kashmir a few years later. It was on this trip that Mansur was instructed to paint a hundred flowers that were in bloom in the Himalayan foothills. It is believed that Mansur was briefed to emulate the style set down in Vallet's book, natural flowers in a formal position juxtaposed against a plain background. Incidentally, the flowers the great artist had been commissioned to paint were the very flowers that had caused Jahangir's forbear to wax lyrical many years ago.

This formalised style of the flowering plant went on to seep into the artistic tradition as something of a legacy, though the originals have been lost to time. It was initially observed in the decorations on the sides of Emperor Jahangir's picture albums that began to transition from depicting arabesques and landscapes to finally portraying flowering plants.

Jahangir was succeeded by his son Shah Jahan, the notoriously pleasure-seeking king. Under his reign flower motifs entered the realm of architecture. Famed for building the white marble tomb the Taj Mahal, it is unsurprising, then, that flowers crept into architecture and flooded the Mughal court, being incorporated by the dozen into tapestries, marblework, carvings, carpets, jewellery, illuminated manuscripts and more. Thus it can be said that the decorative arts were replete with floral patterns in their designs. On keener observation it seemed like the plants, though appearing naturalistic, had been reimagined and perhaps even mated with each other, which would result in puzzling profusions of flowers and leaves, that no botanist would be able to correctly place.

The Mughal emperors had a fondness for travelling across their empire for various purposes, whether it was to hunt or for more politically motivated reasons. Shah Jahan's pleasure-loving nature is witnessed in the fact that he was housed in sumptuous tents during these journeys. One of these tents, now immortalised in a Jodhpur fort incorporates luxurious red velvet furnishings, embroidered in floral motifs framed by dazzling silver and gold threads.

As might be deduced, the courtly style that evolved in the Mughal durbars went on to be the example that regional courts worked to emulate. This included the Bhuj and Gujarat courts. From the artists at the central Mughal courts were evolved the styles which were then passed down to the regional artists. In this way the chain of authority seems to have been quite rigidly delineated.

Flower gardens are very integral to Indian culture as is witnessed in the flourishing tradition in ancient times from the *van* or forest and *upvan* or mini forest to *vatika*s or gardens. These were featured prominently in ancient Indian epics such as the *Ramayana* and the *Mahabharata* and in a plethora of Sanskrit literature and poetry. The leitmotif of the garden then went on to be adopted into murals and miniatures to taking physical form as can

Facing page: 'Three Trees of India', folio from a *Baburnama*, ink, opaque watercolour and gold on paper, 16th century

Below: 'Sunlight Radiance', Maite Delteil, oil on canvas, 2016

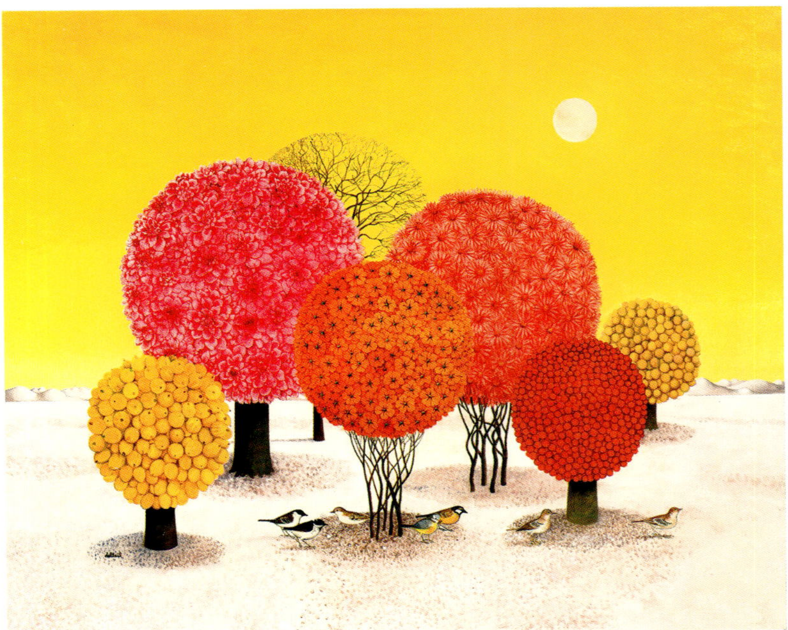

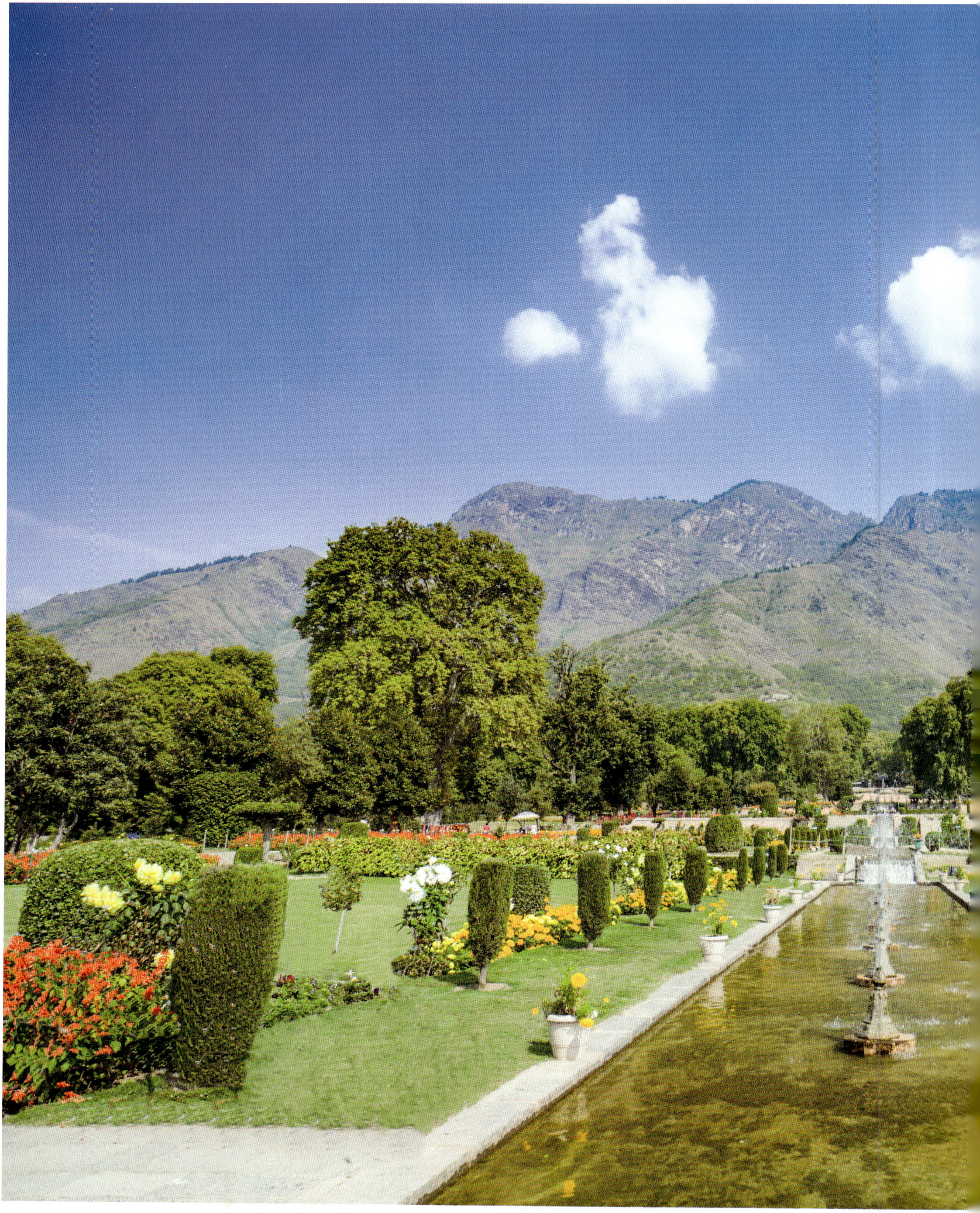

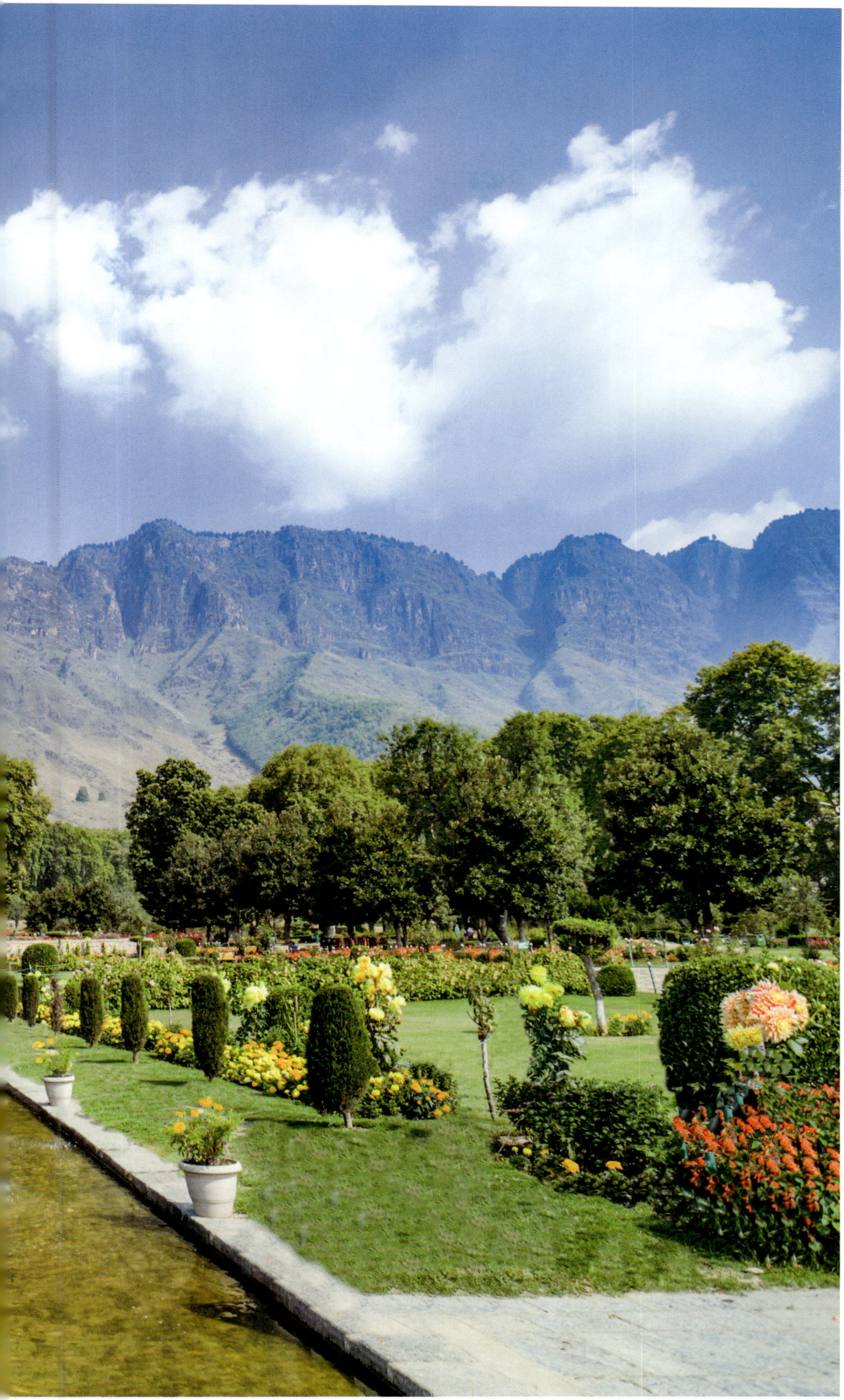

Nishat Bagh is a Mughal garden built on the bank of Dal Lake, Srinagar, Kashmir. Nishat Bagh means 'Garden of Delight' and was built in 1633 by Asif Khan, elder brother of the Empress Nur Jahan

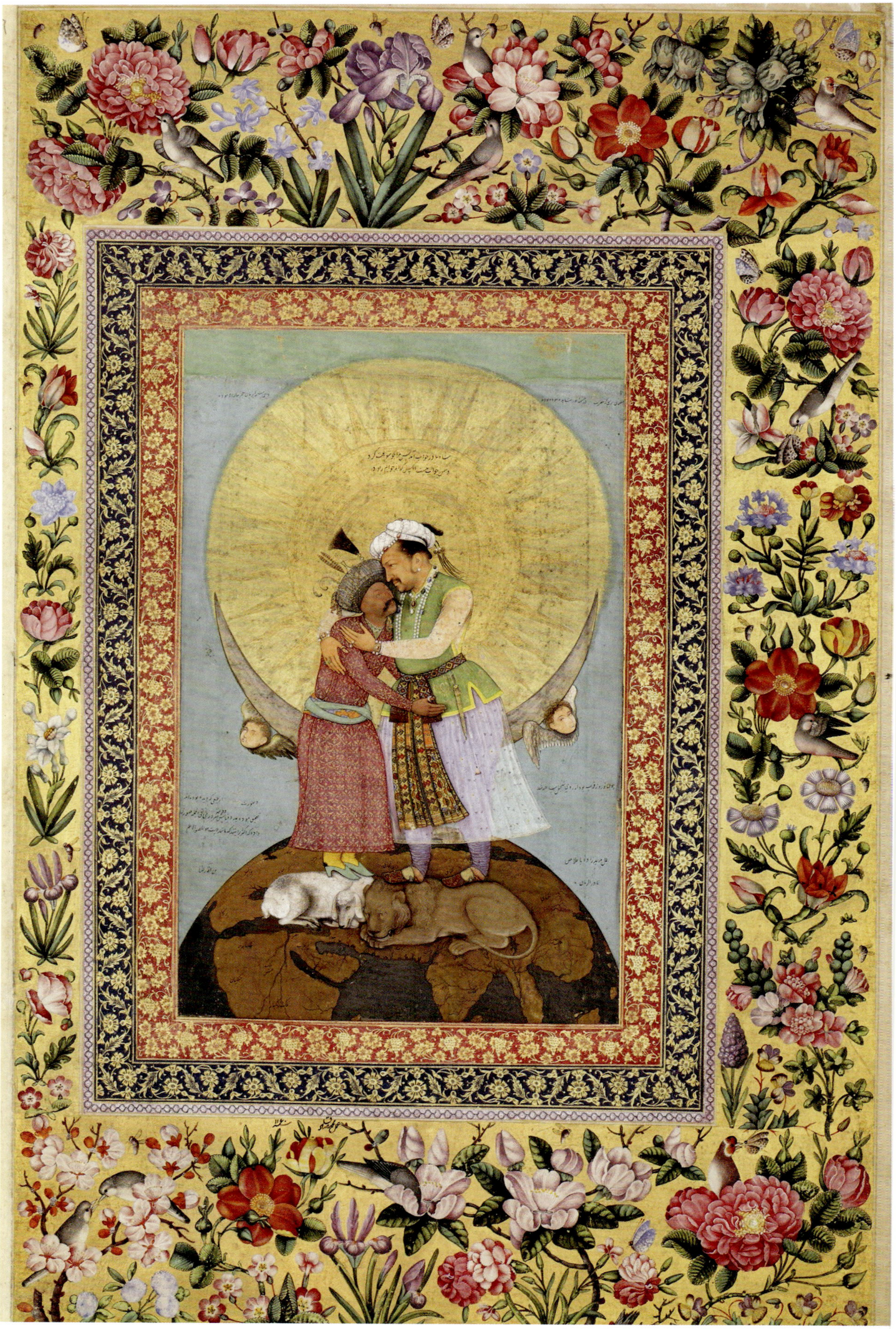

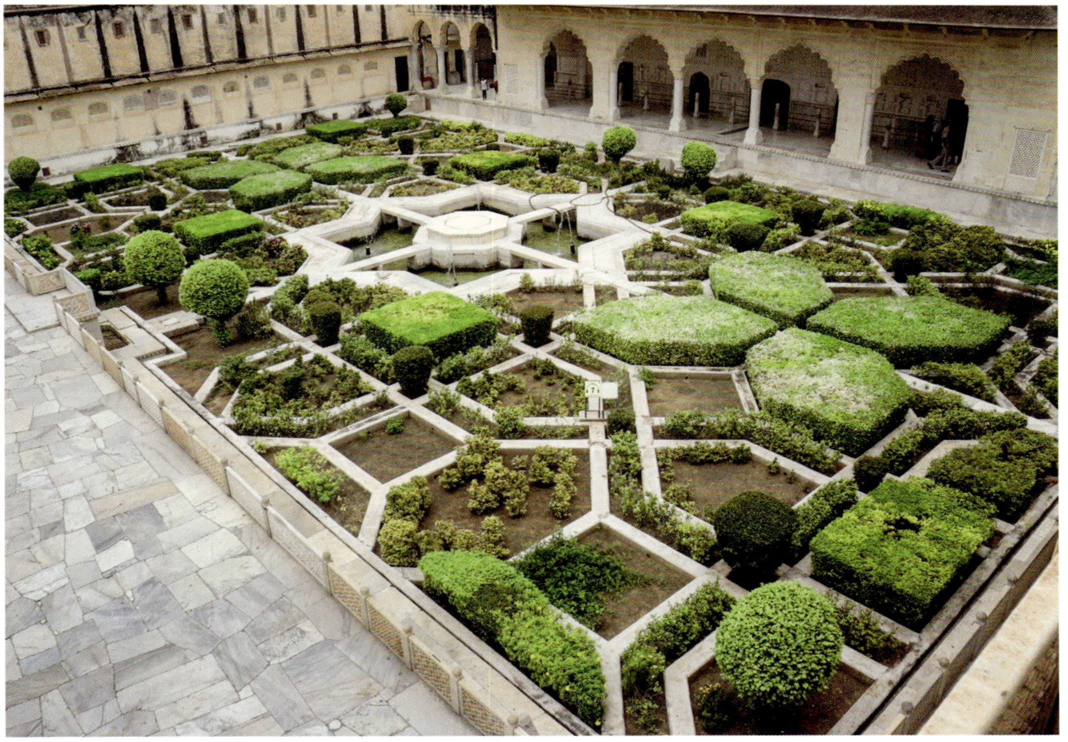

Above: Garden in Amer Fort, Jaipur, Rajasthan

Facing page: 'Allegorical representation of Emperor Jahangir and Shah Abbas of Persia', Abu'l Hasan, opaque watercolour, gold and ink on paper, c. 1618; note the use of flowers as borders in the painting

be seen in the Vrindavan Gardens and the Seva Kunj, a grove that still exists and is much frequented, considered to be one of the sacred spots where Radha and Krishna had their divine rendezvous. There are also the lovely gardens at Golconda fort near Hyderabad in South India, the Shalimar and Nishat Gardens in Kashmir and the crowning glory of the Mughal Empire, the Mughal charbagh, the best examples of which are seen at the gardens of Humayun's tomb in Delhi and the Taj Mahal in Agra. Modern India has its contemporary avatars in every Indian cantonment, as well as Noida's Botanical Garden, the Lodi Gardens in Delhi, the Hanging Gardens in Mumbai, the Lalbagh Gardens in Bengaluru, the Shibpur Botanical Gardens in Kolkata and many others, which carry on and evolve the tradition of fragrant gardens so lovingly and scientifically.

ENDNOTES

1. Chintaman Vinayak Vaidya, *Epic India, Or, India as Described in the Mahabharata and the Ramayana* (Asian Educational Services, 2001), p. 177.
2. Shelly Saguaro, *Garden Plots: The Politics and Poetics of Gardens* (Burlington, VT: Ashgate, 2006).
3. Kālidāsa, *Abhijñānaśākuntalam*, ed. Chintaman Ramchandra Devadhar (New Delhi: Motilal Banarsidass Publishers, 1996), p. 37.
4. Paravur Gopalakrishnan, *The Gitagovinda as the Song Celestial* (New Delhi: Sanbun Publishers, 2008), p. 45.
5. Ruth Vanita, *Gender, Sex, and the City: Urdu Rekhti Poetry in India*, 1780-1870 (Palgrave Macmillan, 2012), p. 46.
6. Calambur Sivaramamurti, *The Art of India* (New York: Harry N. Abrams, Incorporated, 1977), pp. 44–5, 155.
7. Elizabeth Moynihan, 'The Lotus Garden Palace of Zahir Al-Din Muhammad Babur,' in *Muqarnas Volume V: An Annual on Islamic Art and Architecture*, ed. Oleg Grabar. (Leiden: E.J. Brill, 1988).
8. James L. Wescoat Jr and Joachim Wolschke-Bulmahn, eds., *Mughal Gardens: Sources, Places, Representations, and Prospects* (Washington D.C.: Dumbarton Oaks, 1996).

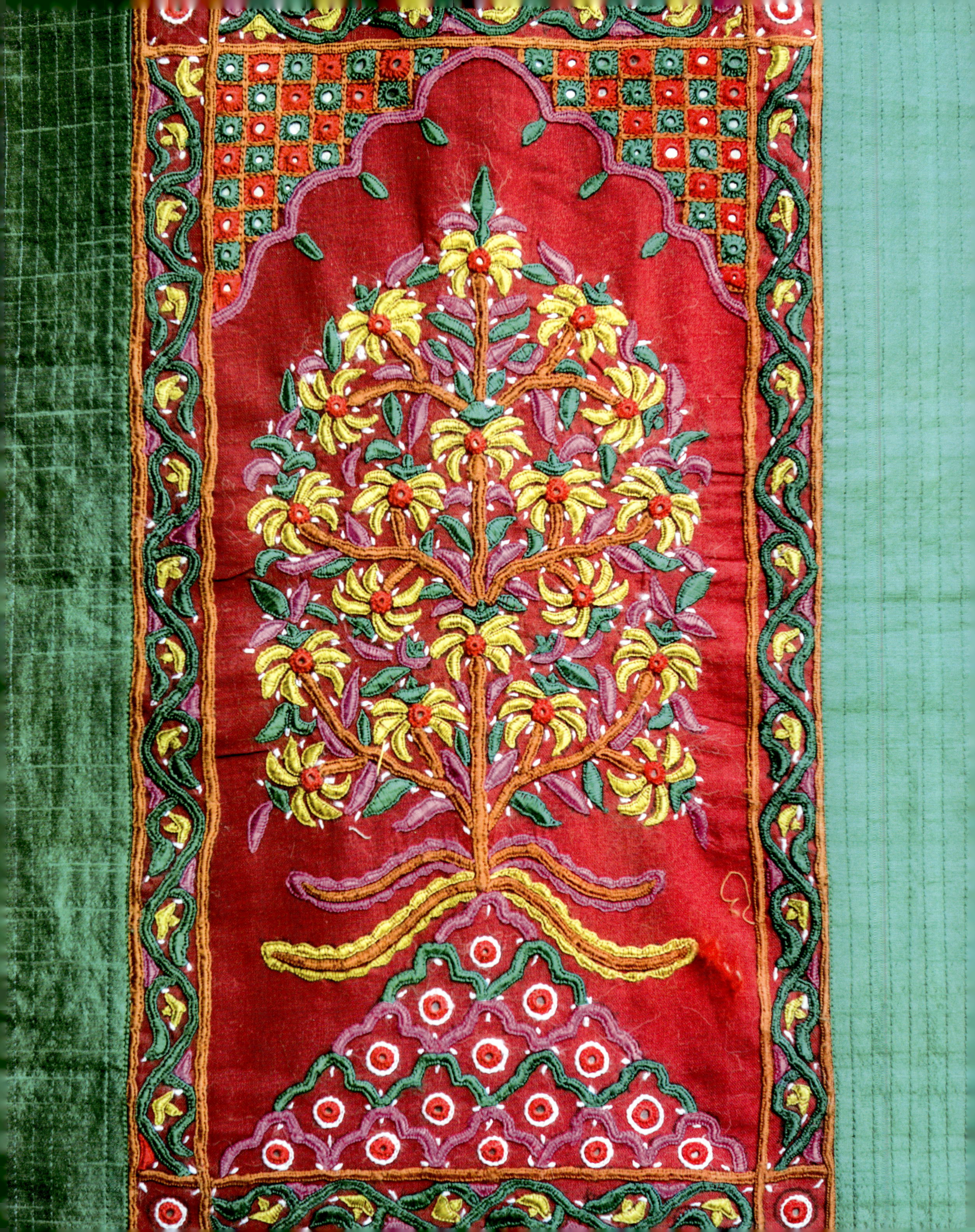

Fabric of Flowers

In bright attire of flowers forged new,
Heavenly of colour, white, red, brown and blue.
—**William Dunbar**

Flowers have been an inspiration for textile patterns since ancient times in India. The floral motifs on the costume of the red male torso from the Indus Valley civilisation show a visible refined version of the jasmine.[1] Numerous Ajanta murals depict people wearing delicate garments with lotus patterns. Over the centuries, a rich tradition of using floral designs in textiles has evolved in various forms all over the country. Among different styles are the kalamkari and ikat traditions of Andhra Pradesh, the Paithani weaves of the Deccan, the embroidered shawls of Kashmir, the patolas of Gujarat, phulkari of Punjab, the Kantha and Jamdani of Bengal and Chamba rumals (handkerchiefs) of present-day Himachal Pradesh, as well as pictorial shawls that have worked images of birds, animals, flowers, plants, landscapes and other natural wonders that can be found across the length and breadth of the subcontinent. Every piece varies in terms of the hues employed, the patterns, the embroidery and its intricacy. The wealth of designs that form the motifs can vary from the symmetrical to the abstract to depictions of the rich flora and fauna of the subcontinent.

Each of the 29 states has its own tradition of weaving, embroidery or block printing and each state has distinctive floral motifs which defines its unique sensibility.

Facing page: *Kalpavriksha* (Tree of Life) motif, finely embroidered on cloth with mirror-work detailing, Gujarat

THE ESSENTIAL MOTIF

Images of flowers and flowering plants form a leitmotif that have run through the textiles of India, whether made by a village woman for her family, by a court artisan, or by a professional weaver for the export trade. Although some of the most common flower motifs (*buta*) and the paisley on Kashmir shawls can also be so stylised that they may appear to be simply abstract forms, their roots are firmly embedded in a love of gardens and flowering plants that dates back at least to the court cultures of the early years of the Safavid Dynasty in Iran (1501–1732) and the Mughal Dynasty in India (1526–1857). They take on many avatars such as embroidered shawls made by artisans for themselves to the *buta* in twill tapestry that were carefully crafted and then sent abroad to enter the realm of the European elite.

Nora Fisher notes[2] that in Gujarat, village women, as well as professional embroiderers from the Mochi community would take the aid of drawings when they were creating their designs at the turn of the nineteenth century.

It is therefore interesting to note that the *buta* that border many shawls came about after various concepts and ideas about floral designs were synthesised from medieval Iranian images, pre-Islamic nature imagery and botanical drawings from France and the Netherlands. The final outcome was so beautiful that it dotted the cultural landscape of the Mughal world and was favoured by peasants and nobles alike.

ASAWALI AND AMARVELL —WESTERN INDIA

The Satavahana Dynasty in 200 BCE made the city of Paithan its capital, alongside the holy Godavari river. The Paithani style is named for this city, renowned as it was for an over two-millennia-old textile tradition that hinged on carefully weaving multi-coloured threads together with gold and silver into a single, magnificent piece. It was but natural that the Satvahana rulers would then go on to give the Paithani style great impetus, being responsible for its spread across the Deccan. So favoured did it become that the style endured under the rule of other kings, the Mughals stamping their mark on it with the addition of floral patterns and the *amarvell* (flowering vine) motif. It even charmed the cold-hearted Mughal monarch, Aurangzeb, under whom it evolved.

Elegant additions such as *asawali* (flower pot with a plant) motif, the peacock, geometrical figures and flowering vines were

Facing page, above: Asawali motif woven into a Paithani sari

Facing page, below: Shawl fragment with paisley and floral motifs, Kashmir, 19th century

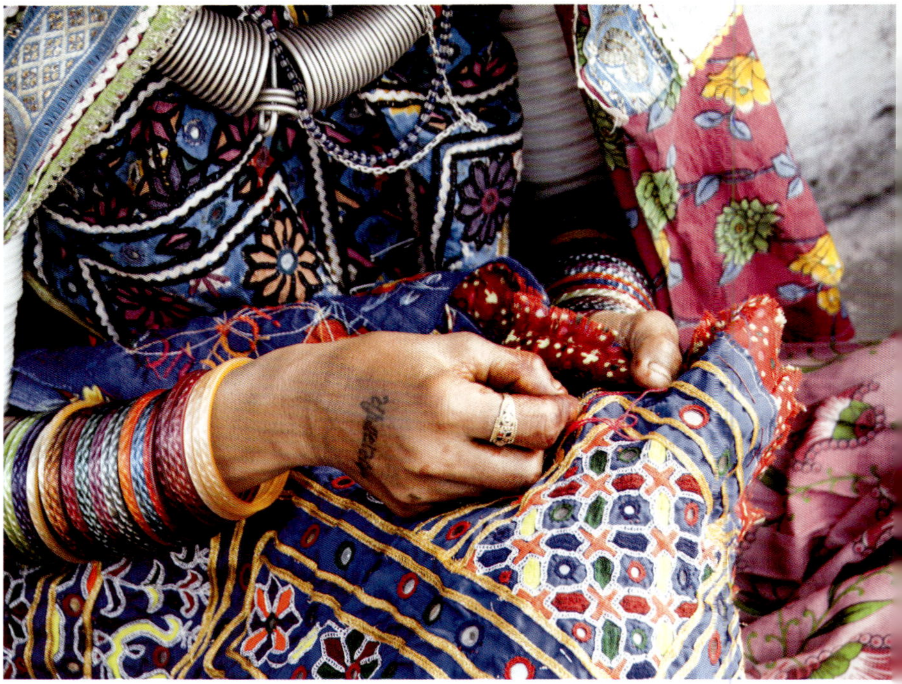

A tribal woman sewing traditional embroidered clothes, helps keep the craft alive and provides income for her home, Gujarat

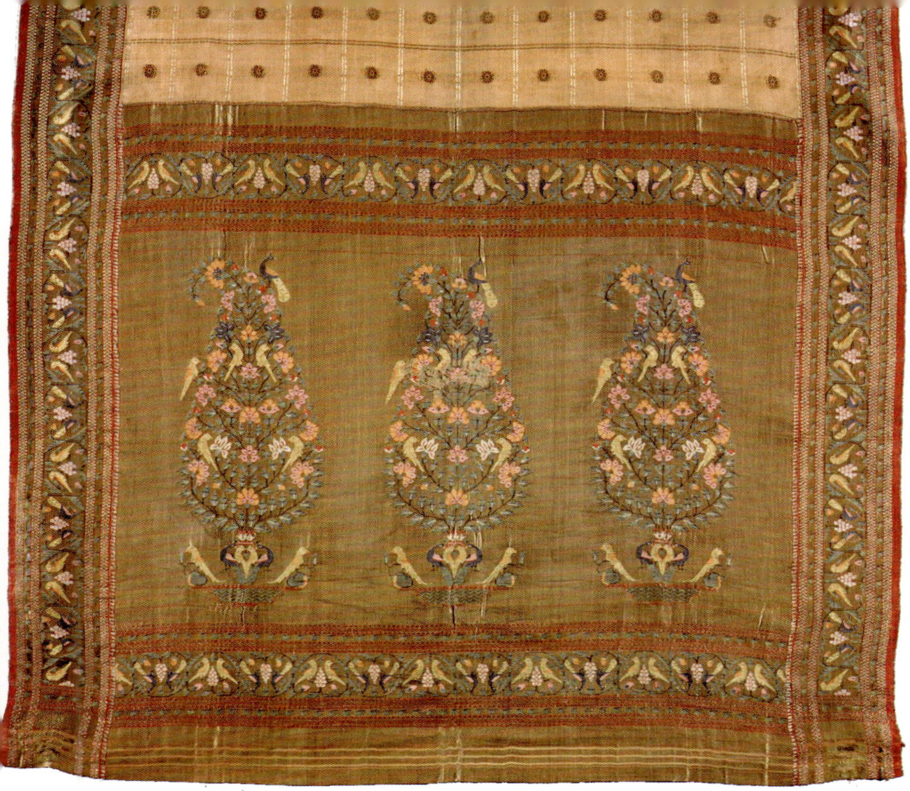

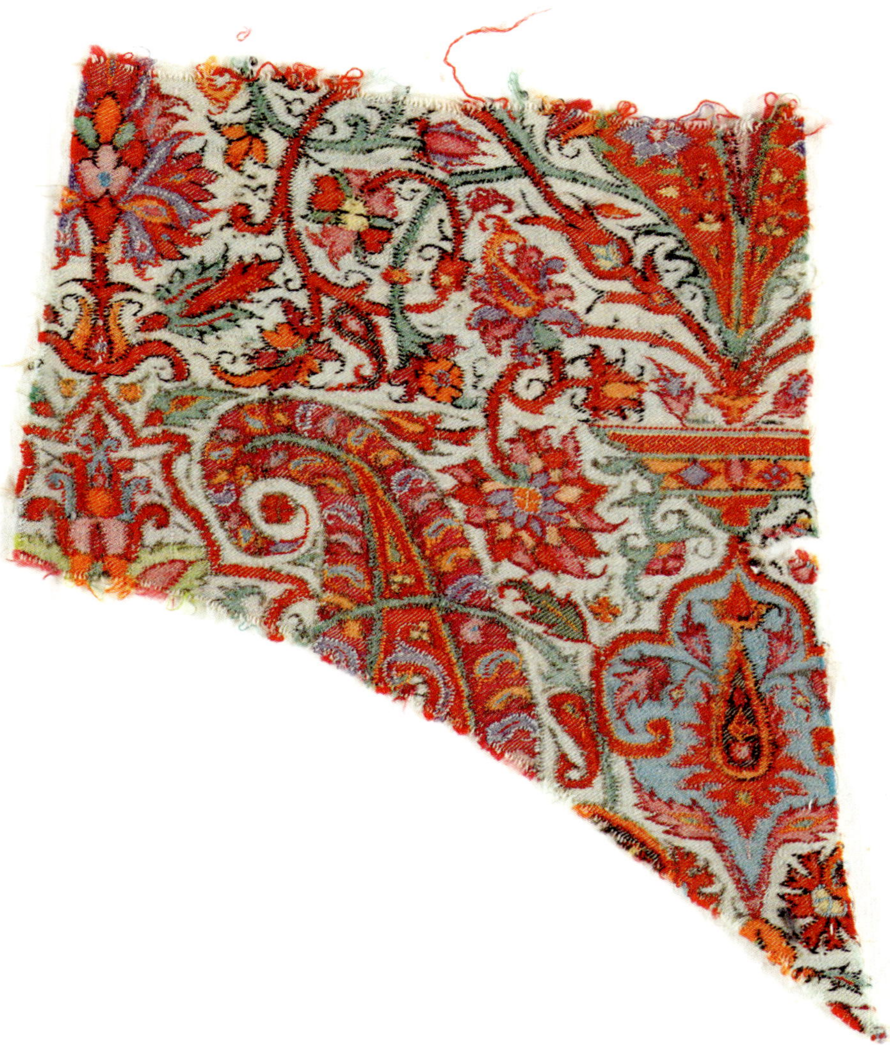

made to the Paithani style as it continued to mature. For example, under the watchful eye of the nature-loving Emperor Jahangir, Paithani grew to acquire many additions in the form of designs that mimicked flowers and other natural wonders. With the rise of the Peshwas and the waning influence of the Mughals, the former became great patrons of the Paithani, the *asawali* or flowering vine being one of their prime contributions in terms of design.

FROM THE LAND OF THE CHINAR—NORTHERN INDIA

Shawls woven in the Kashmir Valley are replete with floral motifs and naturalistic designs incorporating animals, birds and trees. Kashmiri embroidery is known as 'kashida' a vibrant and lovely reflection of the natural beauty of its homeland featuring creepers, leaves of the *chinar*, mangoes and floral motifs. The Kashmiri shawls popularised the 'paisley' design, which—over the years—became the defining motif of the Kashmiri pashminas. The four particularly popular shawls from Kashmir, the pashmina, the kanikar, the dorukha and the jamavar have floral motifs intrinsic to their design.

WEAVES—SILKS AND BROCADES

The Banarasi is one of the important brocades of India. The Banarasi 'naksha,' a Persian device which was similar to the jacquard loom enabled the weavers to weave the undulating floral patterns in intricate brocades and *kinkhwabs* (heavy silk fabric). From realistic roses, to stylised lilies and paisleys, the repertoire of the flowers was expansive, inclusive of the local flowers grown in the region.

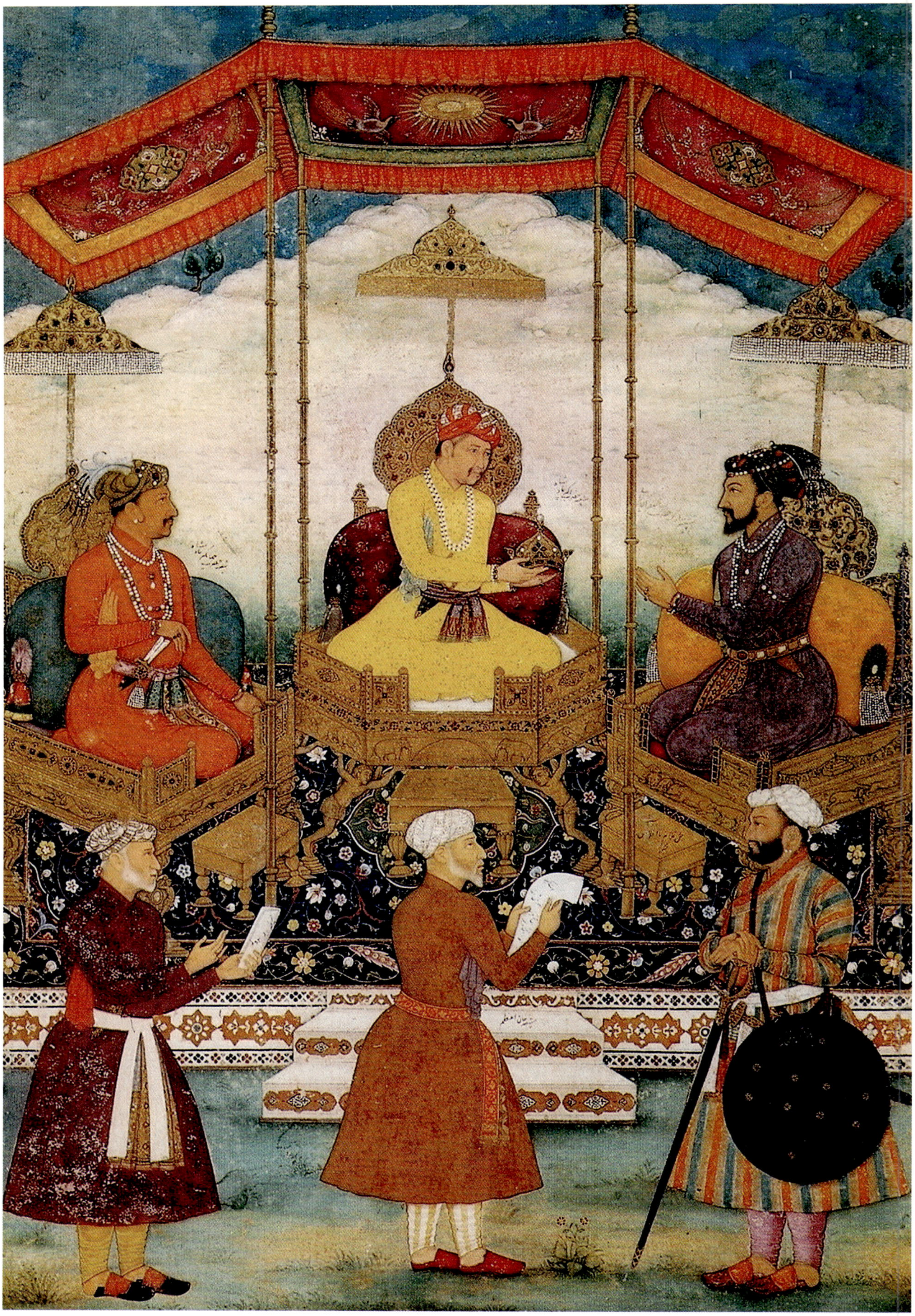

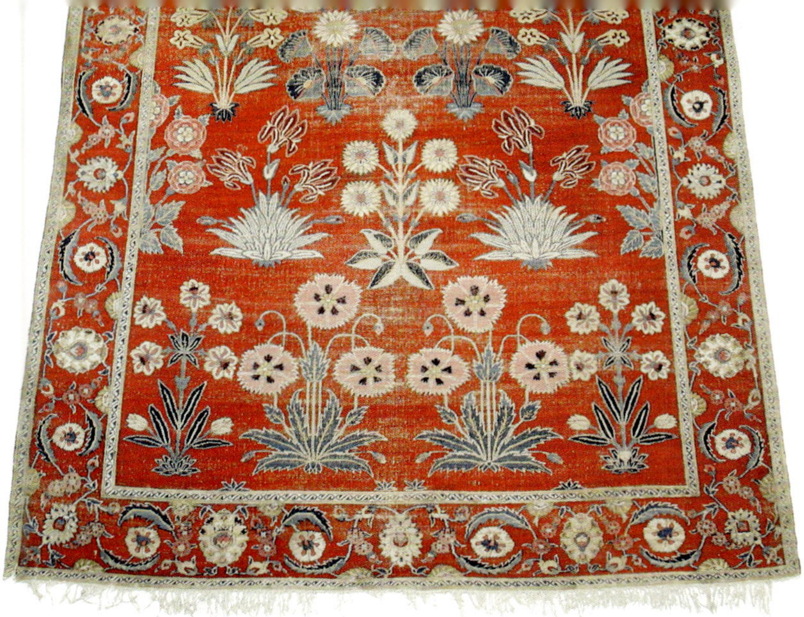

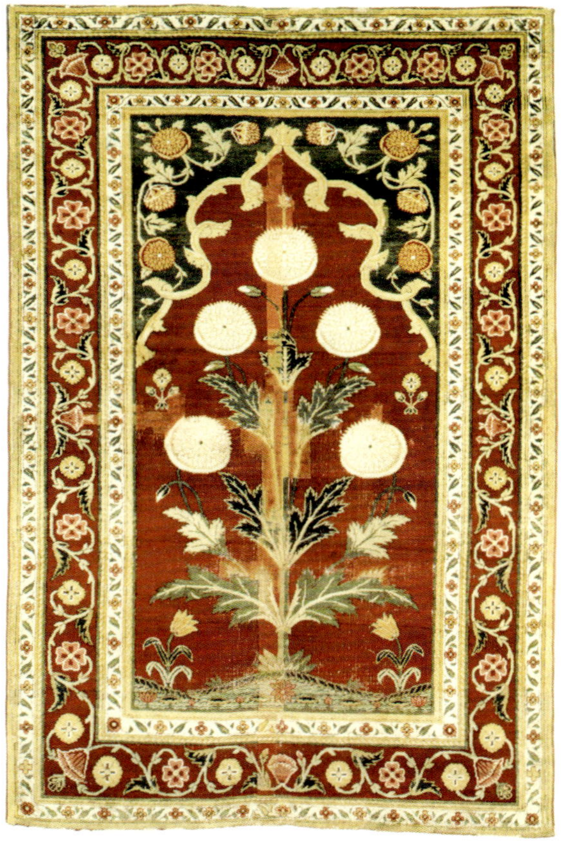

Above: Mughal carpet with motifs of irises, tulips, and other flowering plants, c. 1650

Right: Mughal carpet with niche and flower design, mid-17th century

Facing page: 'Akbar hands his imperial crown to Shah Jahan', Bichitr, 1631; observe the details of the carpet with floral motifs, depicted in the painting

FLORAL MAGNIFICENCE ON THE FLOORS

CARPETS, RUGS AND NAMDAHS, PANJA DARIS

There are numerous references to woven floor coverings in ancient and medieval Indian literature. Some Buddhist texts mention the presence of woollen carpets from as far back as 500 BCE. Pre-modern India saw the presence of woven bamboo mats and *dhurries*. For example they are mentioned in the *Mrcchakatika* of Sudraka. Mats are an age-old cottage industry, references to which are found in the *Atharva Veda* which mentions mats made of grass or *kapisu* and *madur kathi* mats made from *madur kathi* reeds. These were made for priests and Brahmans to sit on, as well as to cover the floors in wealthy households. Woven mats were also found in the Indus Valley Civilisation. All hand-woven, some of them had abstract floral patterns on them.

However, the most luxurious examples are found in the hand-knotted carpets which were introduced to India when the Mughal Emperor Babur made it his home. Babur sorely longed for the luxurious trappings of Persia and it was his grandson Akbar who laid the foundations for the carpet weaving industry at his palace in Agra in 1580. The Mughals not only followed the Persian technique of carpet weaving but were also deeply inspired by the traditional designs and motifs from Persia. The Persian carpets generally had a battle scene in the foreground with exquisite, fine worked borders of vines and flowers. In the Mughal courts, the Persian carpets were recreated with Indian forms and the patterns of these newly birthed Indian carpets varied from patterns featuring creepers and flowers originally with multiple shades of blues and greens on a red and peach base.

Another traditional floor covering of India is the dhurrie. Dhurries can be made of cotton wool, silk or jute and the material depends in the state on which they are made. In Punjab and Rajasthan the dhurries are known as Panja dhurries. Khairabad and Sitapur in Central India are some other centres for dhurrie making. These intricately woven floor coverings often feature geometric patterns which are inspired by the natural vegetation of flowers and vines.

From the states of Himachal Pradesh, Kashmir, Gujarat (Kutch) and parts of Rajasthan come some felted wool carpets known as *namdahs*. This is an Urdu word, brought to India by the Mongols and Mughals. Chain stitch embroidery is used to create beautiful floral motifs, particularly trees and flowers, translated by the artistic imagination of the artisan.

BLOCK PRINTING

Block printing had some prominent centres in Rajasthan, Gujarat (Kutch, Porbandar, Jamanagar, Rajkot), the Coromandel Coast, Punjab and Andhra Pradesh. The British, the Dutch and the Portuguese brought in their respective aesthetics and translated and enriched the Indian floral motifs by introducing flowers from their respective lands. The Indian block printers are exceptional artists. Often they make their own block but usually the block maker is a carpenter and in Rajasthan he is known as a '*bhatt-ghar*'. Flower petals known as '*bootis*' becomes the basis of fabrics. Some of the flower motifs used in block printing are roses, lotuses, rose and lotus buds, sunflowers, lilies, *champas*, *nargis*, marigold, etc. Some local flowers like the *sosam*, *kachnar*, *javakusum* (hibiscus) *jatadari*, lily and *gullala* are used in different forms. Sometimes more than three flowers are fitted in a single *booti* such as a *latkan booti*, where the banana, *sosan* and *saro* tree are all intertwined together. In fact, in the Sanganeri prints of Rajasthan, many flowers do not originate from Rajasthan itself.

PALAMPORES—SOUTH INDIA

The famed Palampores from the Coromandel coast particularly Machilipatnam (previously known as Masulipatnam) in present-day Andhra Pradesh, are hand-printed, mordant-dyed bed covers, which were specially created for the elite classes of England and Europe. Using the 'kalamkari' technique, they depicted a wide variety of plants and flowers, including peacocks, Asian elephants and elegant horses. Each Palampore was special since it was a single, handmade creation which could not be replicated. The stylised flowers and leaves were often hybrid since the craftspersons were also shown patterns and images of flowers which came from England and France.

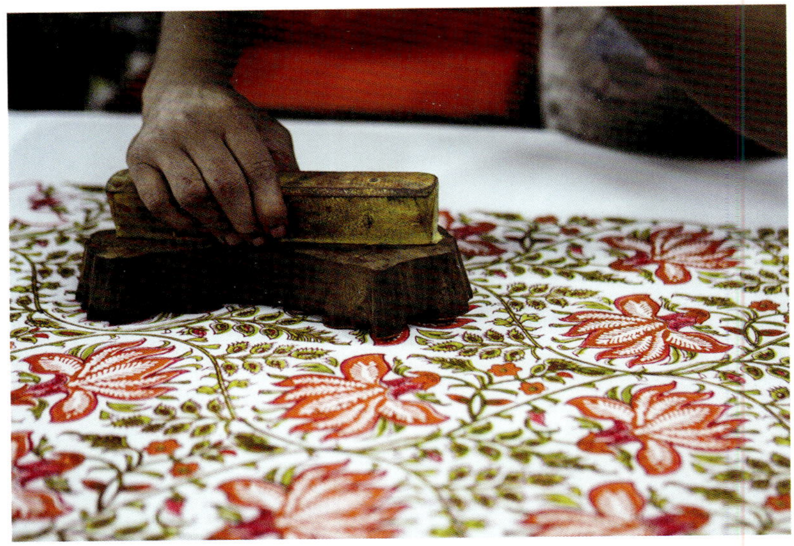

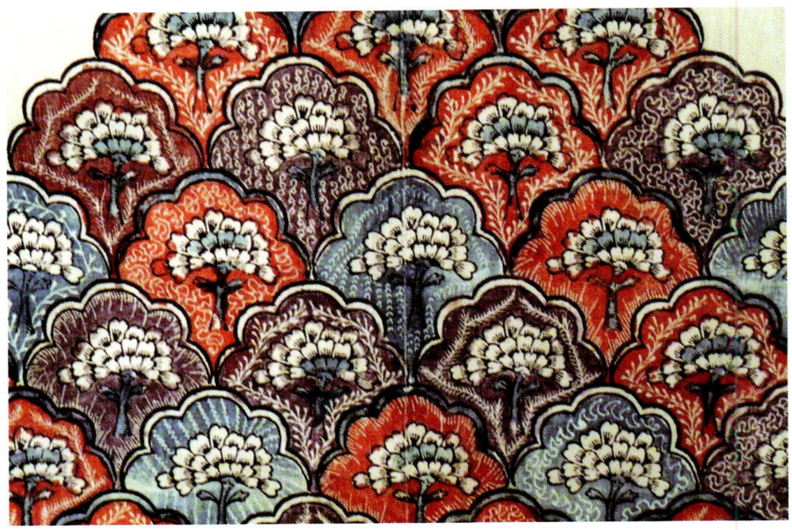

Top: Traditional process of hand block printing, Jaipur, Rajasthan

Above: Palampore with peacock design (detail), Coromandel Coast, late 18th century

KASHIDA THREADING FLOWERS

THE HEIGHTS OF DELICACY

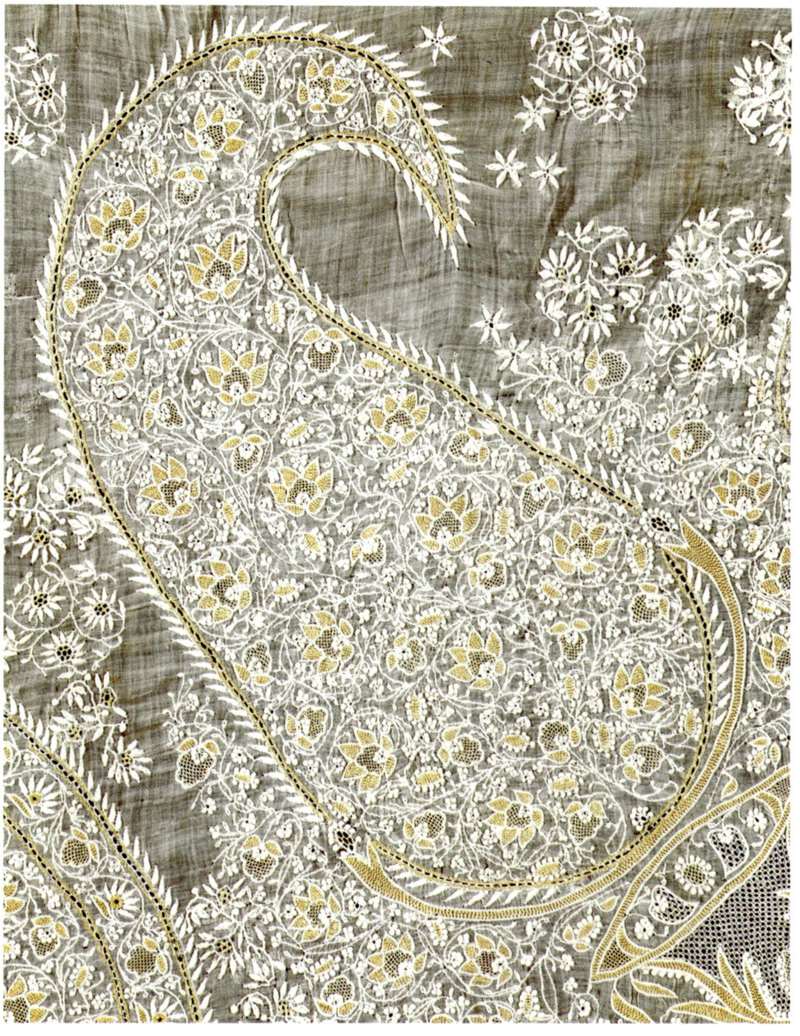

Under the wide umbrella of the Kashida are the embroidery traditions from the different states of India like Chikankari of Lucknow, Kantha of West Bengal and Phulkari of Punjab, which drew its roots from the gulkari of Central Asia. In Sindh too, there is a great tradition of cross-stitch and there are the tribal embroideries from Gujarat with multiple cultural meanings.

Lucknow, in Uttar Pradesh, is one of the most well-known centres of **Chikankari.** The cloth used are delicate chiffons, muslins or georgettes and creeper vines and floral patterns—such as roses, jasmines, paisleys and lotuses—are especially favoured in this form of embroidery.

Traditionally, **Chikankari** is white thread embroidery across white muslin or mulmul (fine muslin).[3] The term 'chikan' is derived from the Persian word 'chakeen', which indicates the art of creating elegant patterns on fabric.[4] Nur Jahan, Emperor Jahangir's refined and politically astute wife, is given the credit for introducing it to India, having been said to be a master at the art herself. Other reports state that muslins featuring delicate floral patterns can be traced back to East Bengal in the year 3 BCE.

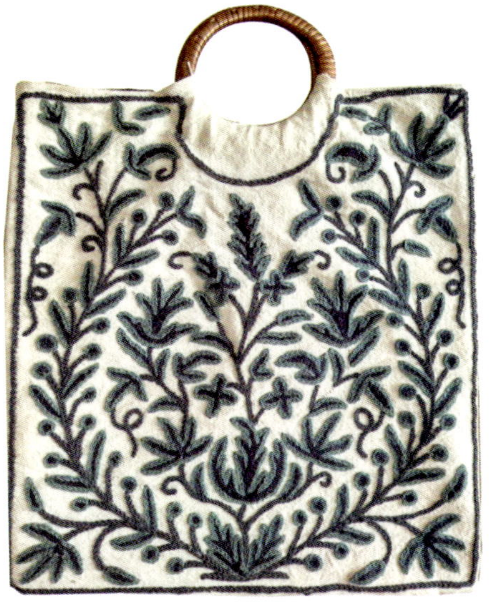

Above: Chikankari embroidery in a paisley design, on a sari, Lucknow, early 19th century

Right: Modern day bag with Kashmiri kashida floral embroidery

Kanthas are the wraps and blankets of Bengal. They comprise of simple materials such as recycled cloth from old saris and dhotis. In stark contrast, the embroidery used to decorate these layered cotton textiles is often a study in intricacy and the designs sophisticated, although a small variety of stitches such as the running stich and darning are most frequently used. Lotus medallions in the middle and paisley or mango patterns called *kalkas* in the corners are elements in many motifs, along with trees. Elaborate designs emphasise the

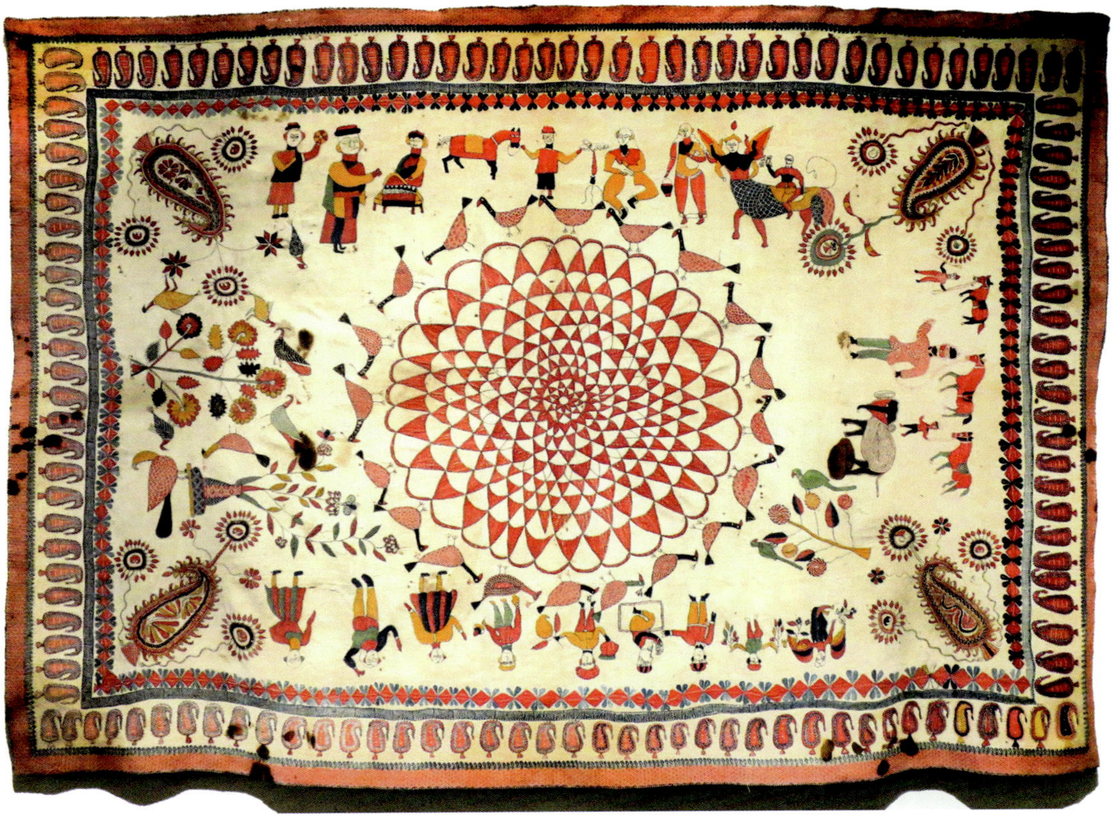

Above: Kantha (bed cover) West Bengal, late 19th–early 20th century

Facing page: Kalamkari, cotton cover depicting a turkey and exotic birds and flowers, western India, early 17th century

borders of the garment. The lotus motifs recall connections to the alpana designs on the threshold floors that Bengalis create, to ensure the well-being of their homes.

THE ART OF KALAMKARI —EASTERN COAST

Kalamkari consists of an array of hand-painted textiles worked in natural dyes from the varied regions of the country. Taking its name from the Persian root *kalam*—an instrument for writing or painting, Kalamkari has held sway over textile traditions for a long time. The Urdu suffix *kari* indicates the craftsmanship inherent in this art form, which reached its height under the Qutub Shahi dynasty (c.1518–1687). It has taken great mastery of skill to create the Kalamkari style of textile painting from medieval times until the present day.

Historically, the art of Kalamkari represented the textile arts practiced along the Coromandel coast, from northern Machalipatnam[5] to places such Srikalahasti, Palakollu and Kakinada in present-day Andhra Pradesh, and Salem, Madurai and Tanjore in Tamil Nadu. Today, it epitomises the technique of painting using natural dyes on textiles. It was favoured by Hindu royalty from the 13th to the 16th century. Mythological stories from the epics like the *Ramayana* and *Mahabharata* alongside episodes from the *Shiv Purana* and *Vishnu Purana* were brought to life in rich and vivid hues. Machilipatnam and Srikalahasti, a holy city, famed as a pilgrimage centre along the River Swarnamurki are renowned for their Kalamkari work.

In Andhra Pradesh, the port town of Masulipatnam (now called Machilipatnam) was a prominent trading site along the Coromandel Coast from where Kalamkari was traded to far-off lands. Many monarchs encouraged the use of Kalamkari textiles. Among them was

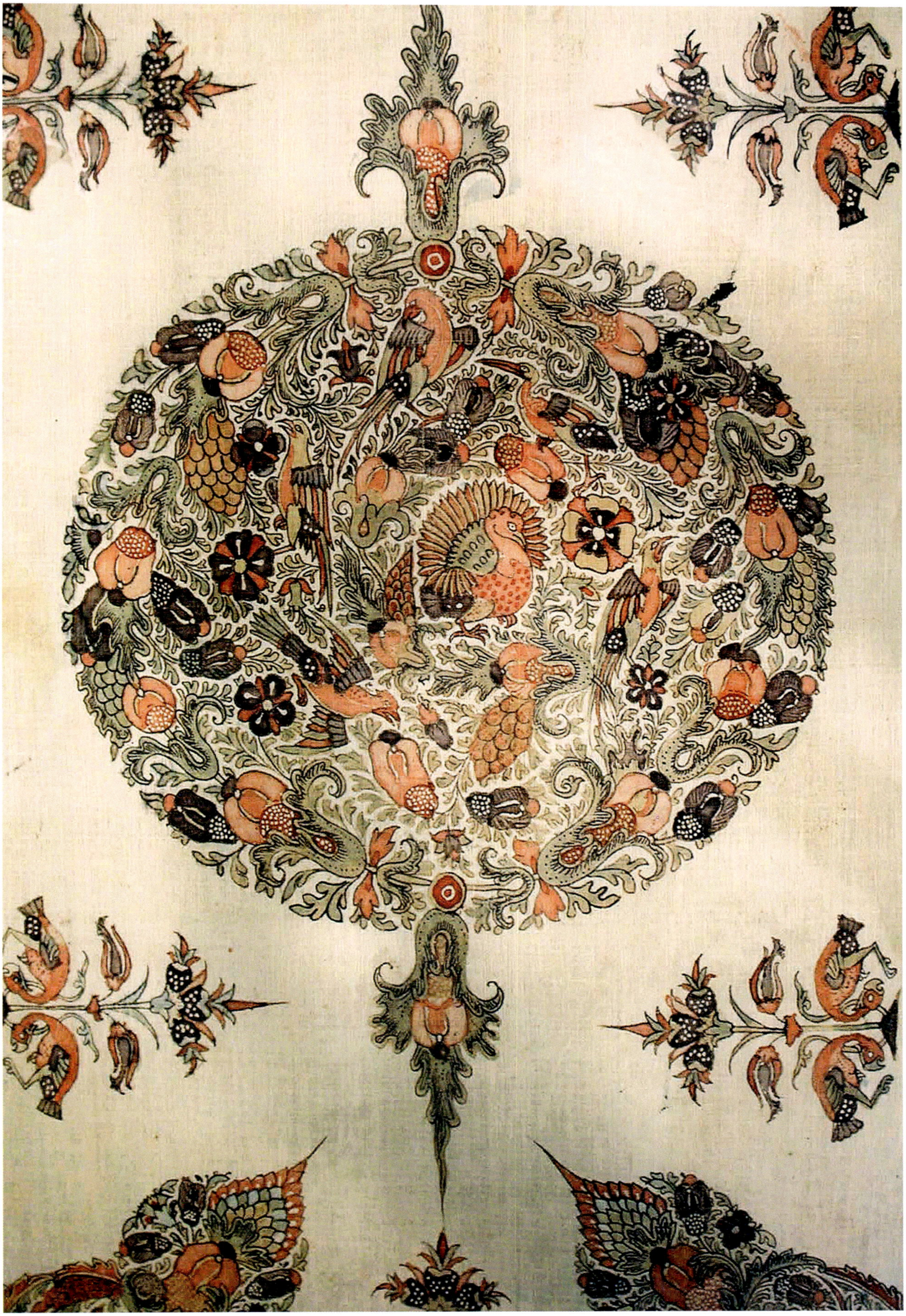

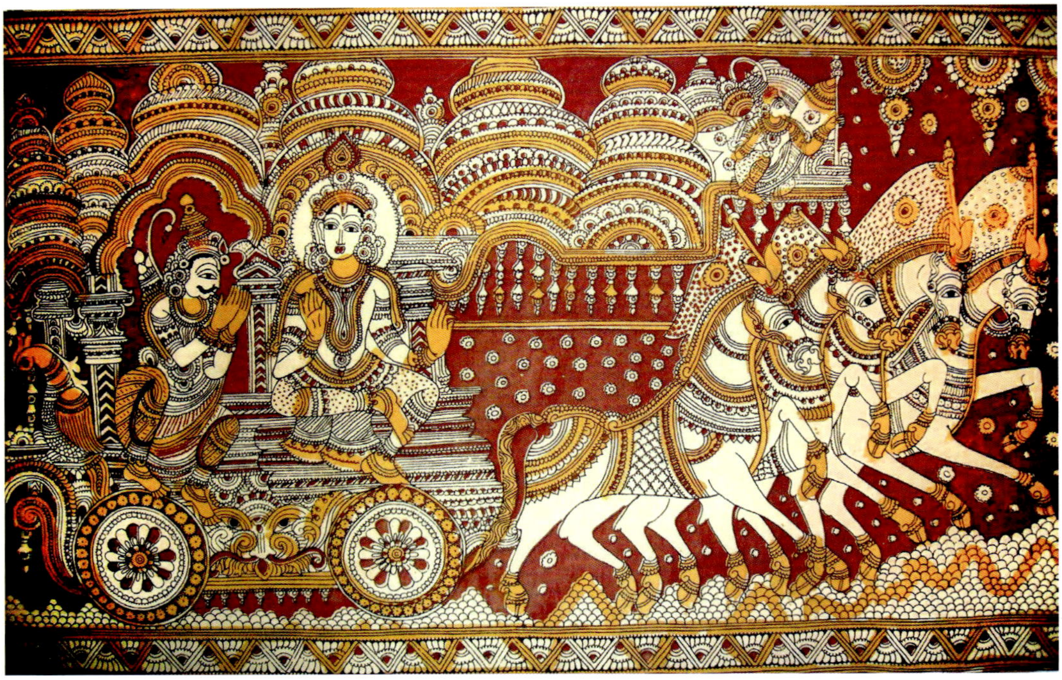

the Qutub Shahi dynasty of Golconda. Trade alliances, such as the ones orchestrated by the Qutub Shahi rulers, saw Kalamkari enter the realm of domestic use during the Persian Safavid Empire, Iran going on to give this art form great encouragement. Floral and natural patterns inspired by the Persian inflections began to seep into the work which took on yet another dimension under the Mughals who preferred portraiture and vignettes of the everyday. In the realm of daily life, sacred tenants, animals, plants, birds and more, dotted the Kalamkari landscape. However, it also seems to have been quite extensively used in temple hangings and ritual textiles. Foreign influences from Europe, especially during the British Raj, went on to cause the art form to be adapted into home decor products like apparel and hand-painted furnishings.

Chintz, which is Indian painted or printed cotton cloth, rose to become a favourite in Europe during the 19th century. Once upon a time the word chintz conjured images of artisans inscribing patterns on cotton with pens and then dyeing them. It was known by different names in various cultures: for example, *sitz* for the Dutch and *pintado* for the Portuguese. The evolution of block printing came about when people began to clamour for it, as it became a popular favourite. The cloth was polished to give the appearance of a sheen. As can be expected, there has been a change in this tradition over the ages. It entered popular culture as well, then went from being associated with the realm of religion to a more secular art that found favour with many rulers including the British and the Mughals.

COMMON BOND

The popular icon of the Tree of Life is found on a variety of traditional textiles like woven on silk in Bengal, woven and embroidered on shawls in Kashmir, painted and dyed on chintzes or Kalamkari from parts of the Coromandel Coast. The motif Tree of

Above: Kalamkari painting on cloth, scene from the Bhagavad Gita—Krishna instructing Arjuna, Andhra Pradesh

Facing page: Palampore, Coromandel Coast, early 18th century

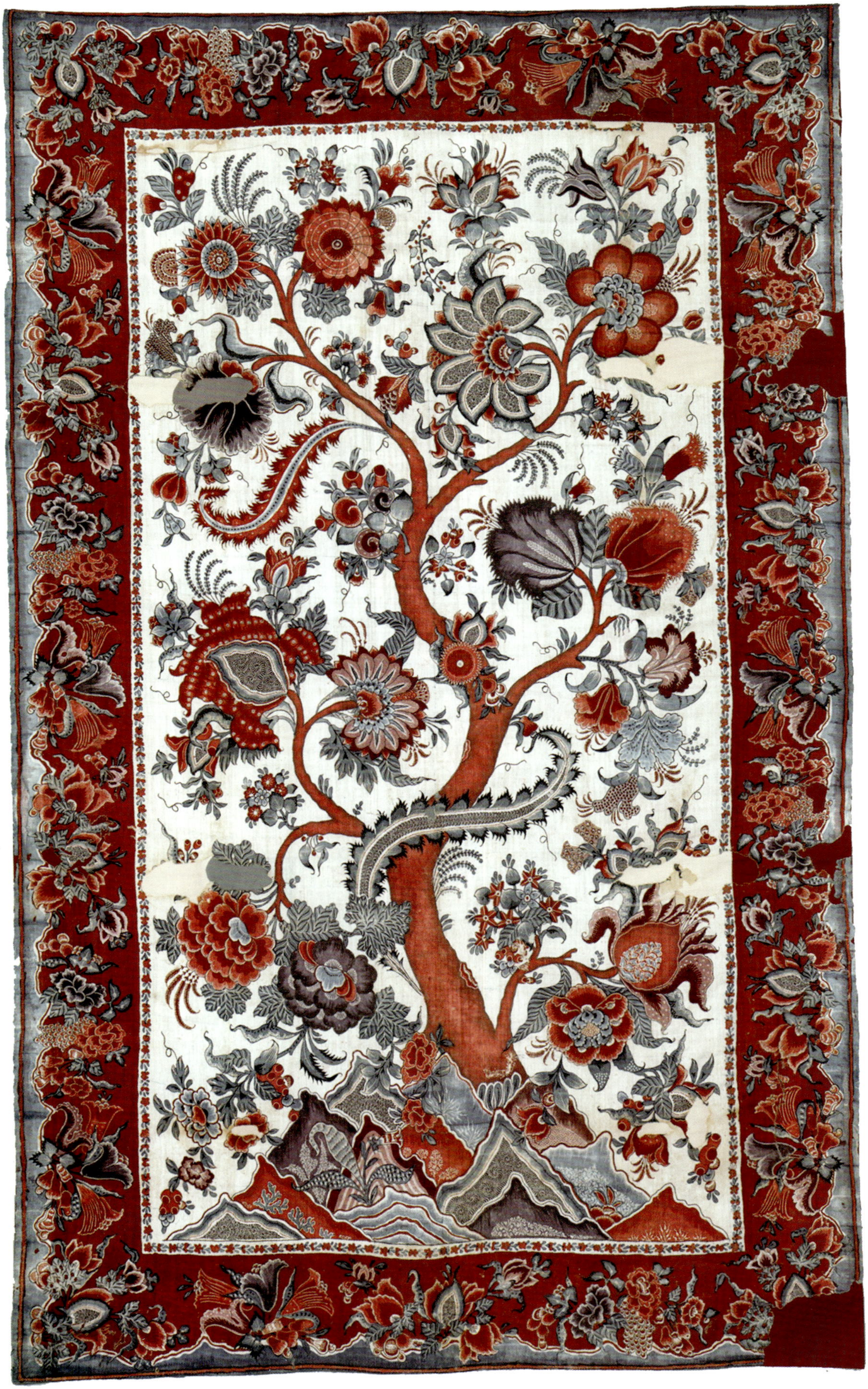

Life is also used in embroidery traditions to highlight ethnic and social identities, for instance Phulkari of Punjab; the *soof, shisha* or mirror work of Gujarat and Rajasthan; *Kasuti* embroidery in Karnataka; *Sujini* work of Bihar and *Kantha* work of West Bengal.

YARDAGE OF FLOWERS

The Indian sari or the six-yard marvel is worn by women of the different states of India in many varied ways. It is draped in a variety of styles and the yardage varies from five to nine yards (as in Maharashtra's *nauvari*). There is also the Gujarati style where women drape the pallu over the left shoulder and display its intricate designs across the front of the body.

A focus on tradition also means that there is an added focus on the aesthetic.[6] Perhaps in no other garment is the history of India's cultural heritage in the realm of textile writ larger than in the sweeping folds of the sari. The elegant weaves, the array of colours and the variety of symbols make for a fine garment, the most splendid silk saris incorporating a floral motif or breath-taking block prints patterned with flowers. With the traditional dyeing techniques the floral designs are, in fact, inked into the cloth, choice motifs being those of vines and flowers.

In this section a few stunning examples of woven flowers have been illustrated—the Paithani and the Banarasi which take their names from their towns of origin.

Under the patronage of the Peshwas the Paithani saris from the Maharashtrian region of Paithan flourished. Characteristic designs are the floral paisley and the geometric and bright flower prints in rich zari on exquisite silk which make these splendid drapes, a favourite of brides. The famed Banarasi saris are among the most sought-after saries in the country, often included as part of the wedding trousseau. The patterns feature multi-coloured floral motifs worked in sumptuous zari along the border.

Colourful Banarsi silk saris with floral motifs

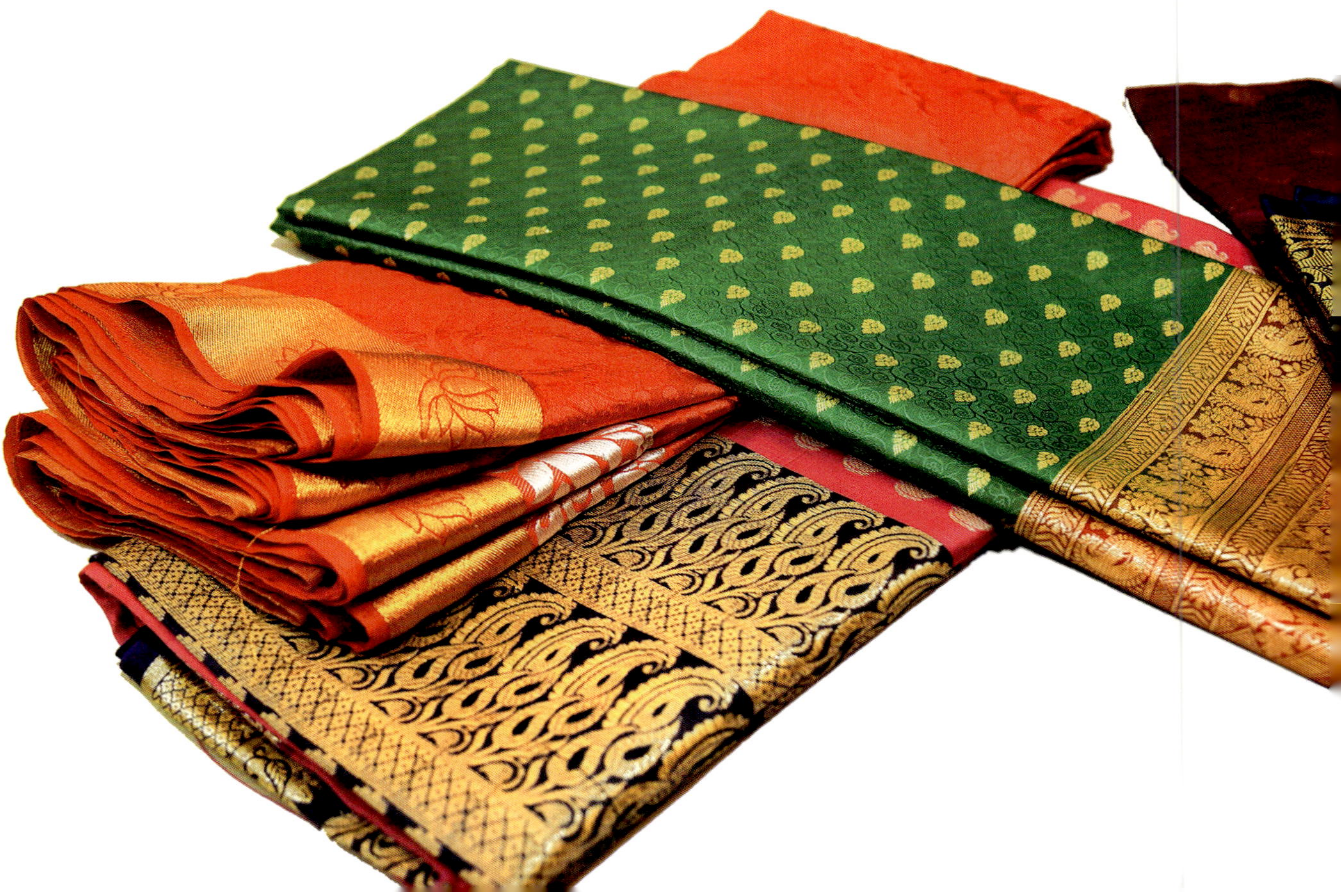

Floral border of a Sambalpuri sari, Odisha

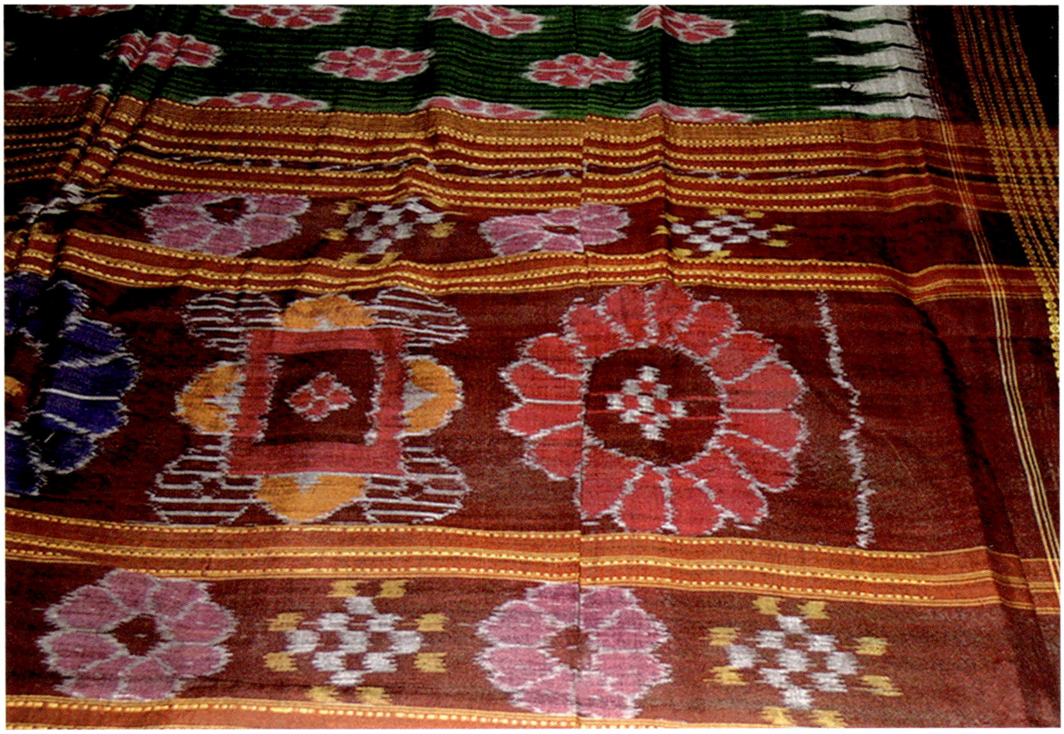

Of the intricate weaves from Eastern India—Baluchari of West Bengal, Bhagalpuri weaves and silks from Bihar and Sambalpuri from Odisha have been picked for their exquisite, elaborate weaves. The elegant woven Sambalpuris, originate from a tie-dye craft style known as *baandha* and are brilliantly hued, featuring flowers or *phul* motifs as well as vines and have recently featured flower pods. The Sambalpuri ikat tie-dye tradition also enhances their beauty and legacy.

These and many others, reflect the environment in which they were fashioned and therefore have as rich a variety of flora as the length and breadth of the subcontinent. In the landscape of Indian textiles spanning from woven mats to delicate embroideries and weaves, to detailed block printing and the finely knotted Kashmiri carpets, the flower in its numerous 'avatars' or manifestations continues to dominate the visual effect of the fabric.

ENDNOTES

1. Kokyo Hatanaka, *Textile arts of India*, Kokyo Hatanaka collection (San Francisco: Chronicle Books. 1996), p. 309-312.
2. Mary M. Dusenbury and Carol Bier, *Flowers, Dragons & Pine Trees: Asian Textiles in the Spencer Museum of Art* (Manchester, Vermont: Hudson Hills, 2004), p. 20.
3. http://www.idc.iitb.ac.in/dsource/resource/chikankari-embroidery-lucknow-ii/introduction (accessed 26 May 2016).4. http://bieap.gov.in/Pdf/FGMPaperIIYR2.pdf (accessed 26 May 2016).
5. Percy Brown, and George Watts, *Arts and Crafts of India: A Descriptive Study* (New Delhi: Cosmo Publications, 1979).
6. Kax Wilson, *A History of Textiles* (Boulder, Colorado: Westview Press. 1979).

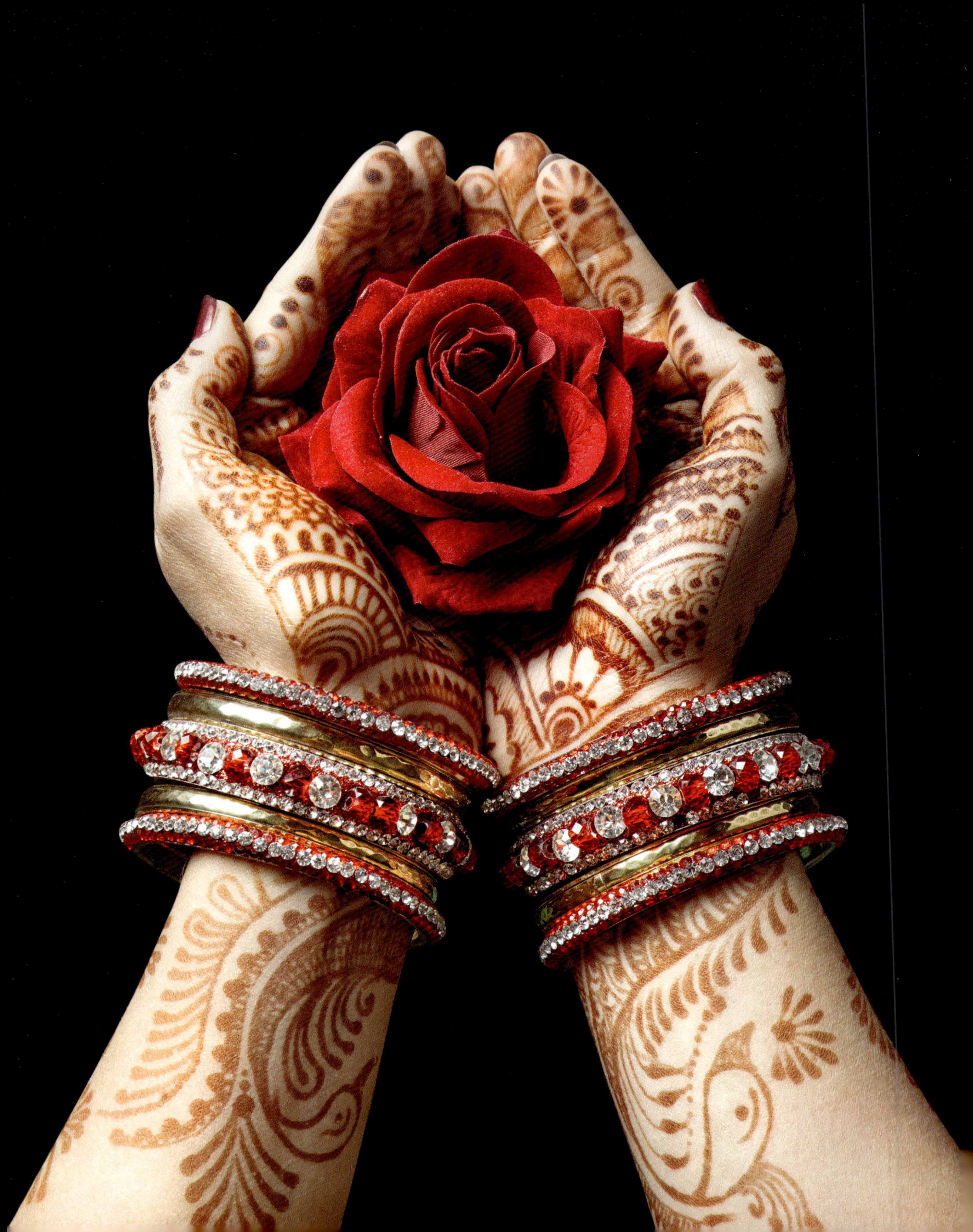

The Queen of Flowers - Rose

Won't you come into the garden?
I would like my roses to see you.
—Richard Brinsley Sheridan

Shrouded within many stories, the hauntingly beautiful, joyous, tragic, spiritual rose is a universal metaphor for love, longing and desire. The origin of the rose in different countries has its own story and time. The first fossilised example of the rose is more than 35 to 40 million years old. Numerous responses of the *patalpushp* or the rose in Sanskrit have been found in the many myths and legends of India. But it was from the medieval period of India that the Indian rose is mentioned time and again in poetry, prose and seen in miniature paintings, particularly the Mughal miniatures, as it was the Islamic influence that brought the roses to the subcontinent. Later innumerable varieties of the rose, particularly from China and some from old Europe, were introduced by the Dutch and the French along with the British colonisers, which added to the repertoire of the already existing indigenous roses, further enhanced by the subsequent roses from Persia and the Islamic world.

The Indian Rose like the land in which it lives, encompasses innumerable, sub-cultures, cross cultures, voyages and plural voices. The Indian Rose thus becomes an icon of the concept of '*vasudhiava kutumbhakam*', one world, one family.

Facing page: Indian woman with hennaed hands proffers a rose as an offering

The rose, because of its beauty and aroma, is perhaps the most popular and sought-after flower in India. And this fascination is not new—the Mughal Emperor Shah Jahan (1592–1666), was symbolised by a red rose. It is popularly believed that this plant did not originate in India and was introduced from Persia by Shah Jahan's great-great-grandfather the Mughal Emperor Babur (1483–1530). However, it has been an integral part of Indian life and religious rituals for nearly a thousand years along with the marigold, the hibiscus and the jasmine. There are references to the rose in ancient Indian texts, such as *Satapatri* or *Suman*, where it was described as a herb and as one of the best offerings to the deities. In the *Mahabharata*, Bhishma Pitamah, on his deathbed, advises the Pandavas to offer this flower to the gods. Descriptions of the beauty of Damayanti compare her face to a rose.

Shakespeare, too, immortalised the rose in many of his sonnets and plays, perhaps the most famous line of which being Juliet's musings in *Romeo and Juliet*, where she wonders, 'What's in a name? That which we call a rose, by any other name would smell as sweet.' The modernist poet and writer Gertrude Stein echoed these sentiments in her poem *Sacred Emily* as she mused, 'A rose is a rose is a rose.'

The flowers of the rose blossom in many different colours, from the well-known red to yellow, white and even purple roses. Grown in North America, Europe, Africa, Oceania as well as Asia, there are over a hundred different species of roses and thousands of cultivars (a plant variety produced by selective breeding). They are usually cultivated, though it is not impossible to grow wild roses in gardens, and the species chosen for gardens have to be carefully selected.

Widely cultivated across rose farms and gardens, the names of the various kinds of roses most commonly derive from the species from which they have descended. If you happen

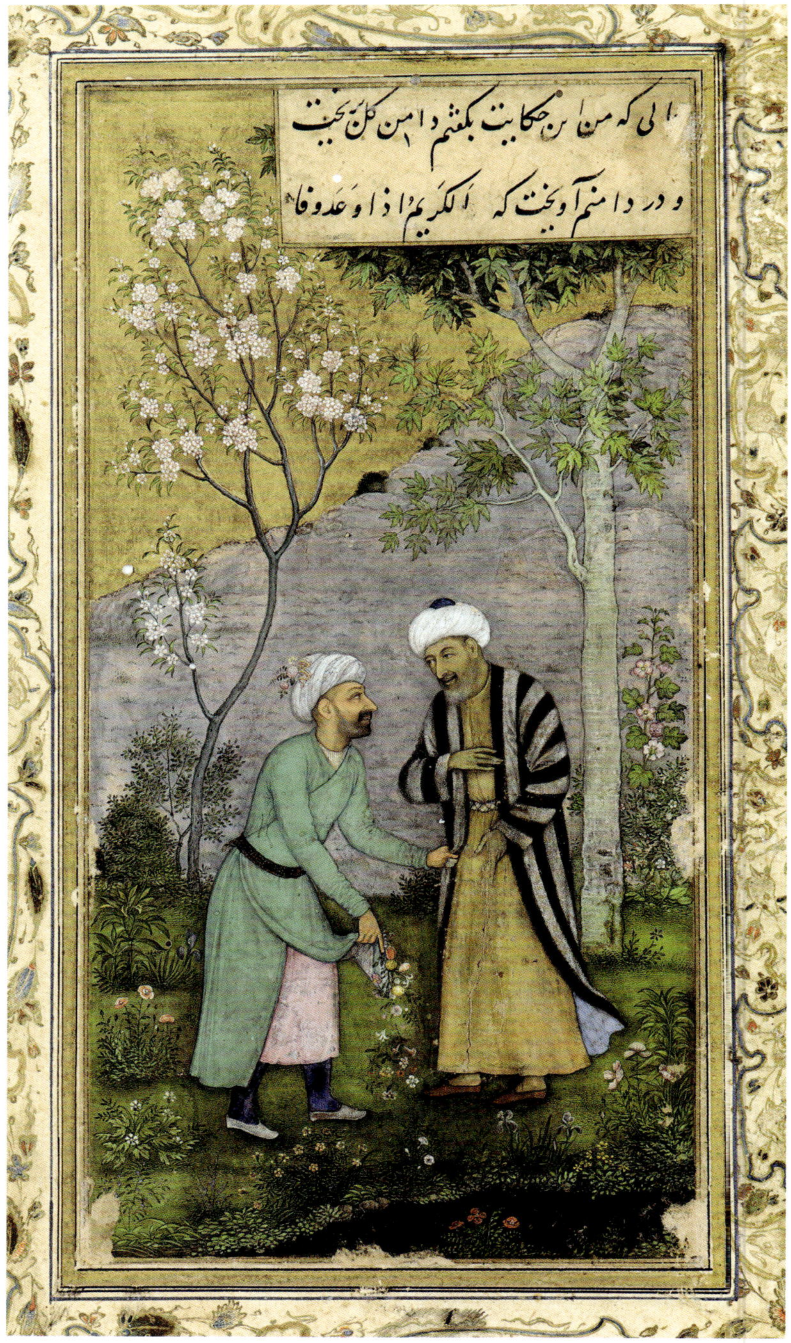

'Sa'di in a Rose garden', from a manuscript of the *Gulistan*, opaque watercolour, ink and gold on paper, attributed to Govardhan, c. 1645

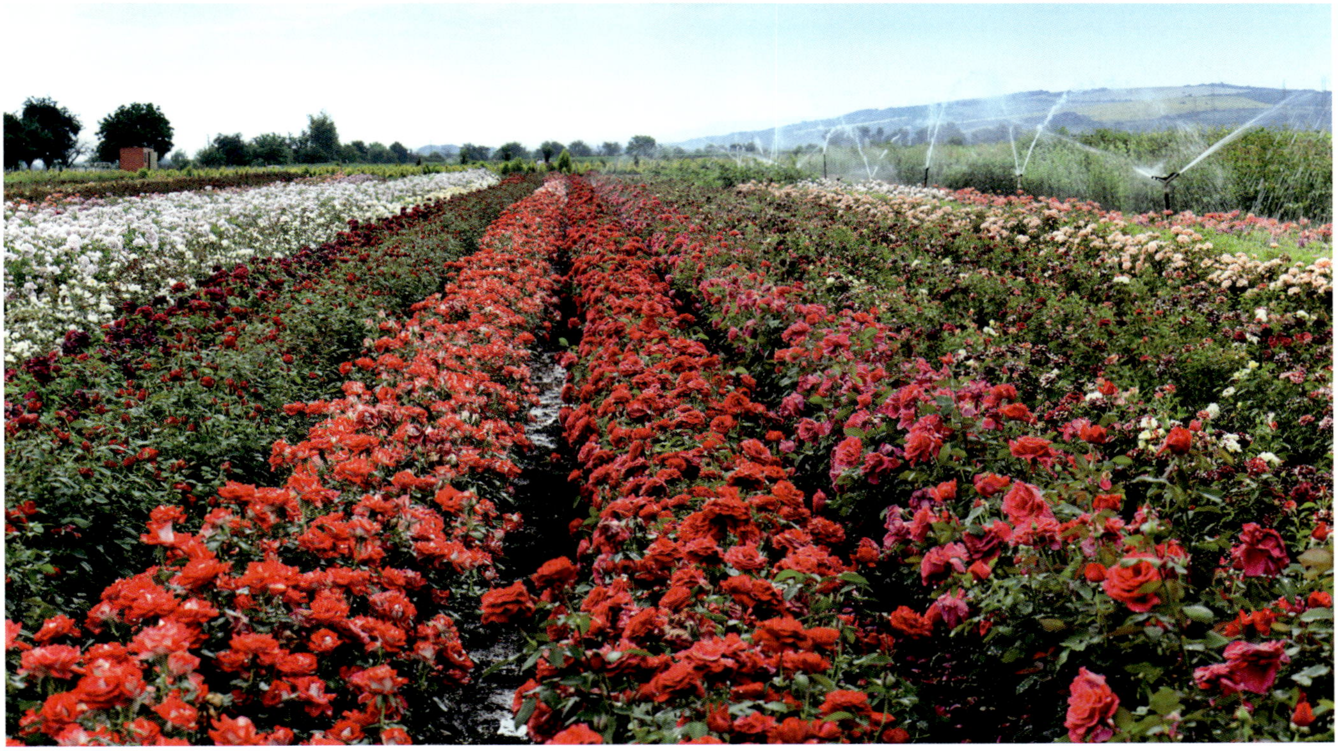

Above: A rose plantation

Below: Dried rose buds for tea stored in a wooden bowl

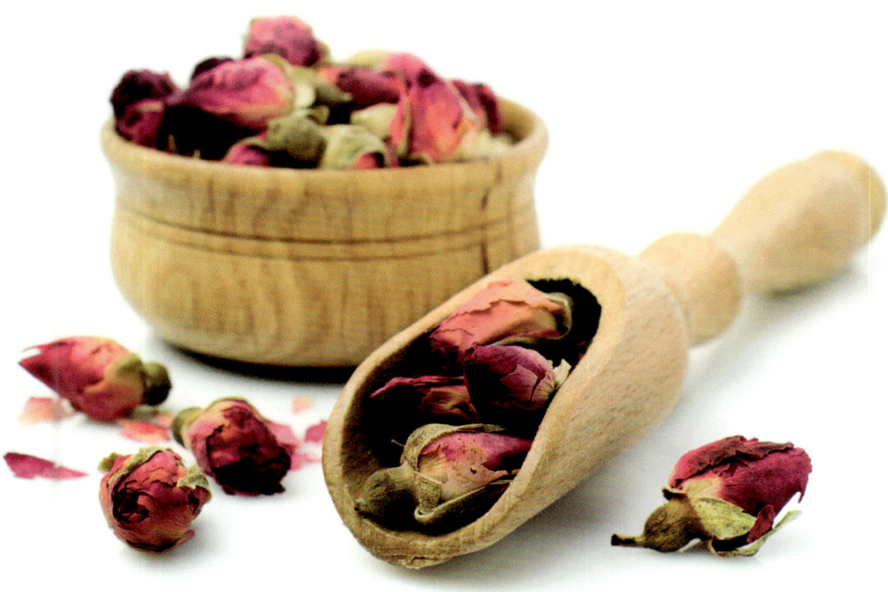

upon a rose garden you might chance upon the well-loved English rose, the recognised hybrid Tea rose, the Floribunda rose, the Chinese rose, the Damask rose, the Climbing rose or Miniature roses. Shrub roses are also becoming quite a common sight these days.

Roses also bear fruit; these are known as rose hips, rose haws or rose heps. Some of these are used for ornamentation and decorative purposes. However, the most common use of the rose the world over is to signal love, passion, sensitivity and beauty. For these reasons it is also a well-beloved name for girl children.

The veritable queen of flowers, innumerable references has been made to the rose from Mughal times. It was said that Nur Jahan, the extraordinarily ambitious and supremely beautiful wife of the Mughal Emperor Jahangir, in 1612, discovered and initiated the process of distillation of rose '*attar*' and ever since the rose or *gulab* has become part of the everyday in the

Indian subcontinent. From '*gulab jal*' or scented rosewater to '*guluband*' or bracelets, the *gulab* is evident in mind, thought, attire and even food.

While the earliest fossil records of the rose can be traced back to thirty-five million years ago, cut flowers and table roses came to Europe from China as late as the 18th century. There is evidence of 5,000 years of rose cultivation in China. There are also innumerable references to the rose in the Roman and Greek cultures. One well-known tale is the story of the ancient Greek poetess Sappho who planted roses in the goddess of love, beauty and sexuality, Aphrodite's shrine, to invoke her, using the flower considered to epitomise the power of love and beauty. Closer home, the lotus inspires similar sentiments, however the rose, in recent times, has inched its way up on the popularity list and has emerged as the queen of flowers. 'If sunshine had a smell this would be it,' as one awestruck perfumer marvelled.

The rose is akin to a chameleon, so many are its incarnations. 'When the Greeks created meanings through their arts, the Romans brought it to the very heart of daily life, making it an essential accompaniment to cooking, gardening, perfuming, feasting, drinking and debauchery, as well as the rites of both living and dying.' [1]

In Roman times, Pliny the Elder, in his encyclopaedic work *Natural History*, described ten different kinds of roses, arranged according to fragrance. Roses also signified luxuriousness and decadence, therefore the dictat of not wearing rose chaplets in times of war.

Roses were revered as the royal flowers of the Byzantines and they subsequently borrowed the idea of exquisite paradisiacal gardens from the conquered Persians. Christendom's red and white roses may have originated from the caliphal gardens of the Muslim Empire or from Persia and then travelled to India.

The highest pedigree of rose cultivation was held be the damasks, the centrifolias, the gallicas and the albas. The Damask, also called the Rose of Castile was said to have been brought from the Middle East by the Crusaders. The Centifolia rose was developed in the 17th century by Dutch rose growers and are sometimes called cabbage roses as they have large blossoms. The Gallicas are considered to sometimes be the oldest of the garden roses. The Alba rose is the famous White Rose of York that is the symbol of the English House of York. The House of Lancaster was symbolised by a red rose. Afterwards, came the Tudor Rose named for the House of Tudor, and a political emblem which united the houses of York and Lancaster, as is symbolised by its red and white petals. A French apothecary claims that the Tudor rose was, in fact, brought back from the Holy Land.

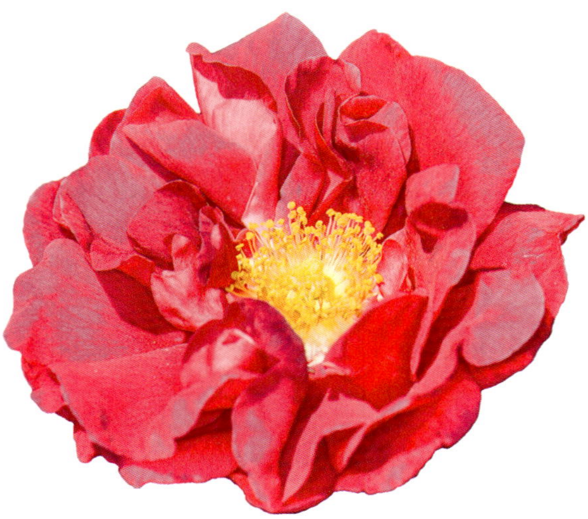

Rosa gallica, also known as red damask rose

Right: *Rosa centifolia* or cabbage rose

Below: The white Alba rose

Bottom: Chinese roses

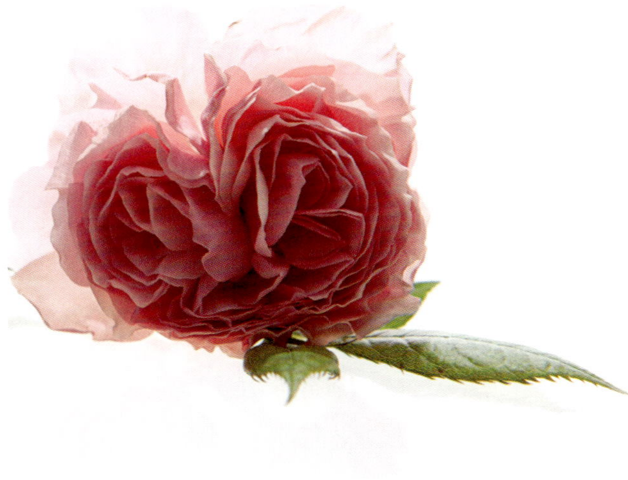

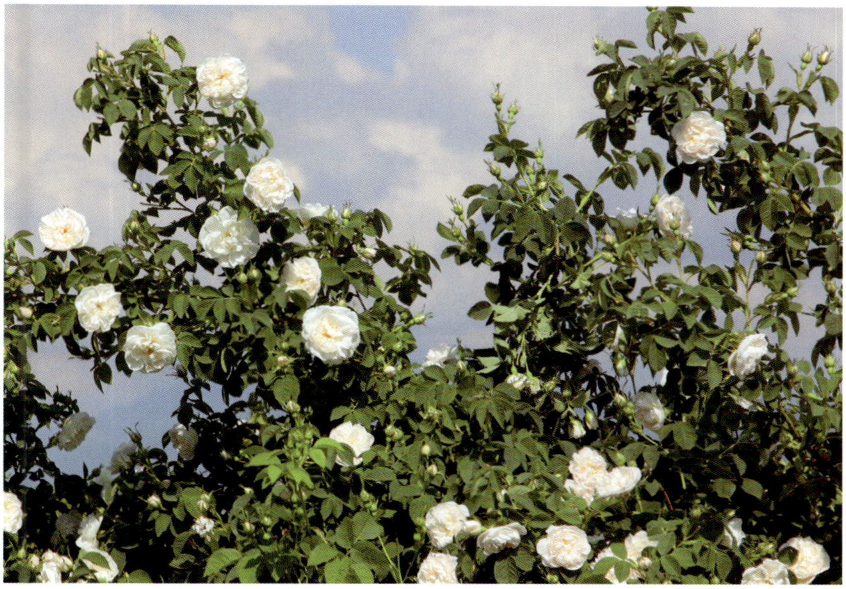

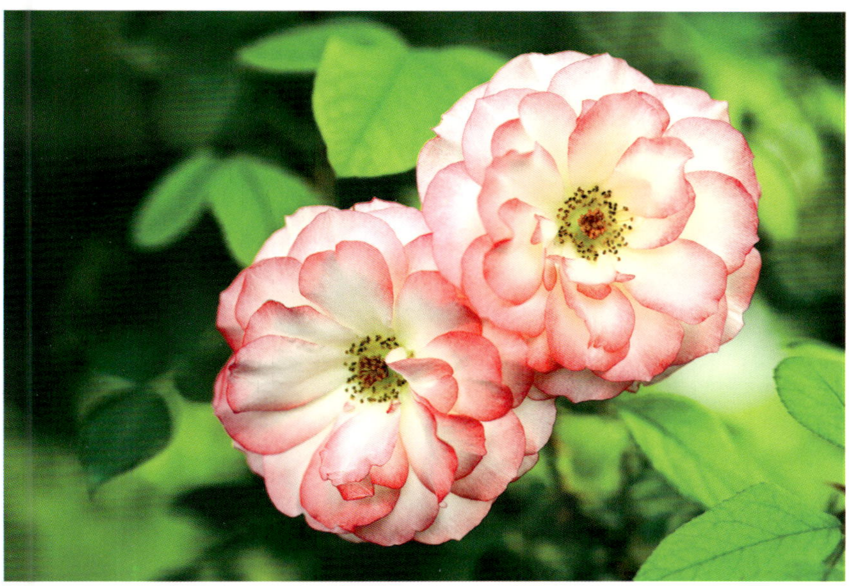

It was China that sparked a frenzy of rose breeding and rosomanie or rose mania, as a result of which roses spread like an epidemic the world over. There are about 150 wild species of roses. The Chinese roses took the British Empire by storm from the 18th century. The Chinese roses were the lightest of silks in stark contrast to the crumpled Damask of Europe. The Chinese roses had repeated flowering in radiant colours of clear reds, scarlets and yellows. The leaves were a thicker green and the buds were elegant. Some of them had a faint sweet smell or a delicate hint of tea. And through the British Empire came a fresh wave of roses from rambling roses to tea roses to Chinese roses into the Indian subcontinent, intermingling with first the indigenous roses of the land, along with the many varieties brought in by the Mughal rulers in India.

In contemporary times, the Mother, from Auroville, the long-time companion of Sri Aurobindo, in her works *The Spiritual Significance of Flowers*, chronicled the unique sacred qualities of nearly 900 blooms, a vast number of which were roses. Her observations were as much about the botanical names of the flowers as well as regarding their spiritual significance and colour. For example, the white-tinged rose with pink and yellow hues signified, to her, love from the

WAR OF THE ROSES

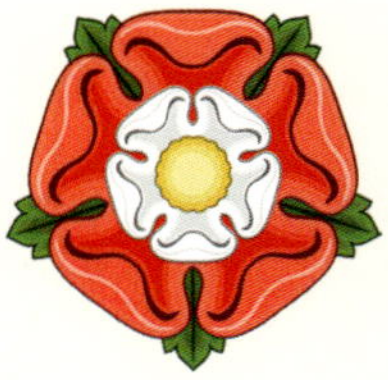

In English history the rose is an important symbol and is best known for its use in the 'War of the Roses' (1455–1487). A civil war broke out in England as the royal families of York and Lancaster each sought to gain control of the English throne during the reign of King Henry VI. The House of York was symbolised by a white rose while the House of Lancaster was symbolised by a red rose. Henry Tudor (of the Lancastrian faction), who ended the War of the Roses by defeating King Richard III and marrying his niece Elizabeth of York, ascended the throne as King Henry VII. He then created a new symbol, the Tudor Rose, which combined the images of both red and white roses.

Divine. On the other hand, the yellow-tinged rose, she posited, signified mental surrender.[2] When a rose was multi-hued it represented, in her opinion, communion with the Divine, while a salmon rose would mean beauty offering itself to the service of the Divine. And, of course, red would represent the universal sentiment that red evokes—passion. However, she posits it is a passion that is changed into love for the Divine.

Thus roses in every shape, size and hue are seen from Kashmir to Kanyakumari replete with their singular myths and legends.

INDIAN POETIC LORE

The relationship between the rose and the nightingale had always been a part of poetic imagery, highlighting the existence of nature with the most powerful of all emotions: love. This bond is witness to all shades of emotion—longing, disappointment, hope, joy—as seen between the birds and the trees, flowers and even celestial bodies such as the moon, the sun and the stars.

The willow tree is shown to be in love with the river and even bends down to touch the flowing water. The white lotus is in love with the sun and droops in sadness during sunset. The blue lotus blooms around the reflection of the moon and, of course, the love of the nightingale is for the rose. The rose is the beloved—or in a symbolic form—it is the beauty and the nightingale is the lover. It is a relationship between nature's two phenomena—a flower and a bird, viewed in poetry from the perspective of human emotions. By looking at it from the philosophical perspective, it is the longing of the finite and the infinite.

The famous poet from Sialkot, Muhammad Iqbal, in his immortal poem on India, writes:

> *'Sare jahan se achcha Hindustan hamara;*
> *Ham bulbulen hain iski, yeh gulistan hamara.'*
>
> 'Better than the whole world is our India;
> We are its nightingales and it is our
> flower garden.'

Though there are other flowers that are praised for their beauty, it is the rose that becomes the symbol of the spring and of beauty. In India, the nightingale has always been regarded as the herald of spring, whose first song assures the arrival of the new season. The nightingale's song is full of sadness when the rose does not respond to her love, though there are carefree and happy moments, too.

Binding of a Divan of Hafiz, original lacquer *'gul-u-bulbul'* (flower-nightingale) motif with gold, red, and black decorative frame, Iran, 1842

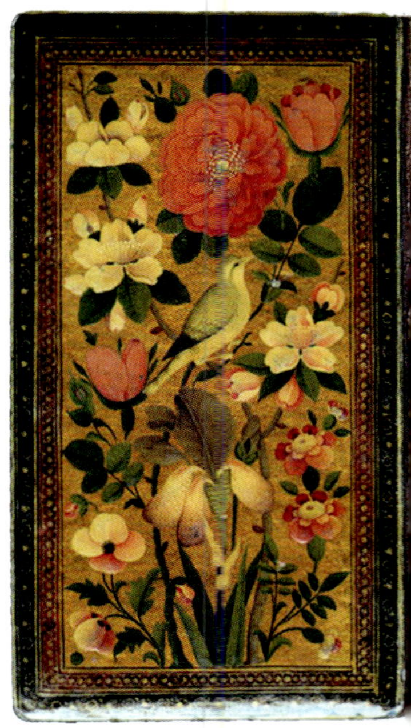

Detail of 'An Allegory of Empress Josephine as Patroness of the Gardens at Malmaison', Baron François Gérard, coll. Metropolitan Museum of Art, New York

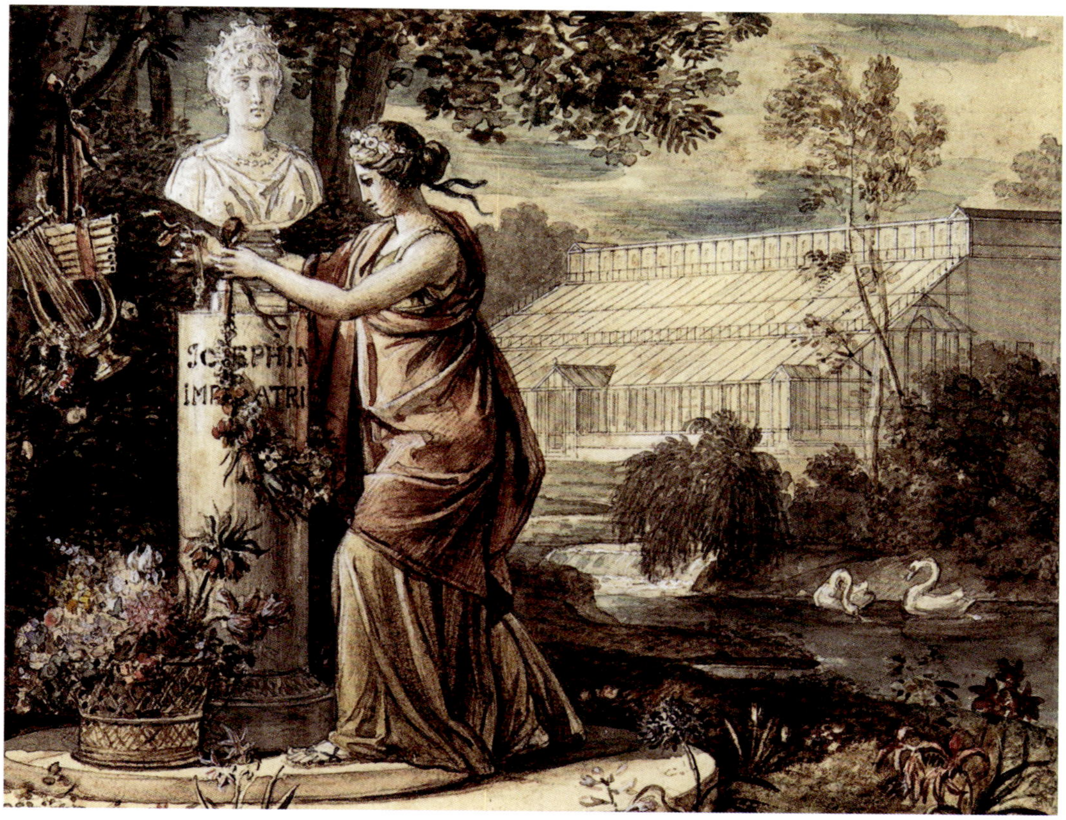

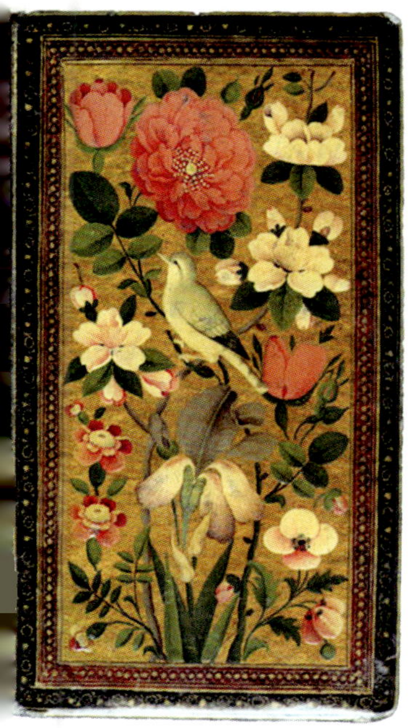

THE ROSE IN LEGENDS

Underscoring the significance of the rose in Indian civilisation, rose breeders and enthusiasts Girija and Viru Viraraghavan, based in Kodaikanal, Tamil Nadu, write:

> Even more frustrating for an Indian rose historian is the fact that there are no written records in India, covering periods where civilisation was otherwise greatly advanced, like the Indus Valley Civilisation (2,500–1,500 BCE). Obviously the association of roses with India's history goes back to well over 5,000 years. But, as earliest Indian history is based on oral tradition and not the written word, one has to depend on traditional myths, legends and folklore.[3]

Only later in the 17th century came the *Hortus Malabaricus*, a Dutch treatise, the compilation of which is attributed to Hendrik van Rheede, then the Governor of Dutch-ruled Malabar. It detailed the medicinal properties of Kerala's flora, touching very briefly upon those of the rose.

One story tells of how Brahma, the Creator, and Vishnu, the Preserver, were discussing which flower was superior. Brahma favoured the lotus, while Vishnu preferred the rose. Brahma had to admit defeat though, when he was brought to Vishnu's splendid rose garden.

As far as creation myths go, there are many about how the gods as we know them came to be. While Lakshmi is often thought to have been brought into existence out of the lotus by Vishnu or is said to have arisen out of the churning of the ocean of milk, there are some legends that believe the flower she was conjured from was, in fact, a rose. This marvellous flower is often said to have been comprised of 108 small petals (108 being an especially sacred number in Hindu philosophy) and 1008 smaller petals.[4] One is inclined to believe the latter story due to the fact that the Himalayas were where the gods were said to reside, the majestic, snowy peaks being home to a host of over 15 kinds of rose. The three ancient civilisations, India, Greece and Rome have their respective mythical association with the Rose. While the Greeks too made a rose oil, references to the rose in India are scanty until the legend which is associated with Nur Jahan who is said to have discovered the art of extracting *attar* or fragrance from the bloom. Yet another interesting legend of the rose is associated with the Empress Josephine, first wife of Napoleon Bonaparte, for whom 50 watercolours of roses were executed on vellum by the famed Belgian artist Pierre-Joseph Redoute. The most expensive painting by Redoute sold for 250,000 pounds and is of a dusky pink autumn damask.[5] Another tale from a distant land is that of the Syrian boy king Helioglobus who was said to have smothered his dinner guests with the perfume of a plenitude of roses.

There are plenty of rose legends in Islam, as the Prophet considered them a representation of the Glory of God. When the Prophet saw a rose he is said to have placed the bloom on his eyes after placing a kiss on it. This sense of the rose's significance is shared by many Islamic mystics.

THE ROSE THROUGH HISTORY

The visual history of the rose in the subcontinent harkens back to the fifth century BCE, during Buddha's times. There are many statues of the Buddha which depict him seated on a lotus in the first and second century Gandhara School of Art. However, in some cases, a five-petalled rose seems to take the place of the lotus. There are also legendary tales of Alexander the Great's invasion of India. He breached the subcontinent's north-west border in approximately 326 BCE and, it

The fragrance of roses refreshes the spirit and its visual beauty promotes peace and calm

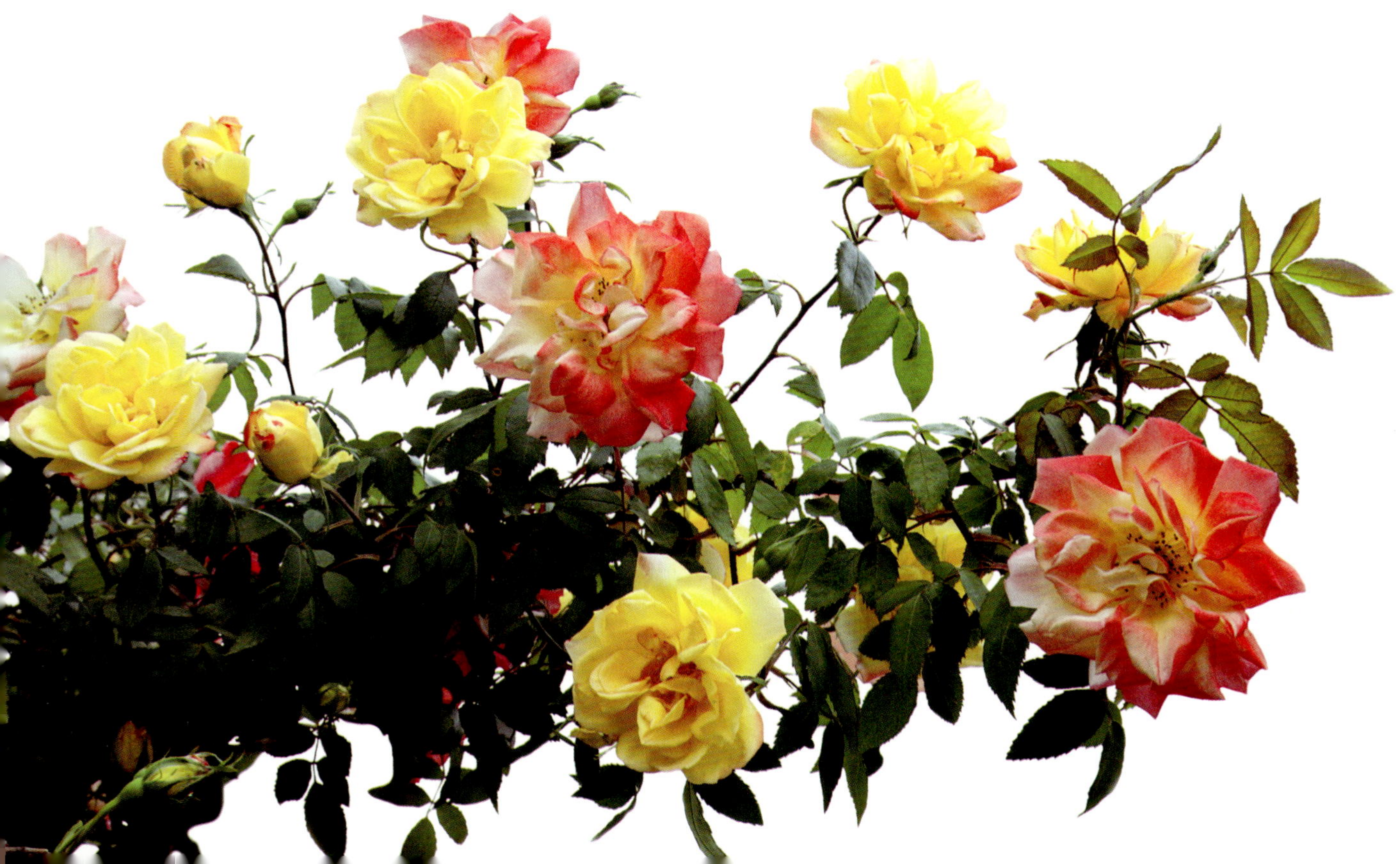

COLOURS TELL A STORY

RED roses are almost universal symbols of love and passion. They can also convey feelings of devotion, sorrow, respect, admiration and regret. A popular practice is to offer a loved one, often a beloved, a dozen red roses, long stemmed ones being preferred.

PINK is a gentler variation of the red rose used to convey feelings of happiness, admiration, respect and perhaps even gratitude.

DEEP PINK roses are things of beauty and elegance, similarly used to express the same sentiments as the lighter variety. They can also be used to express feelings of gratitude, much like their pink counterparts.

LIGHT PINK indicate innocence.

WHITE has a number of connotations most common being peace, purity and innocence, often the reason why brides in the Judeo Christian tradition hold bouquets of white flowers at their weddings. They can be used to signal mourning, which is why white flowers are often used in funerals and on graves.

YELLOW is a warm and bright colour used to express exactly those emotions. Yellow roses can also convey friendship, care, platonic love and joy.

ORANGE blooms are full of passion and vigour and are used to express strong emotions of ardour and fierce joy.

BURGUNDY roses can be employed to signal loveliness.

GREEN is not a common colour for roses but in Ireland green is a symbol of good luck. They can be used to express luck and good wishes, perhaps for someone in ill health.

BLUE roses are another unusual colour and have to be created by humans. Since they are so unusual they can signal things that are enigmatic or beyond one's reach. It could signal that someone cannot put you out of their mind, even though they cannot be with you.

BLACK is not necessarily an auspicious or lucky colour which is why it connotes the ends of things; it can signal the end of life, the end of friendship, the end of a relationship.

PURPLE was considered the colour of royalty during the peak of the Roman Empire. Purple roses can express grandeur and perhaps even protection and deep fondness.

LAVENDER roses can be said to signal falling in love at first sight. They are also used to convey enchantment.

is said, he was astonished at the subcontinent's wealth of natural beauty. The stories tell that he had rose plants, among other gifts from India, sent to the great thinker Aristotle, whom he held in high esteem.

Roses were known in the first century BCE and there are several references to the flower in the treatises of Charaka, a physician, and Susruta, a surgeon, who are believed to be the founders of Ayurveda. Sanskrit had many names for the flower. These names each stood for its curative qualities with regard to a certain illness or ailment, for example as a cure for fever, a remedy for stomach aches, as a laxative, etc.

India had many early texts dedicated to documenting the art of plant growing, for example the 15th century *Vishwavallabha* by Chakrapani Mishra. It states that the presence of roses indicates the presence of underground water. The 800 CE text *Kashyapa's Agricultural Treatise* puts forth that a king should plant roses, among other prescribed flora, in the land over which he reigned.

It was Vatsyayana who wrote the *Kama Sutra* (*Aphorisms of Love*) before the sixth centuary. He offers an enumeration of the duties of a virtuous wife. Among these is to tend the garden in which flowering plants, including the rose is grown. The *Kama Sutra,* an iconic book on judicious behaviour and balance in life between righteousness, pleasure and finance, elaborates in great detail the 64 arts of an aesthetic life. In it, perfumery and aromatherapy are classified as important arts and the rose is mentioned. Several centuries later Kalyana Malla composed the *Ananga Ranga,* in the 12th century. In delineating how the romantic atmosphere could be created in a room in which a couple was to indulge in each other's company, he maintained that rose water should be present to aid the mood for their sensual pleasures.

CROSSING THE FRONTIERS

Beauty gives you wings. And nothing gives more credence to this statement than the voyage of the rose. The physical journey of the rose through the continents in the world is deeply nuanced. The rose has travelled everywhere and in most continents has also been part of its biodiversity. There was a large influx of Chinese scholars who came to India from 1000 BCE onwards since India was the birthplace of the Buddha who went on to have a staggering impact on Chinese thought. Through their accounts (as they visited places of import to the Buddhist faith) we learn that rose garlands were in use in India at the time. Especially noteworthy are the chronicles of the Buddhist pilgrims Fa Hein and Hieun Tsang who visited India in the fifth and seventh centuries, respectively. They reported the Indian gardens were brimming with bountiful roses amidst fountains, which had them spellbound.

A common commercial activity of the time was trade with both eastern and western countries, and India and China did a roaring trade in roses and their by-products. There is evidence that the Kakatiya king, Ganapati, did away with customs duties on Chinese rose water. During his visit to Gujarat, Rashid-ud-Din in the 14th century counted that no less than 70 types of roses were grown there. Another Persian traveller, Abdur Razzak, who visited the kingdom in 1443 as an ambassador of Sultan Shah Rukh of Khurasan, states: 'Roses are sold everywhere. These people could not live without roses and they look upon these as quite as necessary as food.'[6]

One of the founders of Ayurveda, Charaka, mentioned rose water in his texts. The first century Buddhist sage Nagarjuna, also speaks of its distillation in *Arkaprakash.* The early medieval Persian scientist Avicenna is said to be the first to discover how to make rose water through distillation; perfumes were concocted from the precious blooms and then exported. The flower used in those times is probably what came to be known as Rosa damascena (*bussorah*). Rose water is still

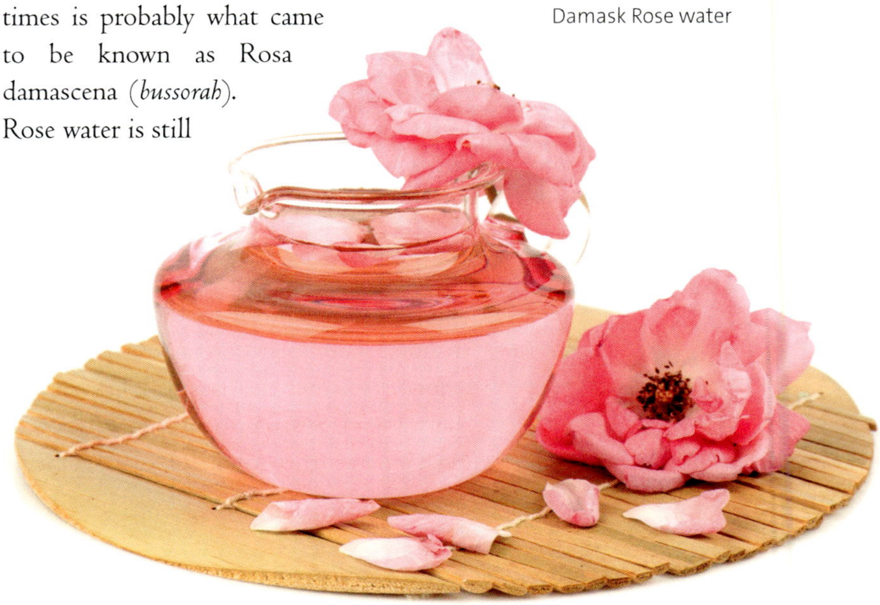

Damask Rose water

'Babur reading in a garden', Bishn Das, watercolour on paper, 16th century, coll. British Library, London, UK

made from this species of rose in India to this day. The Edward rose is another rose used in the creation of rose water and for this reason is referred to as '*panneer*' rose, *panneer* meaning rose water.

ROSES IN THE MEDIEVAL SOUTH

In medieval South India travellers such as the Portuguese Domingo Paes (1520) and Fernaõ Nunes (1535–37), who passed through the Vijayanagara Empire, caught their breaths in surprise as they realised that the luxurious rose blooms formed an important part of everyday life for the local people, who used them as body adornments.

According to them, the exquisite beauty of Hampi on the banks of the Tungabhadra River was enhanced by the many fields of rose bushes, the flowers of which made their way into shops, baskets and garlands.

The most illustrious ruler of the Vijayanagara Empire was Krishna Deva Raya (1471–1529). He was learned in the arts of poetry, literature and warfare. He was described as being attired in snow-white silk robes with intricately worked roses of gold on them. A story pertaining to the king describes how he would rain down white roses upon his courtiers, elephants and horses during his royal audience every day.

It is said that the bedchamber of the monarch had carved stone pillars as well as cross-timber pillars topped with singularly exquisite ivory roses.

An Italian traveller, Ludovico de Varthema, from Bologna, came to the port of Calicut (Kozhkode in present-day Kerala) in 1503 and wrote that he saw innumerable flowers, most notably roses in many hues—red, white and yellow.

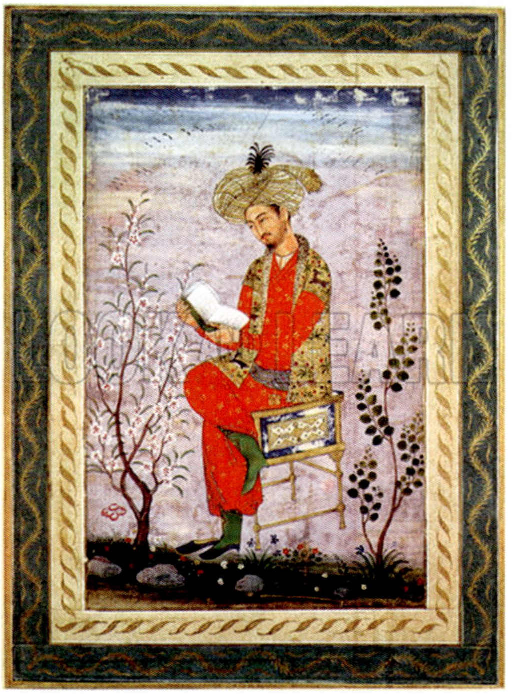

THE MUGHAL LOVE AFFAIR

The arrival of the Mughals[7] ushered in the medieval period of Indian history. Stories are told of Babur, the first Mughal emperor to reign over India, who arrived bringing with him camels laden with loads of roses. Speculation leads historians to believe that these were probably of the musk and damask variety. Babur's reign had centred around the city of Kabul and it was famed for its numerous varieties of flowers. It is believed that Babur thought of the rose and the nargis as among his favourites.

This poem eulogises his fondness for the queen of flowers:

> My heart, like the bud of the red, red rose,
> Lies fold within fold aflame,
> Would the breath of even a myriad Springs
> Blow my heart's bud to a rose?[8]

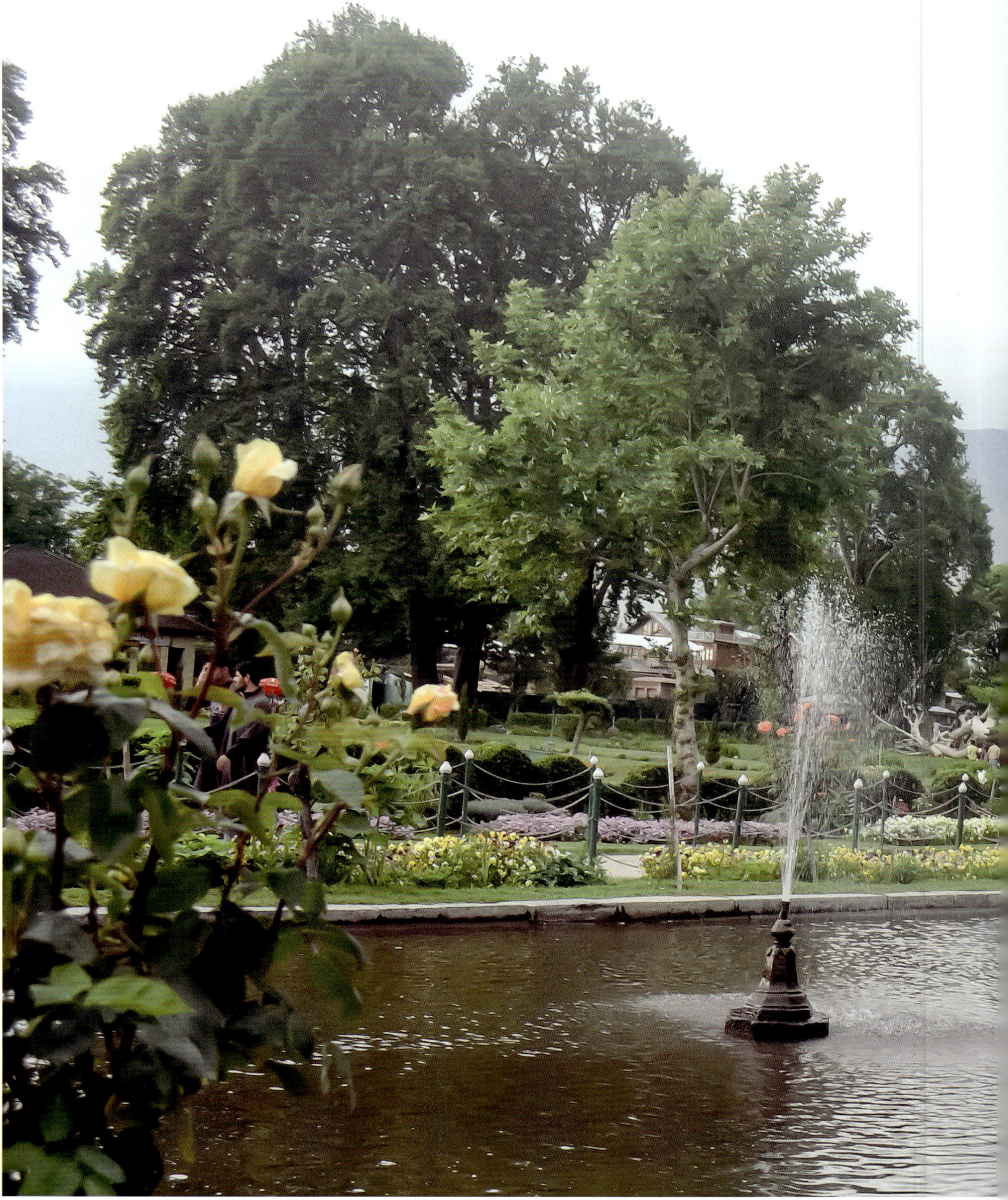

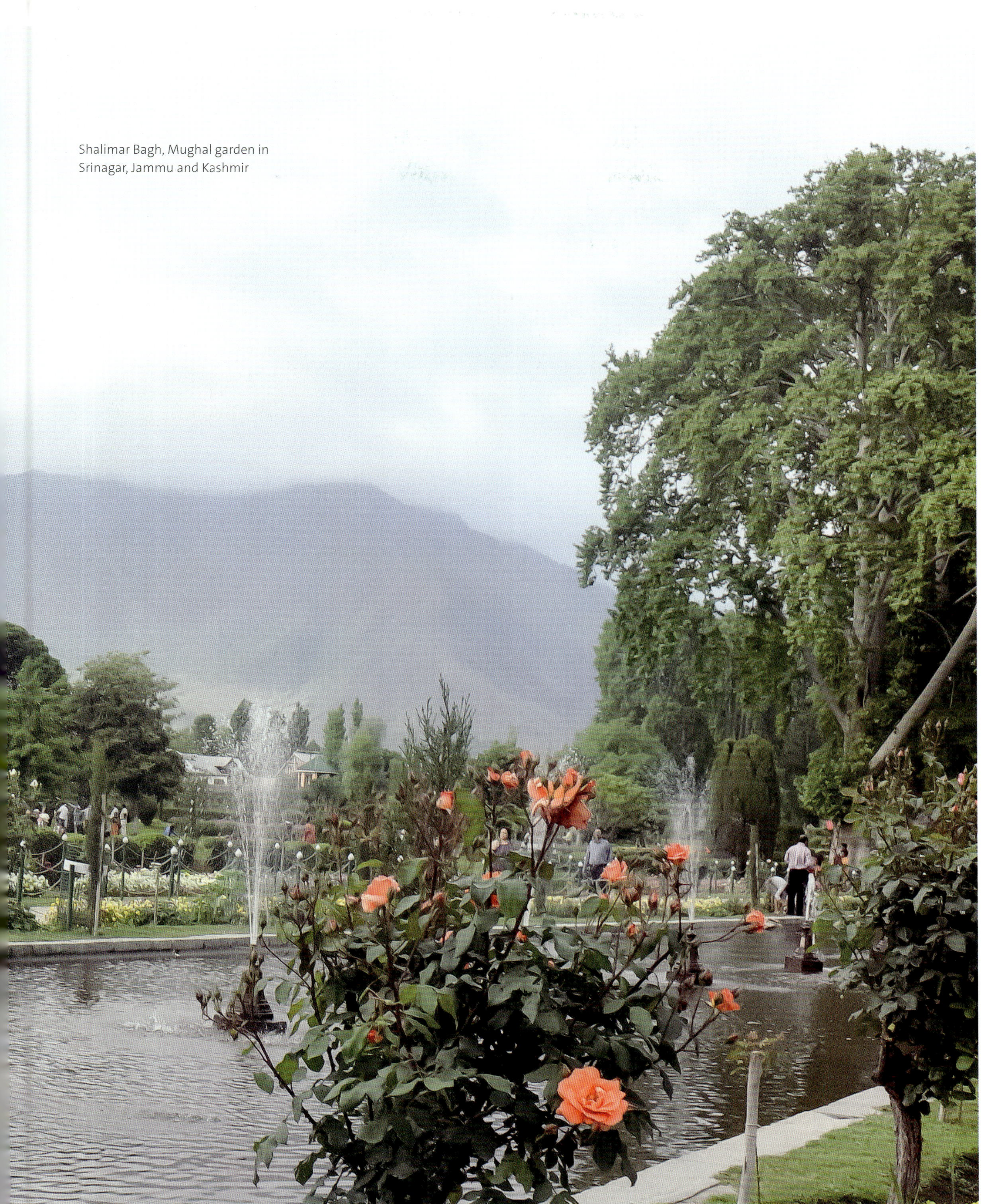

Shalimar Bagh, Mughal garden in Srinagar, Jammu and Kashmir

One of Babur's wives was called Gulrukh, or 'she who is rose faced'. So deeply fond of the flower was Babur that he named each of his daughters for it, as well. One was Gul Chehra or 'she of the rose face'. A second was Gulbadan, 'she with the rose body'. And the third was Gulrang, 'she who is rose coloured'. In Persian, the word 'gul' has the dual meaning of rose and flower.

Gulbadan was fondly known as a 'rose princess' and she had a keen eye for chronicling life at the royal court. It is said that it was Akbar, her nephew, who commissioned her to keep a record of affairs. Her eye for detail was astonishing and one of her descriptions of a feast given by Maham Begum takes note of ewers and sprinklers for rose water.

The geometric 'char-bagh' or quartered plan of the Mughal gardens was conceived by Babur. These gardens were not only stately and sumptuous but also famous for their profusions of roses alongside systematically formulated beds, fountains and pathways that were all extremely intricately planned. Well-known examples are the gardens around the Taj Mahal built by Shah Jahan, the gardens at Humayun's Tomb, built by his wife Bega Begum and the Shalimar Gardens of Kashmir built by Jahangir.

Empress Nur Jahan is well known not only for her love of the scent of roses but also for creating '*attar*,' Persian for rose oil. The stories say that she was fond of bathing in hot water scented with the fragrance of roses. It was during these baths that she noticed the formation of a scum in the bath and she realised the scum carried the scent of roses. From this she went on to fashion her rose water. Said Emperor Jahangir,

> *It was of such strength in perfume that if one drop was rubbed on the palm of the hand it scented an entire assemblage, and it appeared as if many rose buds had bloomed all at once. There is no scent of equal excellence. It restores the heart that has gone and brings back withered souls.*[9]

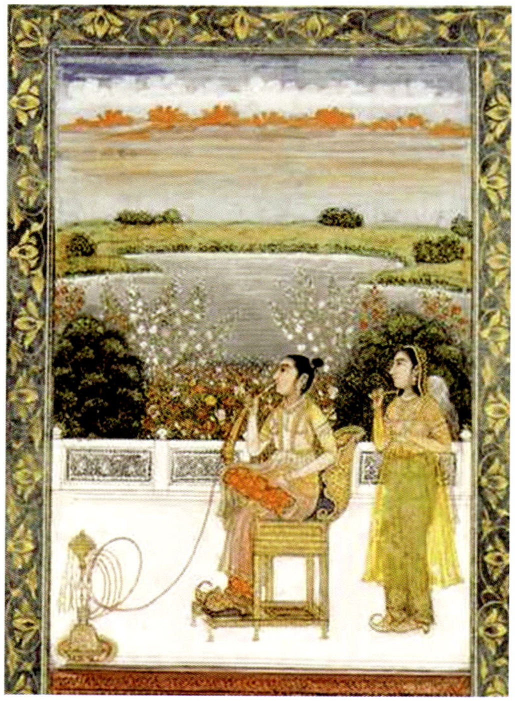

Left: Gulbadan Begum smoking a hukkah

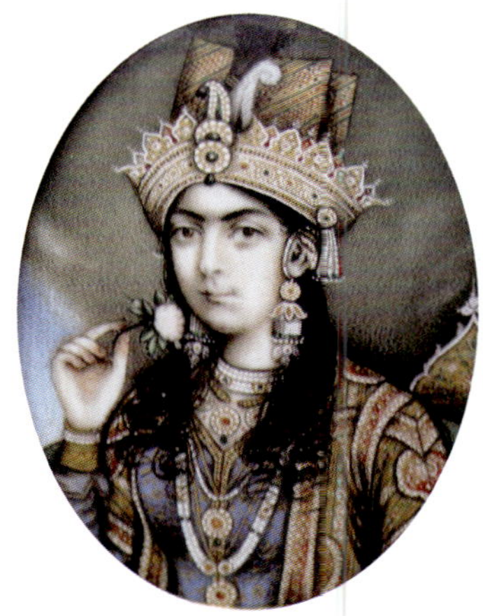

Below: Posthumous portrait of Empress Nur Jahan (1570 – 1627), gouache and gold on ivory, 'Company School', c. 1840

Rose liquor was something of a favourite among the royals. This heady liquid was distilled from an approximate 20,000 rose blooms per bottle, a precious infusion, indeed. It was also almost a regular feature of the subjects of the paintings of this time to be portrayed holding a rose bloom in their hands.

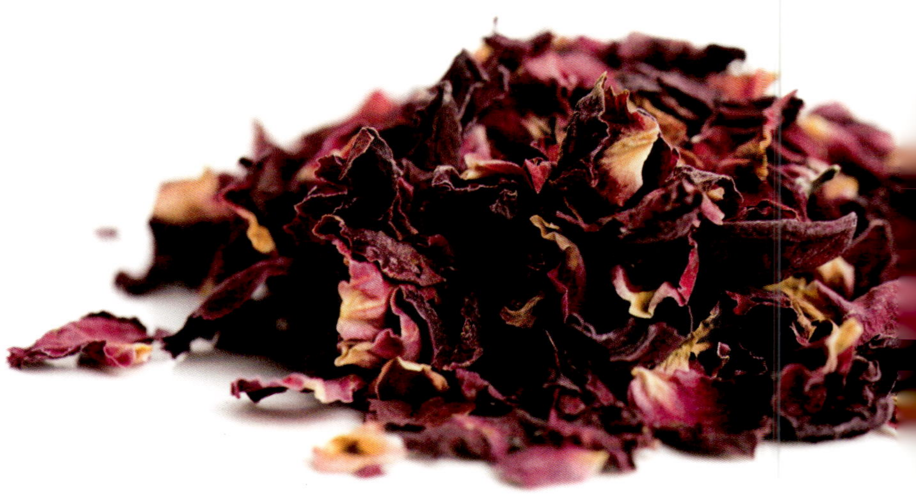

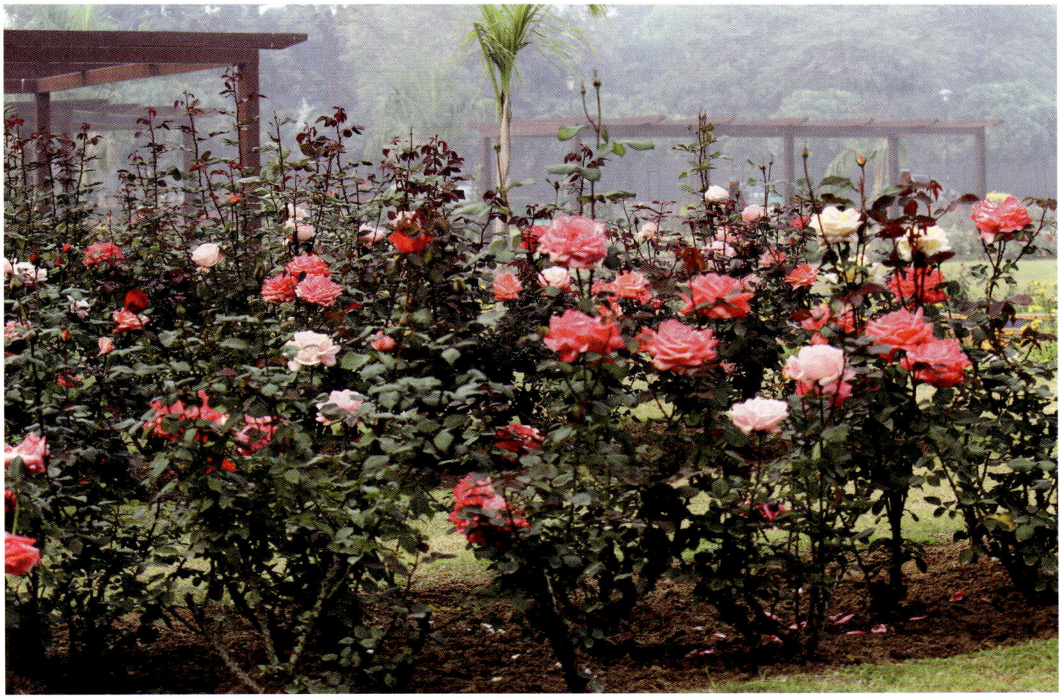

Above: Roses blooming at National Rose Garden, New Delhi

Below left: Dried rose petals are the source of many fragrant products

Manucci, an Italian traveller to India in the 17th century, on arrival found an astonishing variety of fragrant rose water and rose oils. They were not only used by nobles to bathe and scent themselves; they also daubed their horses with it in liberal quantities.

ENGLISH INFUSION

The British did not come to India as invaders or rulers. Their original intention was trade, the activity which they carried out through The East India Company. Through this trading entity, they slowly began to gain a foothold in the country. Their ships travelling from China and the Far East would stop at Calcutta en route to England. It is believed that almost all the ships carried plants—and, of course, roses—as these were becoming a favourite in European countries, such as England and France. The plants were housed in the Howrah Botanical Gardens so they had time to regain their vitality before the long sea journey to Europe was begun anew. The garden was established by Sir William Roxburgh in approximately 1793. Rose bushes from every ship's cargo were planted here, such as Chinas, Early Teas, Noisettes and Double Yellow. To this day, some of these plants can be found in the much older nurseries and gardens, though they might not be recognisable as such, being affected by age, the loss of their names and the ravages of time.

A VERSATILE BLOOM

The rose is a versatile bloom; the lovely flower is used for cosmetic purposes, for its curative benefits and as an essential ingredient in many dishes, as well as in beverages. As India is geographically blessed, the country can engage in a large number of agricultural activities; therefore the fertile lands of northern states

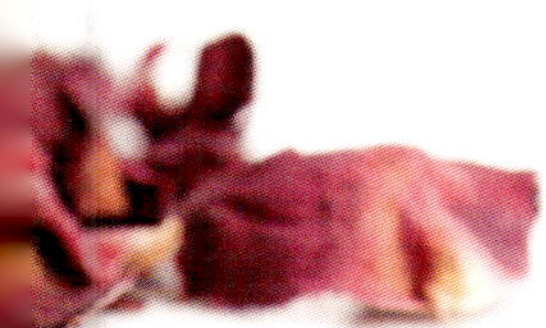

SECRET LORE

An old Roman tradition tells of the practice of marking a room where sensitive information was shared by placing a rose at its door. From this is derived the phrase 'sub rosa', indicating that someone is keeping a secret or keeping something 'under a rose' as they quite literally would, behind a rose-marked closed door in Rome.

like Rajasthan and Uttar Pradesh provide ideal conditions for commercial production of roses, where entire fields are often filled with the perfume of the lovely blooms. These are then harvested for creating rose oil and rose water. *Pankhuri* is fashioned from dried rose petals, a major export from India to the Middle East; there it is employed by the largely Islamic population in traditional cuisine, as well as in medicine. Rose water has a plethora of uses to treat ailments of the stomach, as eye drops and as a soothing relief from the heat. Rose water is said to have great hydrating qualities, thus making it the perfect ingredient for moisturising lotions. Many cosmetics are derived from rose water as it is said to be a wonderful toning agent, resulting in firmer and more supple skin, and it is therefore used in luxurious soaps and creams.

In food, rose water imparts its heady, rich flavour and is used in desserts such as Turkish Delight, gulab jamun, *gulkand* (a sweet preserve made by combining sugar and rose petals), and refreshing sherbet. The fruit of the rose plant called rose hips is not an unusual sight, being sold at markets. Varieties of rose wines and rose liquors can be enjoyed for those who would like to indulge their love for the flower further. Not only is the rose a treat for the taste buds, it is also very nutritious; in it are vitamins A, B, C and K, as well as folic acid. Additionally, it does not lose its nutritive value on being cooked. At weddings it is considered auspicious to greet guests by spraying them with rose water.

Rose oil is distilled from rose water and India is one of the most important manufacturers of this essential oil. Since enormous quantities of rose petals go into the production of minute amounts of this precious perfume, it is extremely expensive, almost on par with the price of gold. Within the country it is the Damascene and Bourbonia varieties that are employed in the creation of the oil.

IMMORTALISED IN VERSE

Springtime has always been received with a special kind of affection by poets. The poet Kalidasa begins his work *Ritusamhara* or the *Cycle of the Seasons* with summer, so that he could conclude with spring, which he called *rituraja* or monarch of the seasons. During spring the new breeze,

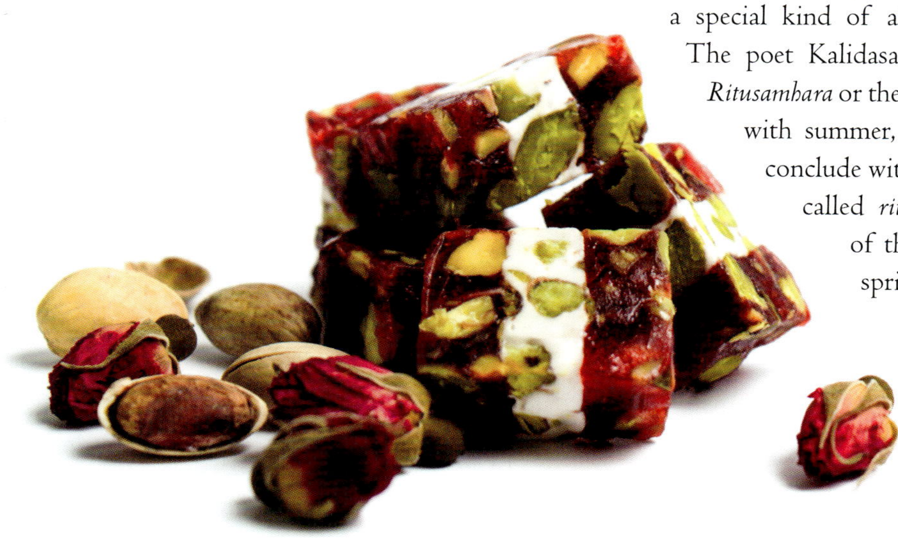

Left: Turkish delights with pistachios and rose buds

Facing page: 'Krishna and Radha conversing', *Gita Govinda*, ink and opaque watercolour on paper, 18th century

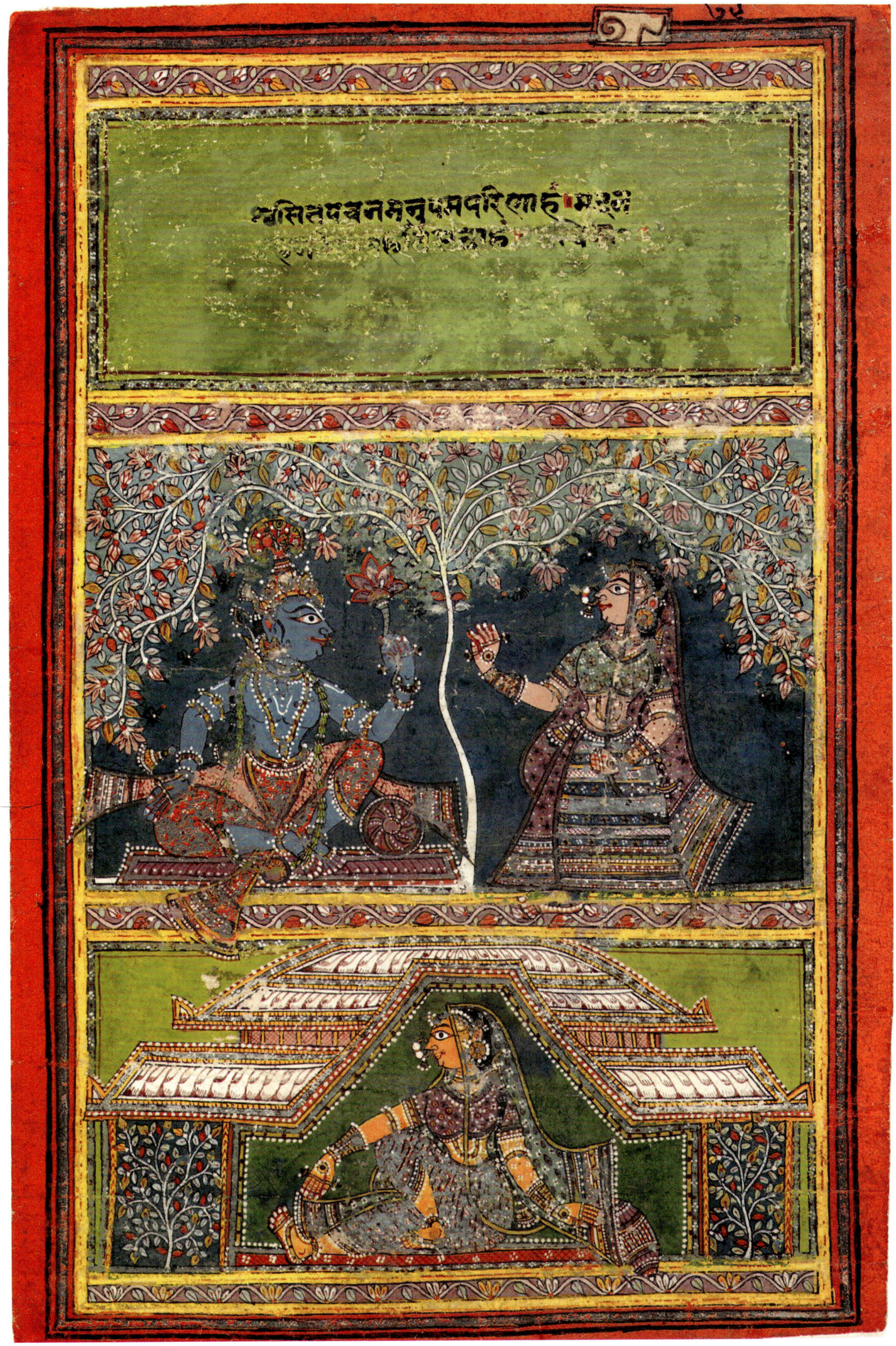

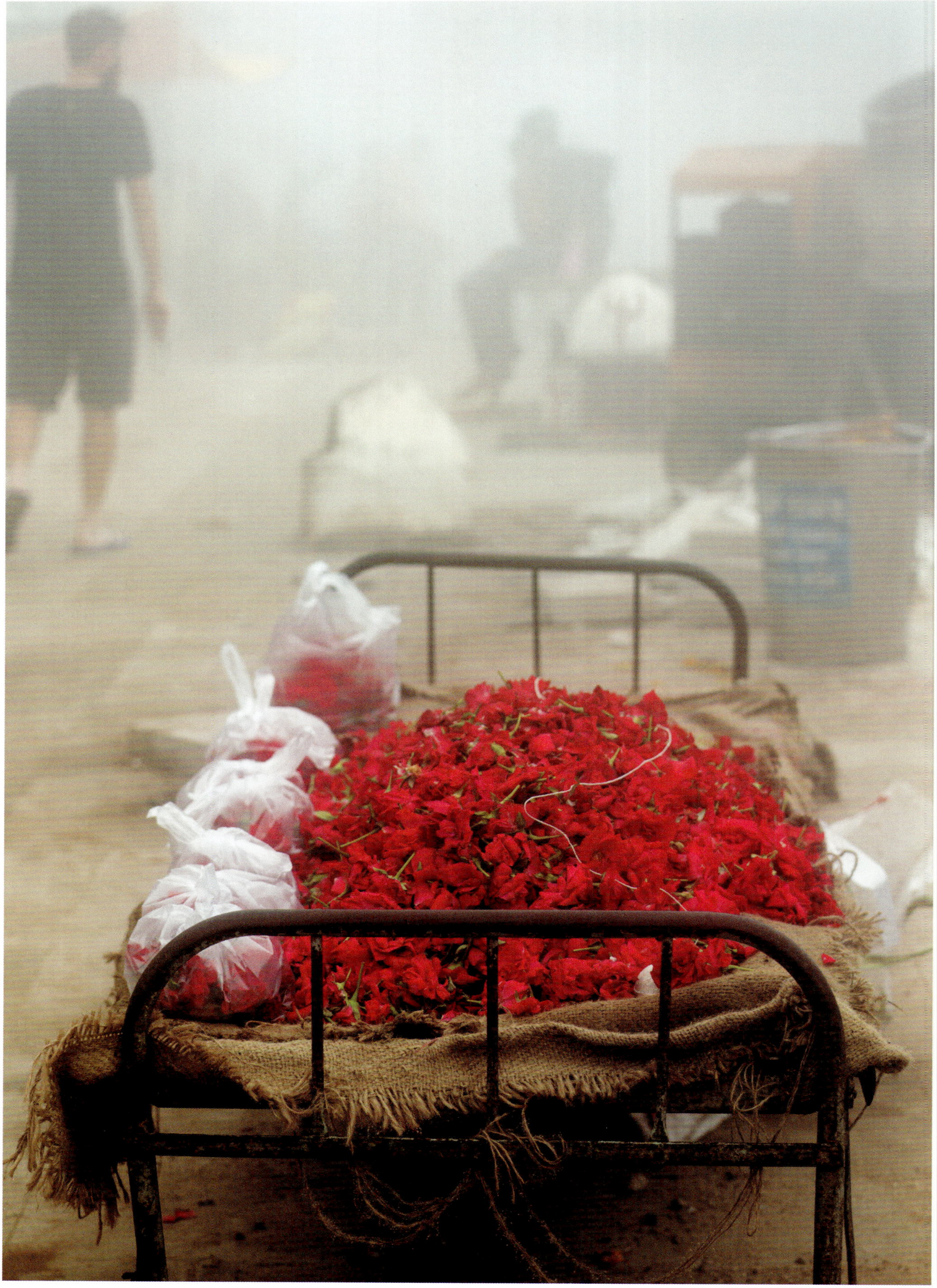

Facing page: Freshly plucked roses in the early morning fog, Champaner, Gujarat

the softness of the flowers and their fragrance, would bring gifts for all the senses. This gaiety of nature was a signal of celebration as it was the 'time for wine and roses.' Jayadeva's *Gita Govinda* (written in the 12th century CE) is full of references to the rose.

The ghazal form, which was perfected during the medieval period under Mughal rule, sang praises of the splendour of the rose, which was said to be why the nightingale sang its lament. These are images that were very popular in Urdu poetry, especially in the works of Hafez. Much like the Bhakti movement, Sufism also used the metaphor of the lover to speak of divine love. In this tradition there were poets like Ibn Arabi who likened their beloved's cheeks flushed with blood, to the rose, which he also associated with the names of the divine. Sufism is brimming with such instances to demonstrate the reverence its followers had for the rose. For instance, Sheikh Abdul Qadir al-Jilani, an eminent Sufi personage was called 'The Rose of Baghdad'. The story goes that when he returned to the city of Baghdad after gaining spiritual insights and a large following of disciples, the scholars of the city sent him a vessel filled to the brim with water to represent the city of Baghdad, which was full of holy men. Sheikh Abdul Qadir al-Jilani placed a rose upon the surface of the water to indicate he would only add the fragrance of his spiritual wisdom to the city. The holy men accepted his greatness and named him 'The Rose of Baghdad'. Qadiriyya, the order he established, went on to adopt the flower as their symbol.

Thus, we see, from the tender bud to the fully blossomed flower, the rose has enchanted and inspired people through the ages. No wonder then that the rose is called the 'Queen among Flowers.'

ENDNOTES

1. Jennifer Potter, *Seven Flowers And How They Shaped Our World* (London: Atlantic Books, 2013), p. 135.
2. The Mother, *The Spiritual Significance of Flowers – Part 1* (Sri Aurobindo Ashram Publications Department, 2000).
3. http://www.helpmefind.com/gardeningezinephp?publicationID=926&js=0 (accessed 6 December, 2015).
4. Frankie Hutton, *Rose Lore: Essays in Semiotics and Cultural History* (Lanham, MD: Lexington Books, 1955), p. 56.
5. Jennifer Potter, *Seven Flowers And How They Shaped Our World* (London: Atlantic Books, 2013).
6. Bangalore Suryanarain Row, *A History of Vijayanagar: The Never to be Forgotten Empire, Part 1* (New Delhi: Asian Educational Services, 1905), p. 317.
7. http://www.indianrosefederation.com/aboutus.html#about(accessed 6 December, 2015).
8. Som Prakash Verma, *The Illustrated Baburnama* (London: Routledge. 2016), p. 278.
9. Nidhi Dugar Kundalia, *The Lost Generation: Chronicling India's Dying Professions* (New Delhi: Random House India, 2015).

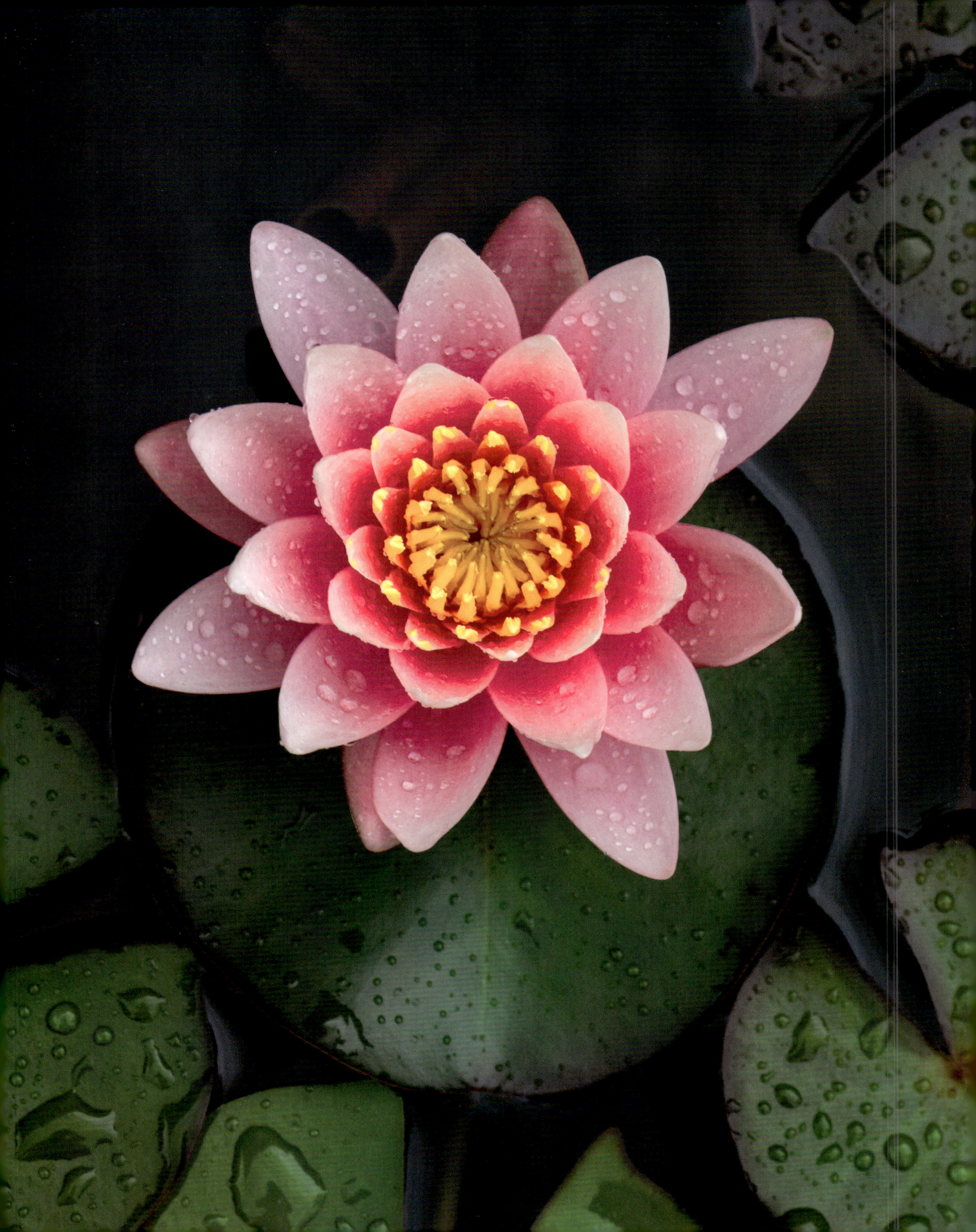

The Emperor of Flowers - Lotus

As a lotus flower is born in water, grows in water and
rises out of water to stand above it unsoiled,
so I, born in the world, raised in the world, having overcome the world,
live unsoiled by the world.
—**The Buddha** [1]

The lotus has had a deep association with the history, culture, religion, and arts and crafts of India. It has been described in various mythological legends, epics, scriptures, literary works and historical records. Due to this rich heritage, it has been selected as the national flower of India. The lotus is native to India, China and Japan, but is now widely distributed from tropical East Africa to Northeast Australia. It was introduced to Egypt by the Romans about 50 BCE. It is found growing in wetland habitats including floodplains, lagoons, marshes, swamps, rivers, lakes, ponds, pools and the backwaters of reservoirs. This aquatic perennial plant (*Nelumbo nucifera*) belongs to the family Nelumbonaceae. It is also called Sacred Lotus and Indian Lotus. In Sanskrit, the lotus is known by various names like pundarika, padma, kamala and utpala.

Facing page: A pink lotus flower bathed in dew drops rises resplendently above its surroundings, symbolising beauty and perfection

UNIVERSAL AURA

Frequently associated with pre-modern Hindu and Buddhist mythologies, the lotus has deep roots in Eastern history. The life cycle of this aquatic perennial has been connected symbolically to the concept of spiritual enlightenment in Buddhism. In Hinduism, a creation myth includes the unfolding of a lotus to represent the origin of the universe as it is dreamed into existence. In Egypt, the mythology has a similar creation myth, though the flower in this mythology is the blue lotus. The aromatherapy practitioners attempt to connect similar ideas of spiritual enlightenment to lotus essential oil. It is frequently used in meditation, where the combination of deep breaths and the perceived effects of relaxation from the scent increase tranquillity.

The lotus has inspired poets, artists and sculptors of both the ancient and present periods. Its shape has also been used as inspiration in architecture. A perfect, contemporary example of this is the Baha'i House of Worship in New Delhi, popularly known as the Lotus Temple due to its lotus-like shape. It was built in 1986.

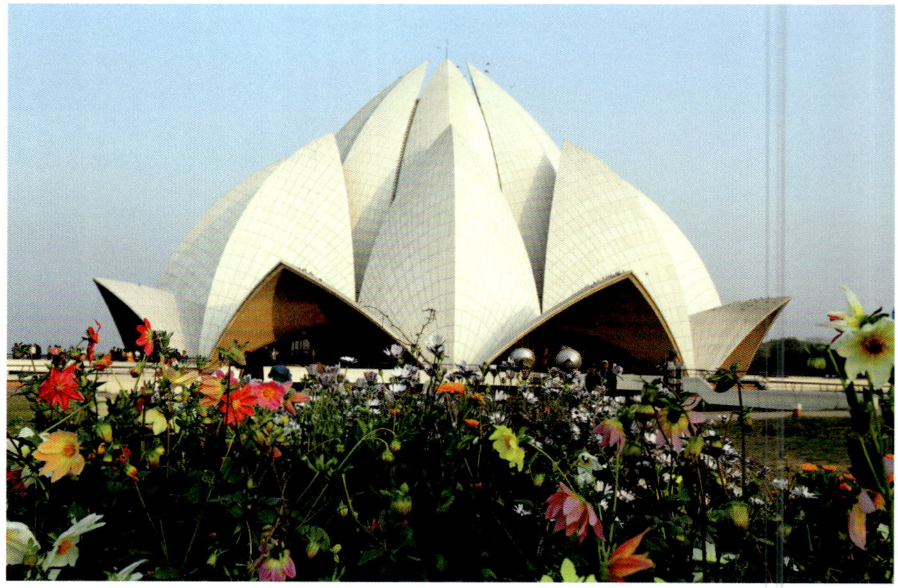

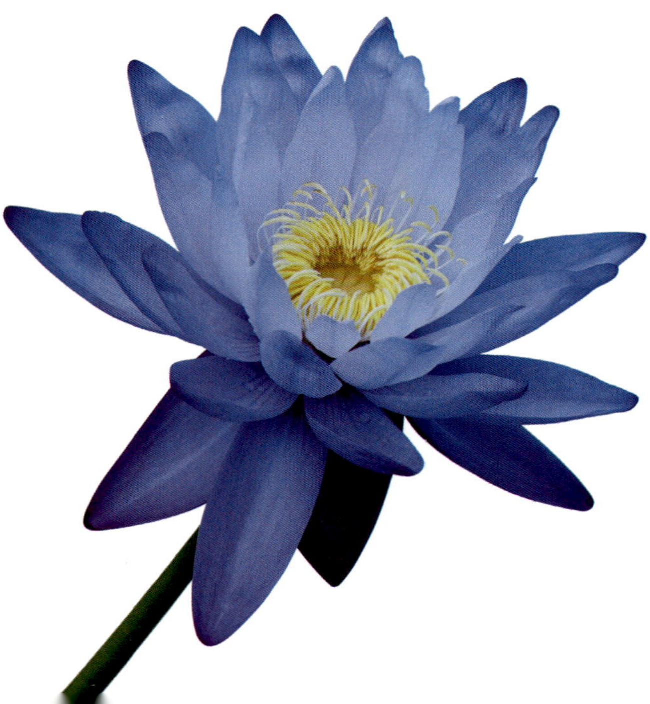

Lotus motifs are also seen in textile designs and pottery, and it is referred to in Indian folk songs and dances. It symbolises divineness, purity, beauty of the eyes and face, kind-heartedness, fragrance, coolness, fertility and prosperity. In Asian classical literature, especially poems and songs, the lotus represents the ideal feminine attributes such as elegance, beauty, perfection, purity and grace.

Above: Lotus Temple, New Delhi

Below left: A blue lotus

Facing page: God Upulvan is identified with Lord Vishnu

Below: A statue of Lotus-headed Fertility Goddess Lajja Gauri, sandstone, Madhya Pradesh, c. 6th century

ANCIENT PRESENCE

Before the fall of Mohenjo-daro and Harappa, a bare-breasted goddess with lotuses decking her hair seemed to have been worshipped. This could have been an early version of the goddess Lakshmi, whose prominence diminished as the Aryans conquered the subcontinent. One of this ancient goddess's names was Padma, meaning lotus. After the decline of the Indus Valley civilisation, the lotus-headed goddess came to be known as Lajja Gauri, or sometimes, Aditi. The Lajja Gauri is depicted with her legs open to display her genitalia. It is interesting to note that the lotus also symbolises the female genitalia, another reason why

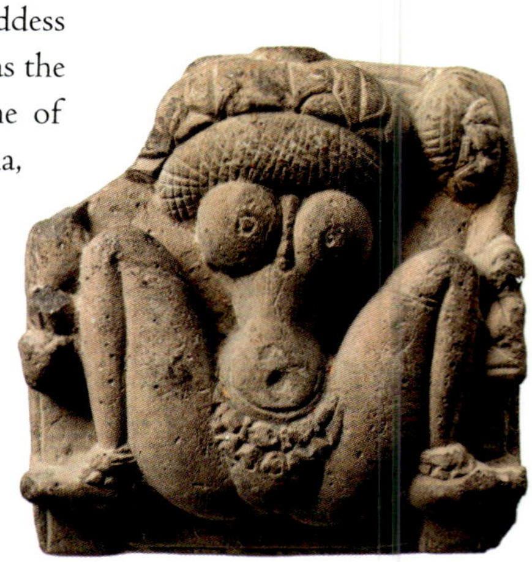

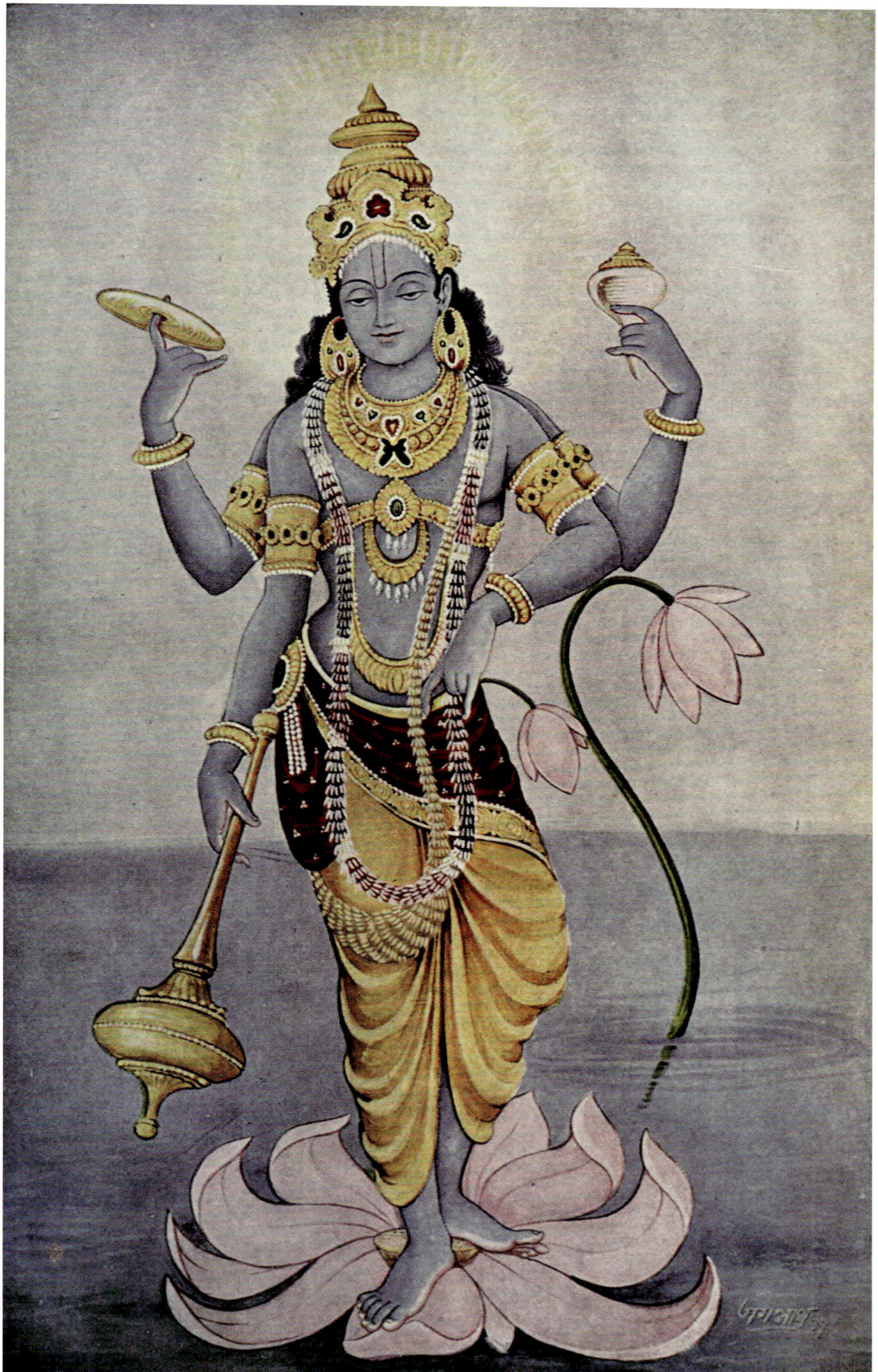

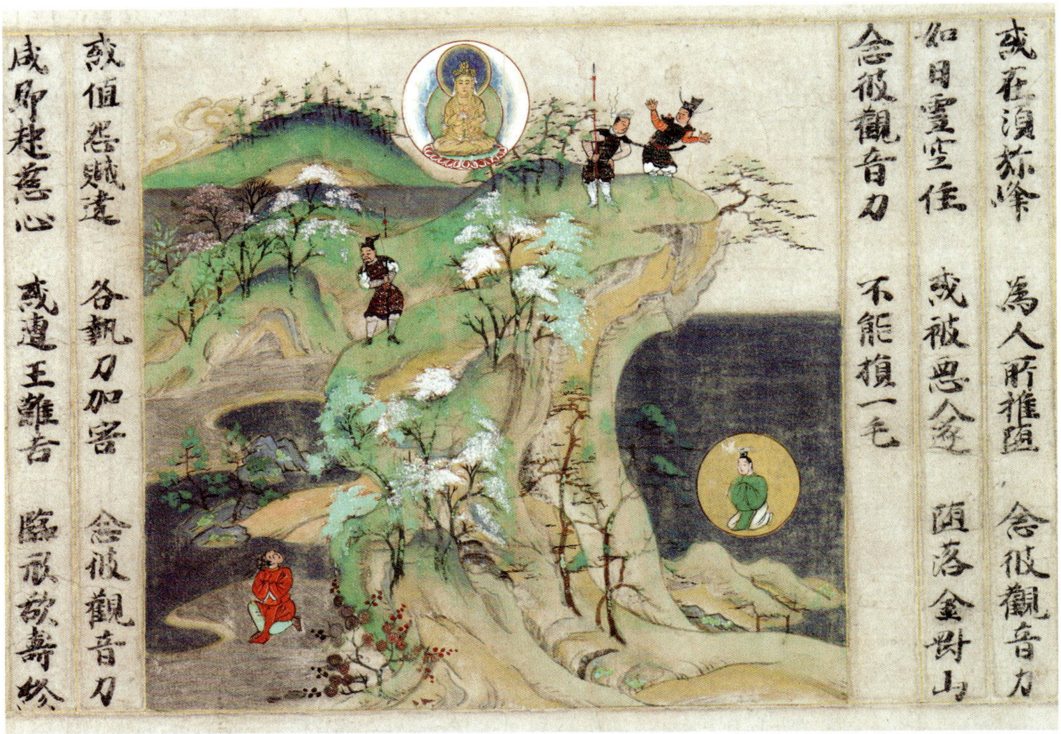

Left: 'Universal Gateway', detail from the *Lotus Sutra*, Chapter 25; Calligrapher: Sugawara Mitsushige; ink, colour, and gold on paper, Japan, 1257

Facing page: 'Guanyin as the Nine-Lotus Bodhisattva', ink and colour on silk, Ming dynasty, 1593

Below: Pink lotus flowers symbolise elegance and charm

these goddesses may have been called lotus goddesses. In Hinduism, the three Brahmanical deities—Brahma (the Creator), Vishnu (the Protector) and Shiva (the Destroyer)—are all associated with the lotus. Vishnu's four arms hold a conch, wheel, club and a lotus—which is probably the symbol of the created universe. The Brahmana part of the *Taittiriya Upanishad* (6th century BCE) describes the lotus in association with Prajapati (Brahma) in cosmogonist myths. When Brahma was sitting by the side of a pool of waste water, he saw a solitary lotus leaf floating in the moving waters. He dived into the water, in the form of a boar, to find the earth underneath and broke off its fragment. Brahma brought it to the surface and spread it out on the leaf—and that was the beginning of life on planet earth.

All the major deities of Hinduism (Brahma, Vishnu, Shiva, Lakshmi, Saraswati, Parvati, Durga, Rama and Surya), Buddhism (Buddha and Bodhisattva) and Jainism (Mahavir and others) are associated with the lotus. They are often depicted sitting on the divine lotus or holding a lotus in the hand. One of the most important Buddhist texts of the Mahayana sect, is known as the *Lotus Sutra*. This sutra depicts the events and circumstances that occur in a cosmic world of vast dimensions.

In 100 CE, Asvaghosha, one of the earliest Sanskrit poets, writing in the *Saundarananda*, describes in Canto 3 the beauty of the Buddha's half-brother's wife, Nanda Sundari, and compares her to a lotus pond, with her laughter for the swans, her eyes for the bees, and her swelling breasts for the uprising lotus buds. The 'lalita' meter in the *Natya Shastra* has been personified as a beautiful woman and is thus described: 'O lady, hurriedly but gracefully moving, wearing beautiful clothes and with delicate hands and having a blooming lotus-like face, you look charming after the fatigue of love sports.'[2]

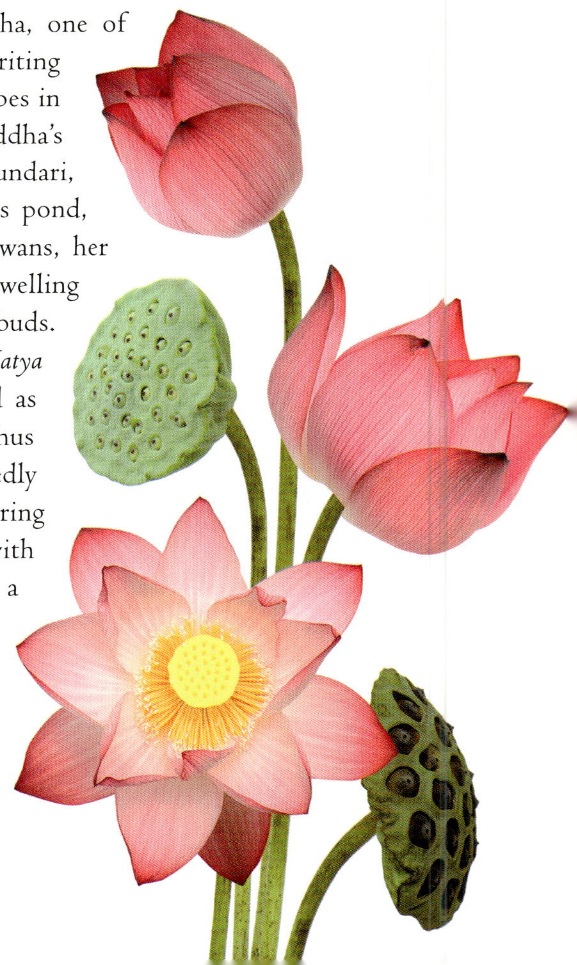

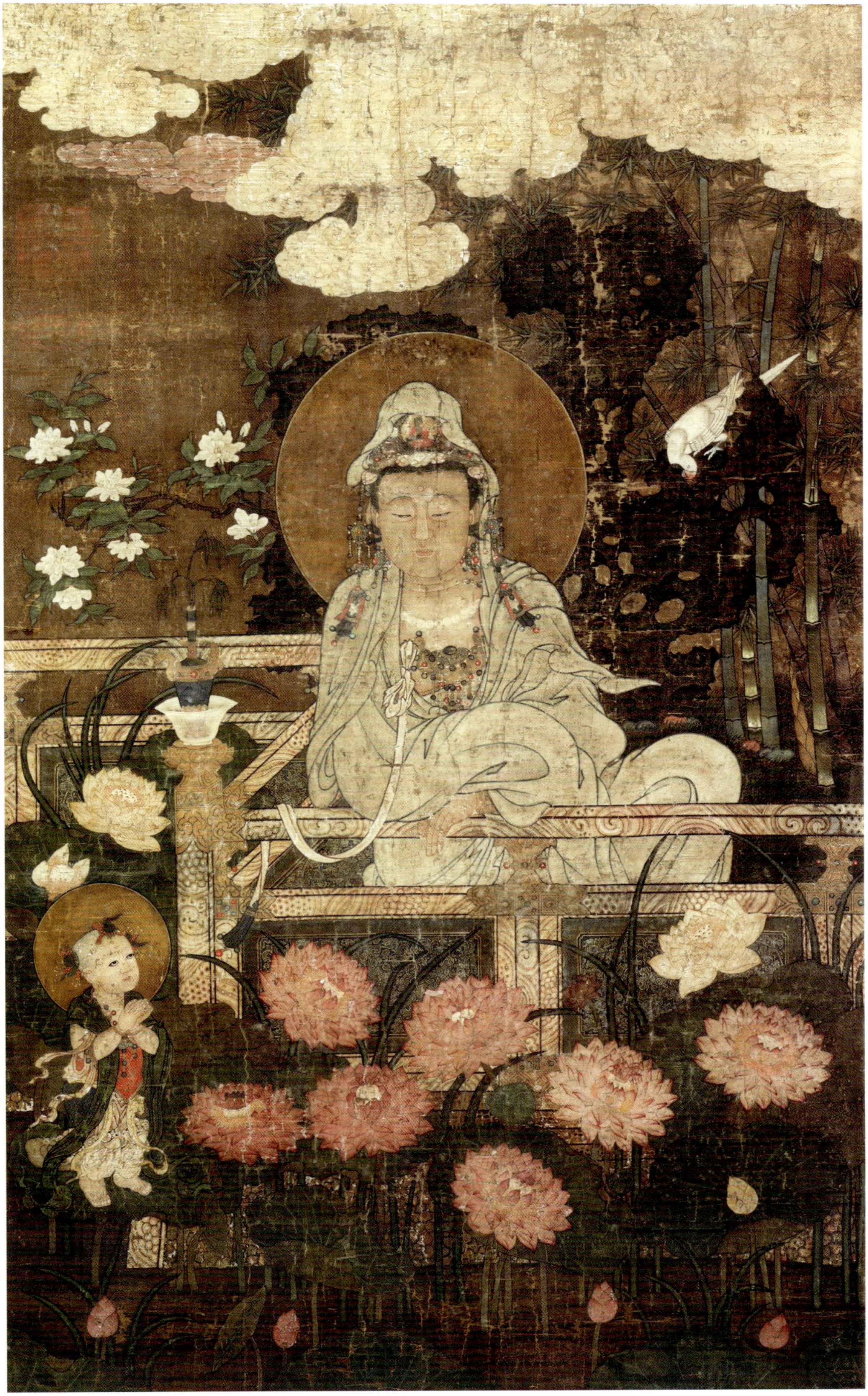

NATURE'S MYSTICAL CREATION

The lotus flower stands tall and undefiled in murky waters, and lends itself to powerful religious interpretation. Purity, divine beauty, resurrection, enlightenment and faithfulness are all contained in this symbol of the lotus flower. Surrounded by various myths and legends, the lotus is considered divine and one of nature's mystical creations that represent the purity of the mind and soul. Its metaphysical analogy is its perennial rise to faultless beauty from a muddy environment—symbolic of the evolution of man's consciousness. In Buddhism and Hinduism, the lotus is the embodiment of perfection. It defines the capacity of the soul to expand and embrace the beauty of life with a clear conscience. The lotus flower sheds its seeds and blooms at the same time. In Buddhism, it is said to signify the norm of cause and effect. The magnificent flower grows from the dirty and unclean part of the pond, but still remains unscathed and unblemished by the dirt and mud surrounding it. This makes the flower highlight the purity of spirit, rebirth, beauty, fertility, transcendence and resurrection.

DIVINE BEAUTY

India is home to a host of religions that share a common ancestry. For this reason there are many similarities between Hinduism, Buddhism and Jainism. One tie that binds them together is that all three consider the lotus to be sacred and hold it up not just as an ideal of beauty but as an ideal of divine beauty. The 16th century prayer 'Sri Ram Chandra Kripalu' by Tulsidas contains the following lines.

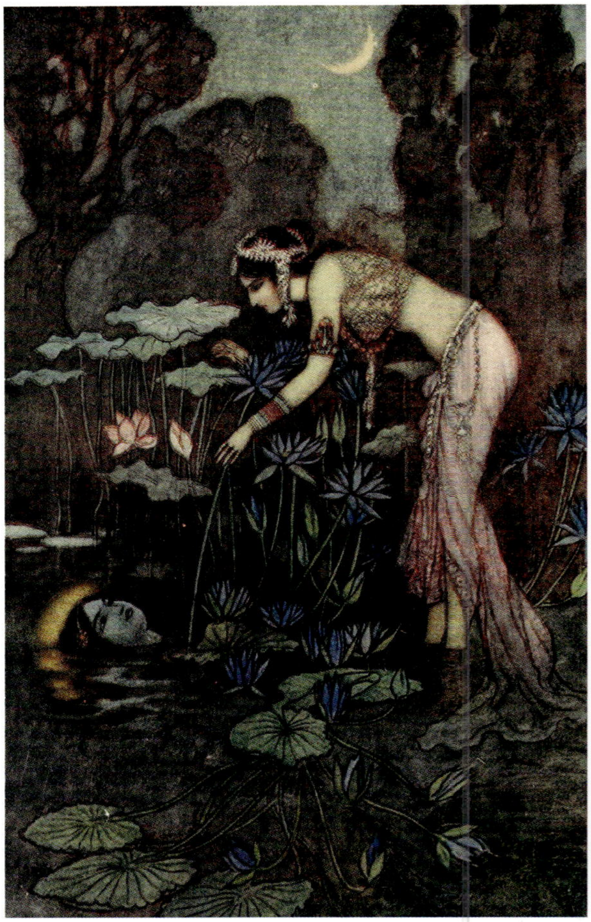

Above: 'Sita finds Rama among lotus blooms', Donald Alexander Mackenzie, 1913

Facing page above: 'Lotus-clad Radha and Krishna', Basohli, opaque watercolour on paper, c. 1730

Facing page below: View of wall art of the God Brahma seated on a lotus flower

Shri Ramacandra kripalu bhaju, mana, harana bhava bhaya darunam,navakanja-lochana kanja-mukha, kara-kanja pada-kanjarunam.[3]

O mind! Revere the benign Shri Ramachandra, who removes sorrow, pain, fear, and scarcity or poverty,
Who has fresh lotus eyes, a lotus face, lotus hands, and feet like the lotus and the rising sun.

Yet another set of lines from the *Bala Mukundashtakam* by Sage Bilvamangala, also known as Lila Sukha reads:

Kara aravindena pada aravindam
Mukha aravinde vi niveshay antam

My mind remembers that beautiful Bala Mukundam, Who with His lotus-like hands holds His lotus-like feet, And puts the toe in His Lotus-like mouth.[4]

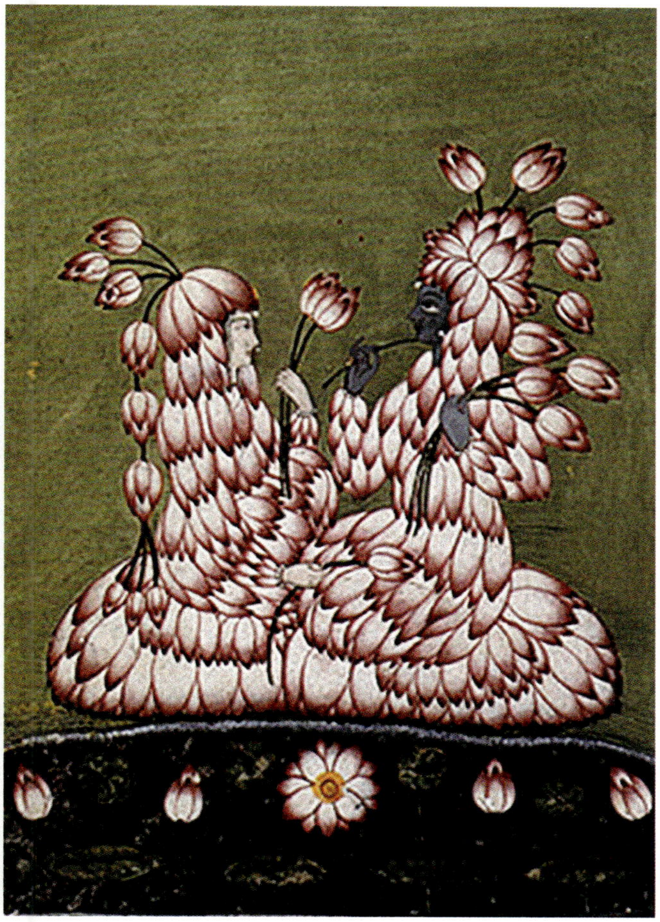

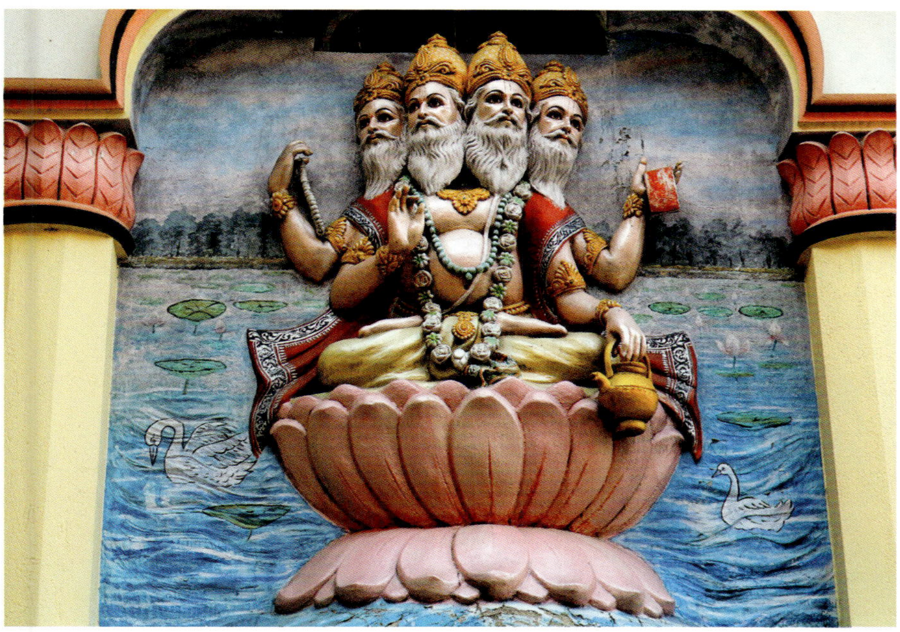

In both examples, Lord Ram and Lord Krishna's beauty is extolled by juxtaposing their lovely bodies with the lotus. Here it becomes apparent, that the lotus was an ideal of divine beauty in mortal form. Similarly, the emperor of flowers is a recurring leitmotif in Buddhist and Jain iconography, where the serenity and beauty of the enlightened ones is equated with the lotus.

THE MANY SYMBOLS OF THE LOTUS

According to Sehdev Kumar,

> In an eighteenth-century watercolour from Thailand, the Buddha is not represented as a human being but as a golden lotus bud placed in the otherwise empty pavilion above the Earth goddess. In a nineteenth-century Japanese woodblock, the Buddha Sakyamuni is depicted with his feet resting on lotuses, and he appears to float in a starry atmosphere. In another illustration, the child Buddha stands naked on a large lotus-shaped basin as two mythical dragons bathe the divine child from above.[5]

The lotus symbol, like all other Indian symbols, has a metaphysical or subjective as well as an objective significance, given that the beautiful blooms blossom with their roots in the mud and yet respond to light and close up against the darkness when night falls.

> The lotus symbol, which originally gave birth to beings and existences in unending succession, now carries the powerful wisdom of Nirvana: the word that puts end to all individual existence, whether in heaven or on earth. —Heinrich Zimmer [6]

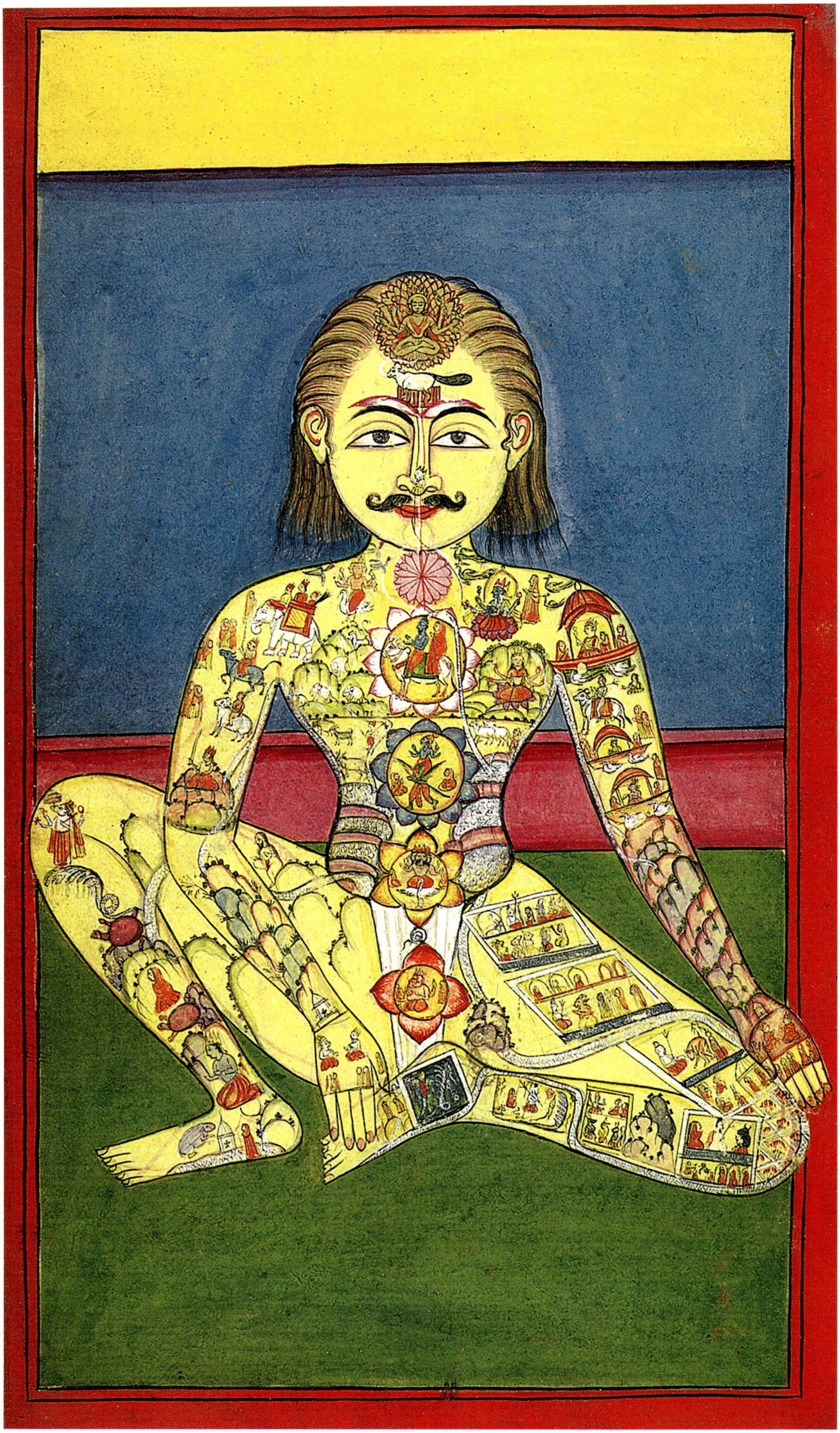

Facing page: Sapta Chakra, from a Yoga manuscript in Braj Bhasa, 1899

Below: Lotus motif carved on the roof of the Ellora Buddhist Caves, Aurangabad, Maharashtra

MYSTIC INTERPRETATION OF THE LOTUS

The *Mahanirvana Tantra* gives the mystic meaning of lotus symbolism. The root, it says, is Brahman, the Unknowable, from which all creation springs. The stalk is *Maya,* the unreality of world phenomena. The flower is the world itself. The fruit is *Moksha,* the soul liberation when it is released from worldly desires. The bell-shaped fruit was the mystic *Hiranyagarbha,* the womb of the Universe, holding the germ of worlds enumerable and still unborn.

From the earliest times the lotus has remained a unique metaphor for the heart of the devotee: *hrd-puskara; hrd-padma,* the primary abode of the deity. The lotus-heart is further likened to space, *akasa;* its eight petals representing four directions and four intermediate points of the compass. The eight-petalled lotus is basic to the *mandala* drawings, which later came to play an important role in all three important spiritual traditions—Buddhist, Jain and Hindu. One of the earliest uses of the lotus with this symbolic meaning is represented in the ceilings of the Gupta-period Buddhist cave temples at Ajanta.

To further explore the concept of the flower in Tantric and Kundalini Yoga, Sehdev Kumar says,

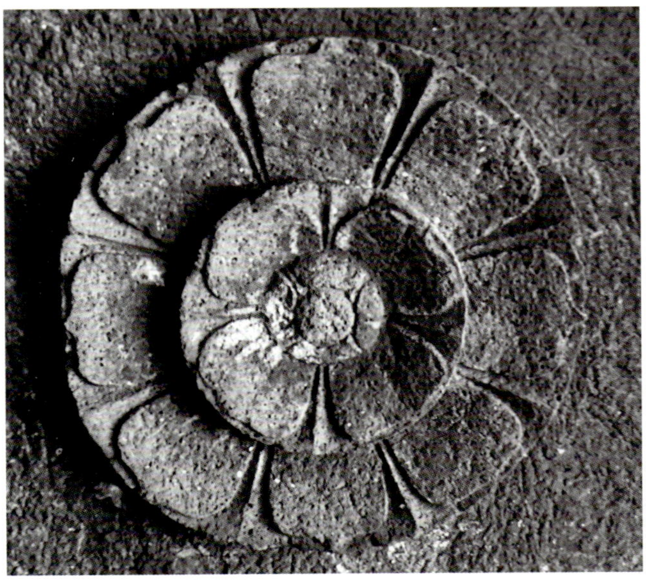

the representation of the human body as a microcosm and its special geography is of supreme importance. The main axis of the human spinal cord is considered the pivot of the universe; it is the meru-danda compared to Mount Meru. Along this axis are a series of chakras in the shape of lotuses, with different number of petals. Significantly, Tantric spiritual praxis requires the practitioner to pass through several stages or processes of cosmogenesis, the final plane located at the summit of the skull known as 'lotus-head', usnisa kamala, to the Buddhists and 'the thousand-petalled lotus'—sahasrara padma or chakra to the Hindus and the Jains. A very early representation of this idea occurs in a mid-6th-century Narashima, an avatar of Vishnu, where a lotus grows out of the god's head.[7]

PURANIC ASSOCIATIONS

The lotus motif, though the most visible, is the least understood image of Indian art. But if the student fails to follow the inner working of the Indian mind he or she is not likely to enjoy or appreciate the aesthetics of Indian art. Puranic ideas are very strongly emphasised in the symbolism of Indian art. Since Puranic ideas are so deep-set in India's spiritual thought, the Indian no longer seeks corroboration from references of their source or feels the need

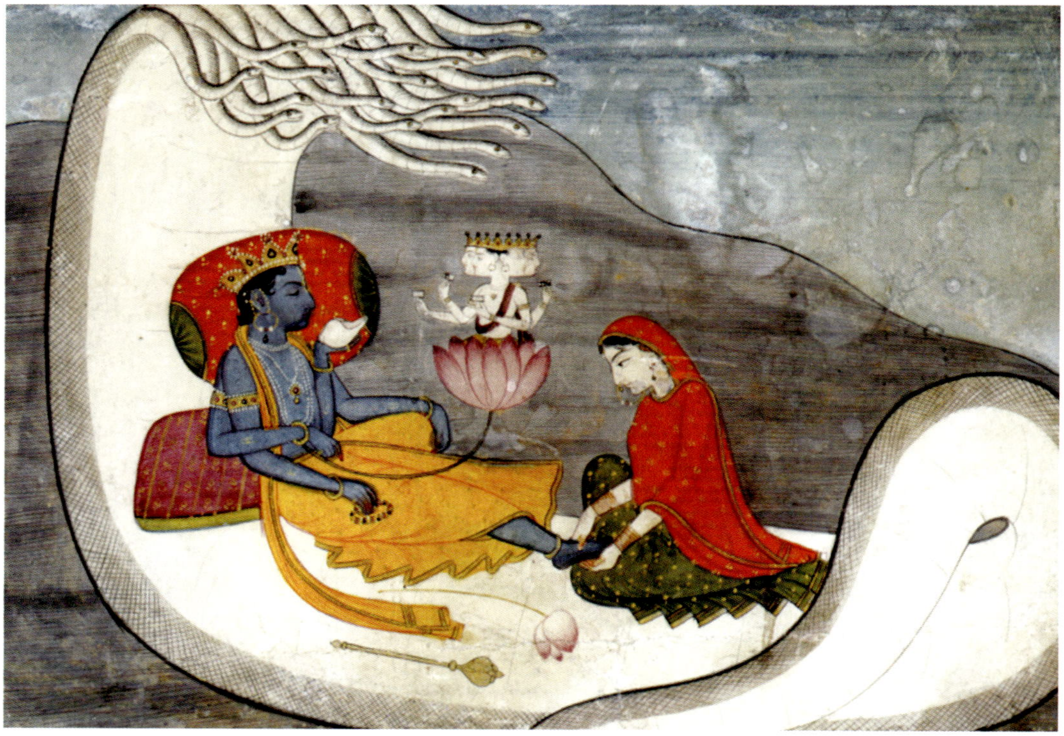

'Vishnu and Lakshmi on Shesha Naga', Kangra, opaque watercolour on paper, c. 1870

to do so. But, for the student of Indian art, absorbing these impressions is a significant process. Imagine and retain in your mind's eye the Puranic description of the Himalayas as the centre of the World Lotus with the deep blue waters of Lake Manasarovar mirroring the Creator's mind. The seed vessel or pod is Brahma's holy city near Mt. Kailash and the Himalayan snow peaks are the glittering upturned Lotus petals. The plains of India, together with the sub Himalayan slopes, form the southern one of its four great petals turned down upon the stalk, which springs from the navel of Brahma the Creator, the Eternal Spirit, reposing on the bed of the cosmic ocean.

Another version of this tale that can be traced back to the Taittiriya Brahmana, is that, in his quest to create the universe, Brahma brings forth a splendid lotus with a thousand petals. This was considered a way to reach the epicentre of the universe itself. In other versions of this tale, Brahma is born from within this radiant lotus. It can thus be noted that the lotus is associated with the act of creation in Hindu thought.

The lotus is the stool and footstool of the Gods, the symbol of the material universe and of the heavenly spheres above it. Its colours—red, white and blue—are symbolic of the Trimurti, the three aspects of one. Red for Brahma the Creator; white for Shiva, the Divine Spirit and Destroyer; and blue for Vishnu, the Preserver and upholder of the Universe.

MANY HUES

Rooted deep in the mud of a lake or river and pushing its way gradually upwards through the water until its beautiful flower blossoms in the light of heaven, the lotus was Nature's own symbol of the spiritual process by which the human mind won liberation or *nirvana*. It seemed a perfect

Trays of colourful lotus, marigold and rose flowers presented as offerings to the deities at a Hindu temple

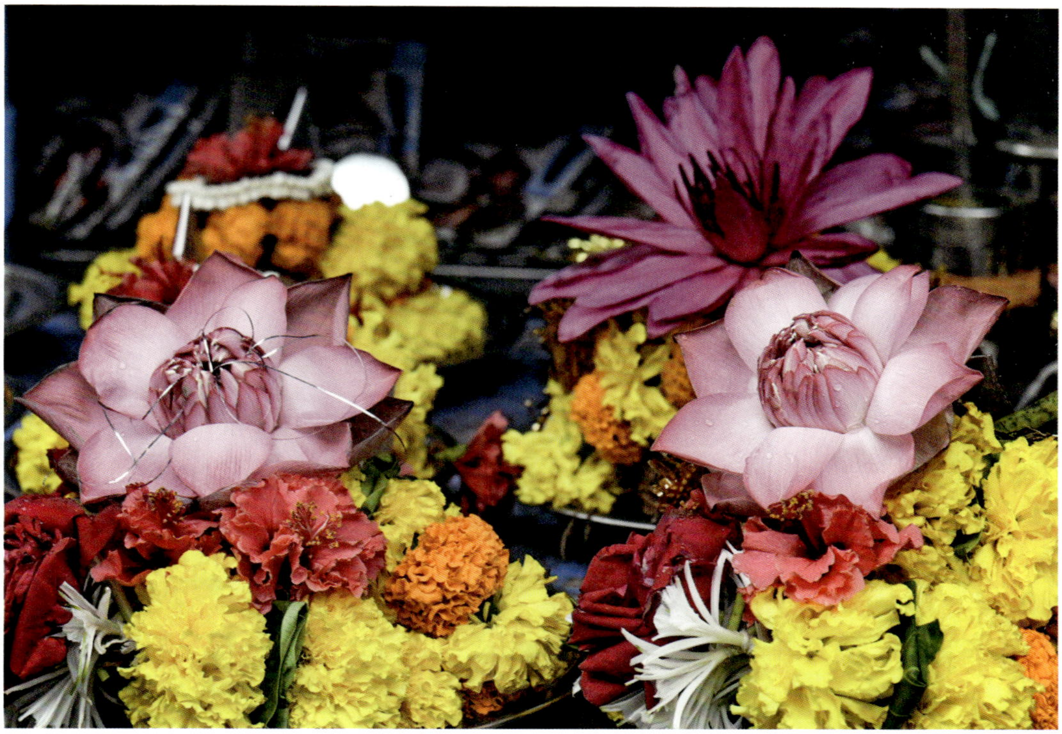

symbol of creation, of divine purity and the beauty of the cosmos evolved from the dark void of chaos and sustained in equilibrium by the cosmic ether, *akash*.

It has often been noted that both Buddhism and Sufism use unfolding flowers, like the lotus and the rose, as symbols of the opening of the mind. The parallels between the two faiths cannot be identical, but the two traditions have crossed each other's paths in certain ways. Buddhism is often associated with light, while Sufism is associated with the heart.

Every civilisation has its symbols, which may be religious, such as the cross in Christianity, the swastika in Hinduism or the crescent in the Islamic tradition. They bring about religious sentiments, as well as a sense of solidarity and unity. In Indian culture, philosophy, poetry and mythology are closely intermeshed and symbols have occupied a significant position. In the modern day, Mahatma Gandhi adopted the spinning wheel (*charkha*) to signify quiet dignity, dedication and simplicity. In its form of self-confidence, the wheel was sometimes associated with the lotus. In Yogic meditations and healing practices, the energy travels through many centres known as *chakra*s, where each *chakra* is associated with a lotus of a specific colour and a certain number of petals. The highest *chakra* is known as *sahasrara*, the '*Lotus with a thousand petals*'. Thus we can see the grandeur in the design of the halo that is behind the Buddha's head. It is brought out in the combination of the qualities of all three: the illumination of the sun, the movement of the wheel and the creative power of the lotus.

One of the most common metaphysical analogies compares the Lotus' perennial rise to faultless beauty from a miry environment to the evolution of man's consciousness—from instinctive impulses to spiritual liberation.

DIVINE ATTRIBUTES

According to Indian culture, the lotus flower denotes prosperity, knowledge and learning, fruitfulness and illumination. The lotus flower, regarded as sacred, has religious connotations. The flower represents the energy centres of our body, known as *chakras*, and is connected with beauty, prosperity, knowledge, fertility, and above all, eternity and spirituality.

The sacred lotus illustrates the metaphor for growth and illumination. The self-cleansing lotus never gets dirty, even in the muddiest of ponds, exemplifying an empowering narrative of integrity, remaining pure and unaffected amidst the impurity of events. In Asia, the lotus (*padma* in Sanskrit) holds a position of great cultural and spiritual significance. In the ancient practice of yoga, the cross-legged 'lotus pose', or the '*padmasana*', universally symbolises wisdom and serenity.

Heinrich Zimmer, in talking about the lotus as a metaphor, says: 'Rising from the depths of water and expanding its petals on the surface, the lotus (*kamala padma*) is the most beautiful evidence offered to the eye of the self-engendering fertility of the bottom. Through its appearance, it gives proof of the life-supporting power of the all-nourishing abyss.[8]'

As per Jennifer Potter in *Seven Flowers*, the *Brihadaranyaka Upanishad* states that when the universal spirit is visible, one of its forms is as a white lotus. Potter recounts how at the end of an elaborate ritual to seek greatness and after making offerings and chanting incantations, the individual undertaking the ritual must address the morning sun and cry, 'You are the one lotus among the quarters! May I become the one lotus among men!' When the woman 'he has created' is to give birth, he is instructed to sprinkle water on her and say, 'As from all sides

Facing page: Floral decorations on the stone platforms of the cloister walk in the Mahabodhi Temple Complex in Bodh Gaya

Below left: Goddess Lakshmi—the Hindu Goddess of wealth and prosperity, stands on a lotus pedestal

Below: Lotus flowers placed as offerings at a Buddhist temple

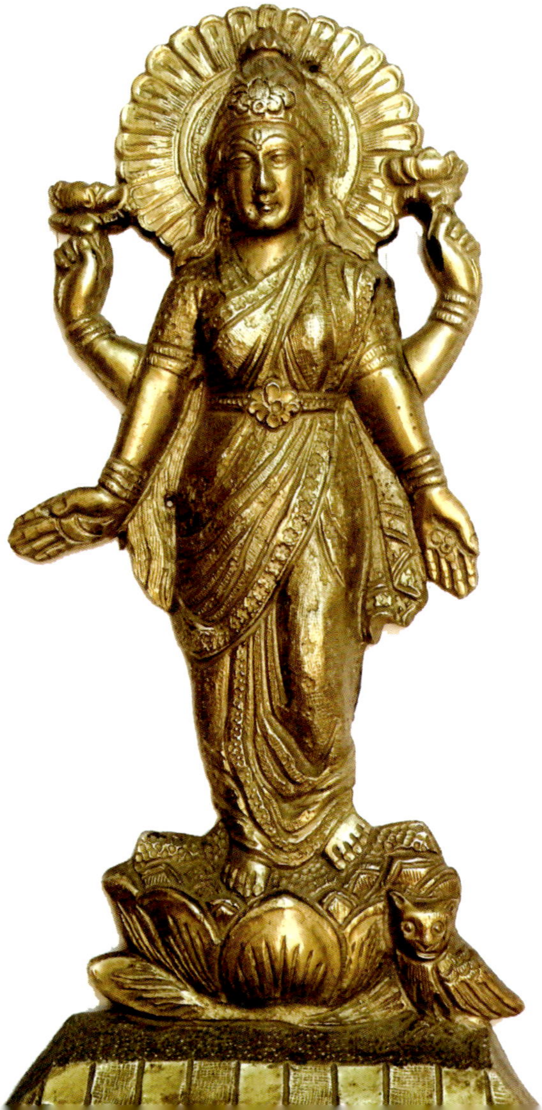

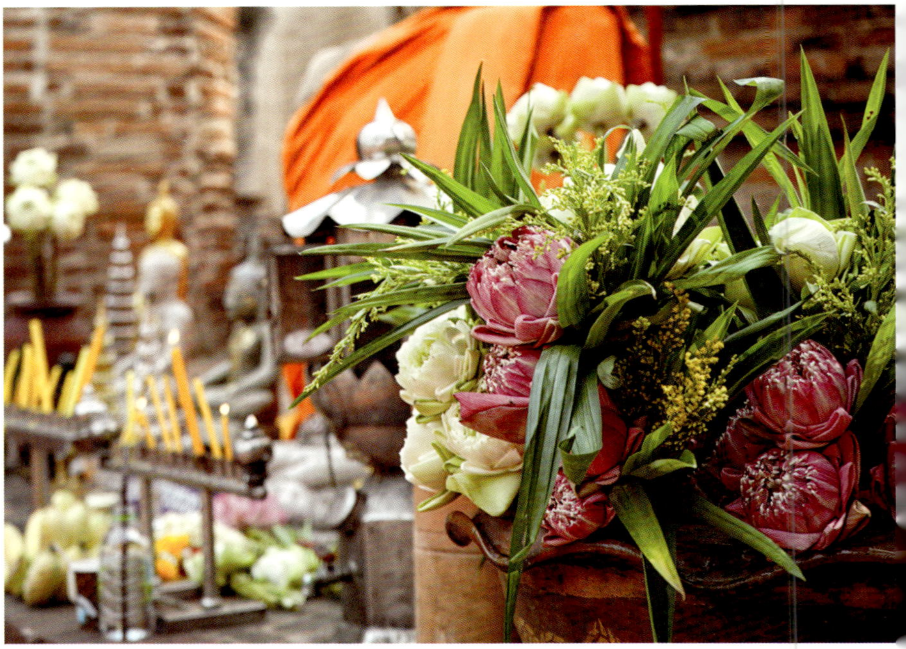

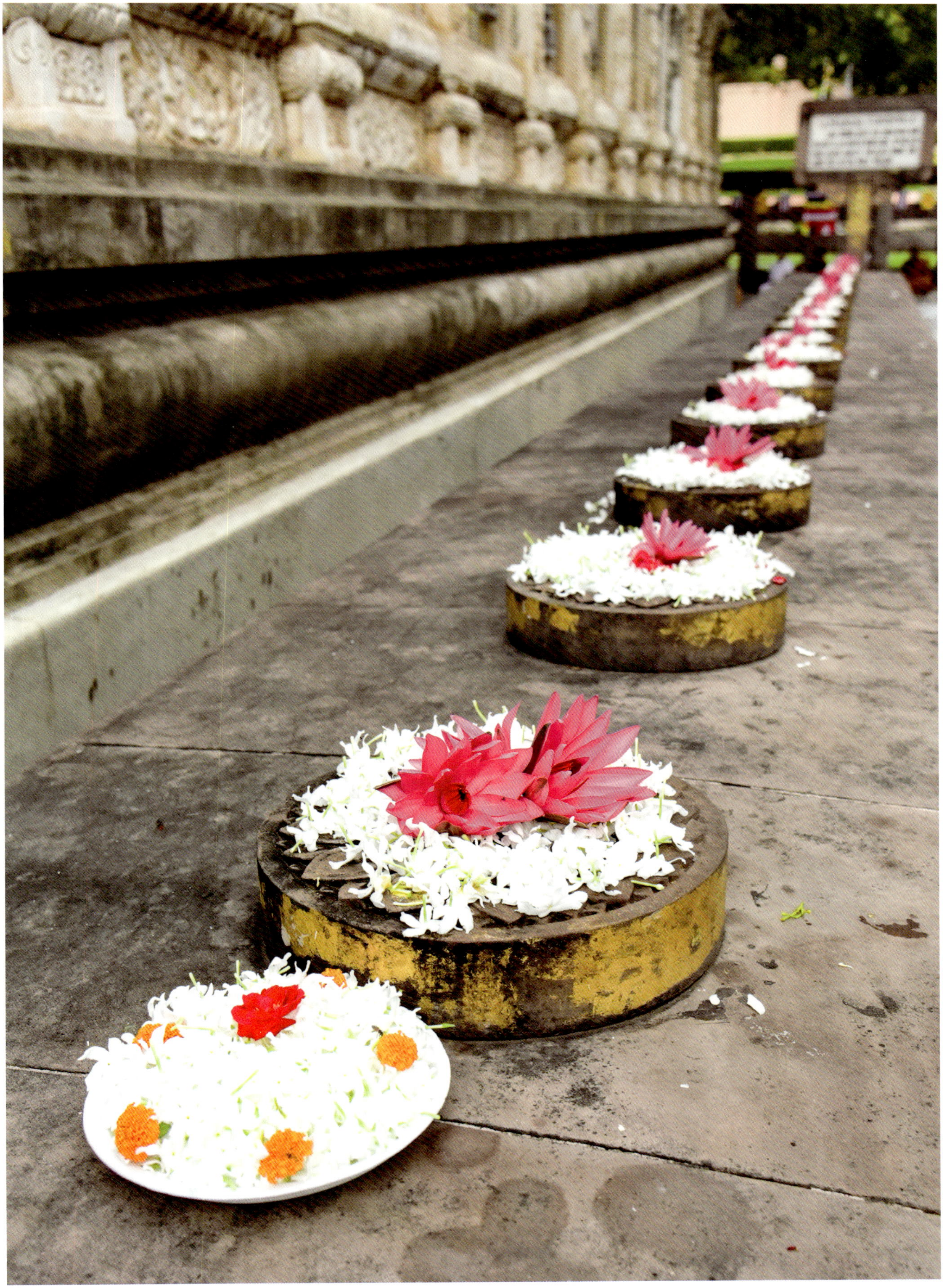

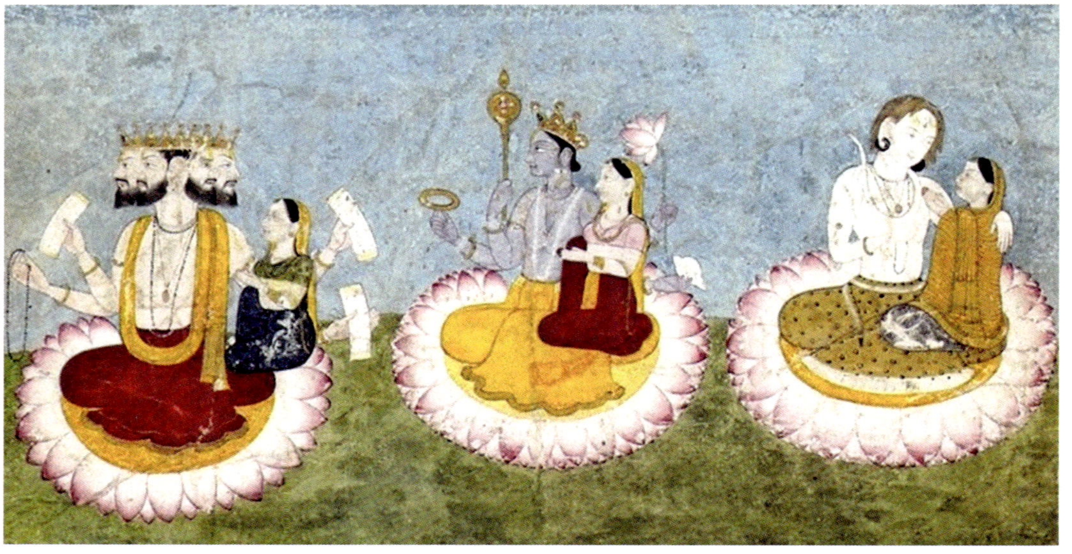

the wind churns a lotus pond so may your foetus stir and come out with the afterbirth.'

Another beautiful example from the *Chandogya Upanishad,* mentioned in Potter's book, seems to attribute to the lotus the status of the space in the heart that contains everything. 'Now here in this fort of *brahman,* there is a small lotus, a dwelling place, and within it, a small space...What belongs here to this space around us, as well as what does not—all that is contained within it.'[9]

From as early as the 2nd century BCE, the lotus was the principal attribute of Lakshmi as well as her seat.

During the Gupta period (between 320 and 550 CE), the lotus had become the principal support for most deities of the various religious systems, and in many instances the lotus was the main decorative motif on their halos. It was considered to be the most beautiful flower in India and thus played a special role in divine worship as it came to be held by the sun god Surya, Lord Vishnu, the Buddha and several Jain deities. It was no doubt the most frequently used metaphor for beauty, where Lakshmi was described as lotus-eyed, lotus-complexioned, lotus-thighed, decked in lotus garlands and lotus-faced. The feet of certain gurus, gods and other higher beings worthy of adoration were referred to as lotus feet, their hands as lotus hands and the devotee's heart, where god dwells, was described as the lotus heart. Thus the lotus was a metaphor for purity and of wisdom to those who remain untouched by the sorrows of life.

At night, the lotus flower closes; at day it blossoms to the glory of the sun. Ancient Indian cultures consider this to be analogous to sun worship, thus giving rise to varied creation myths. The flower blooming every morning and closing its petals each evening also symbolises the endless cycle of life and death. It is iconographic of the sun god who signified the cosmic cycle of the birth and dissolution of the universe. The lotus halo behind the head of deities, as well as a lotus support below their feet, represents the possibility of a yonder world of heaven and earth. This is taken further in the following passage from an ancient Vedic text, the *Satapatha Brahmana*: 'Fire is verily the lotus of this Earth, the Sun the lotus of yonder Sky.'

The lotus also signifies the firmament of middle space (*antariksa*). Vishnu's navel,

Above left: 'Brahma, Vishnu and Shiva seated on lotuses with their consorts, Saraswati, Lakshmi and Paravati, respectively', Guler, c. 1770

Facing page: 'Lord Vishnu holding lotus flowers', Mankot-Basohli, Pahari School of painting, c. 1750, coll. National Museum, New Delhi

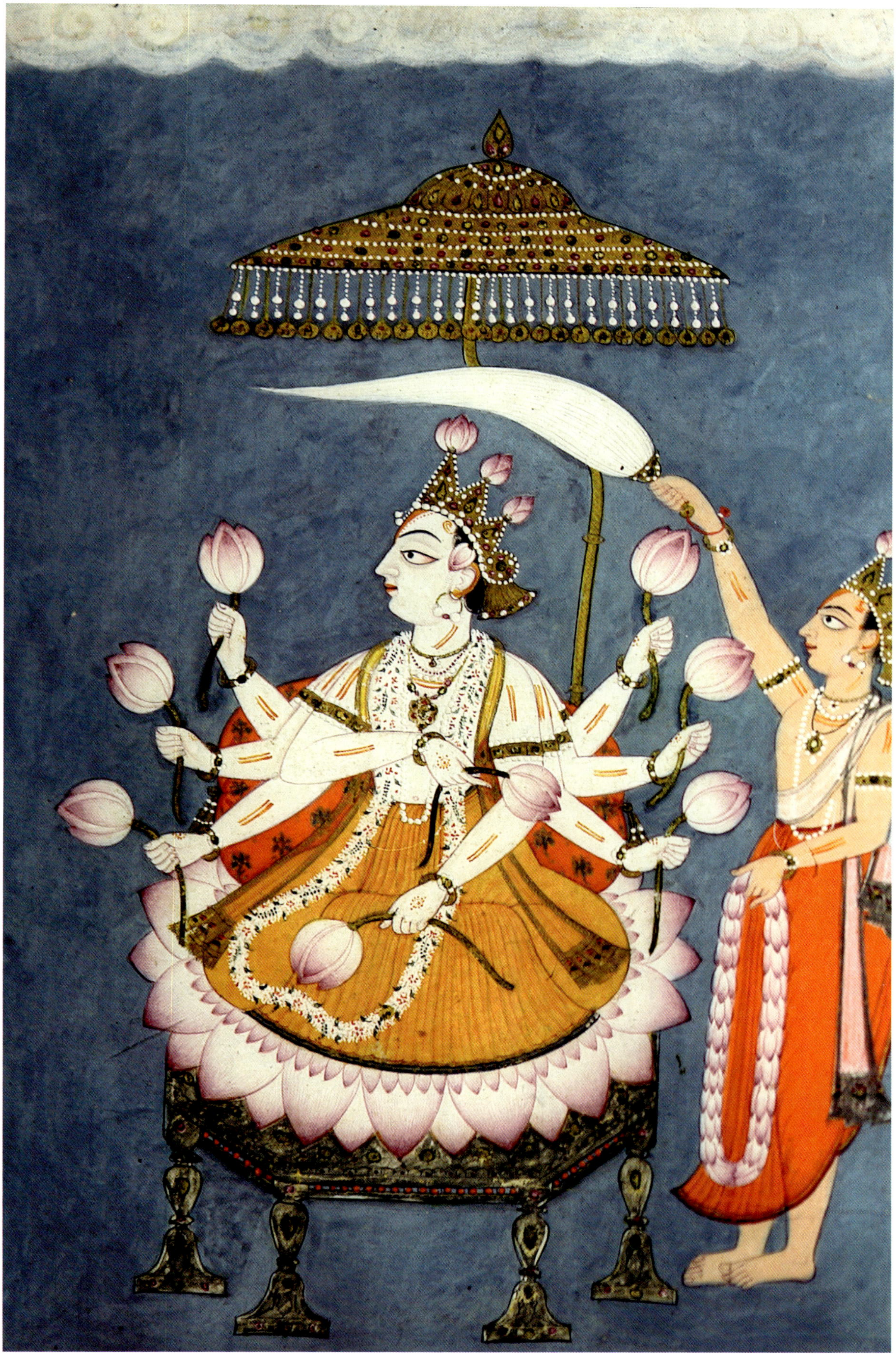

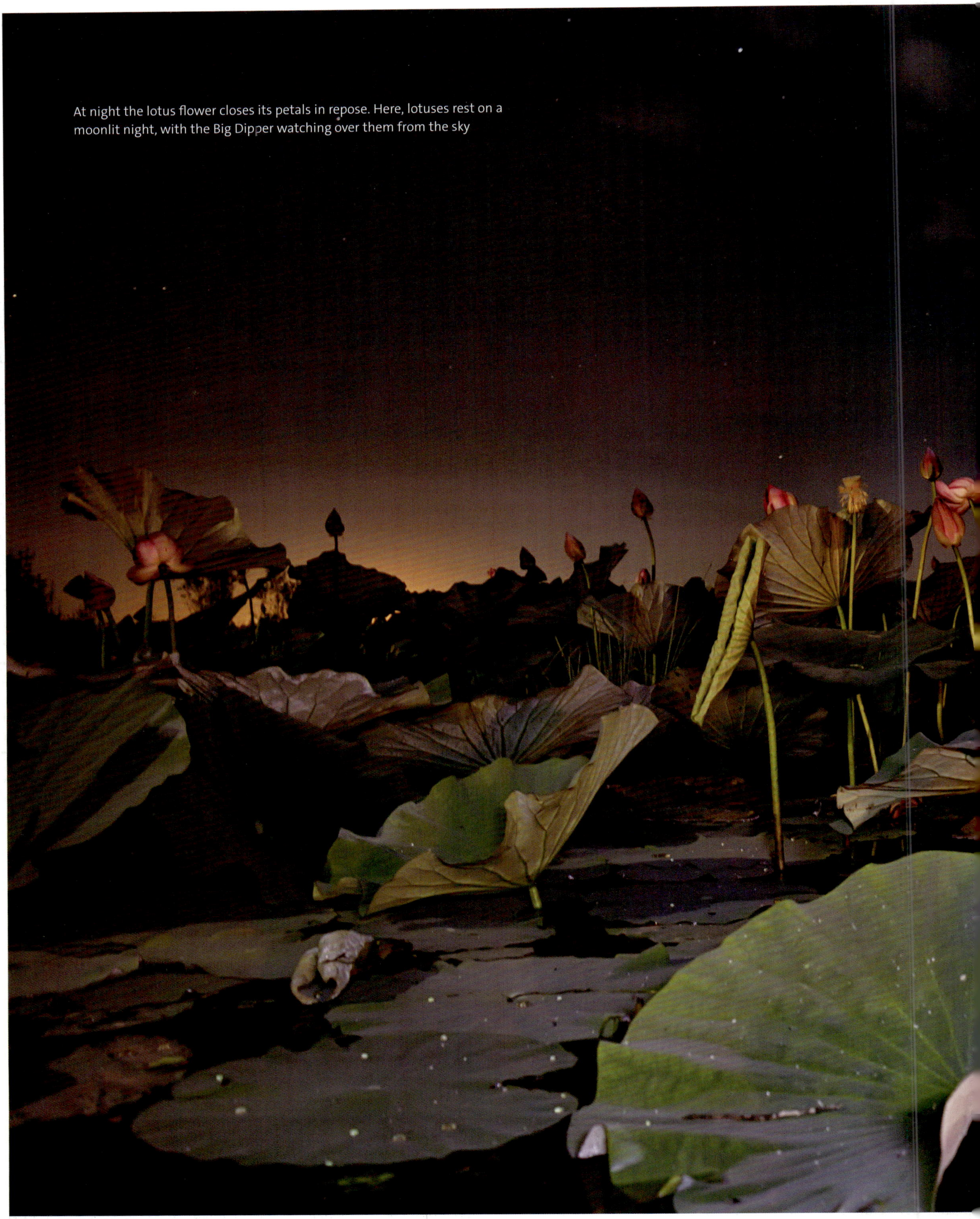

At night the lotus flower closes its petals in repose. Here, lotuses rest on a moonlit night, with the Big Dipper watching over them from the sky

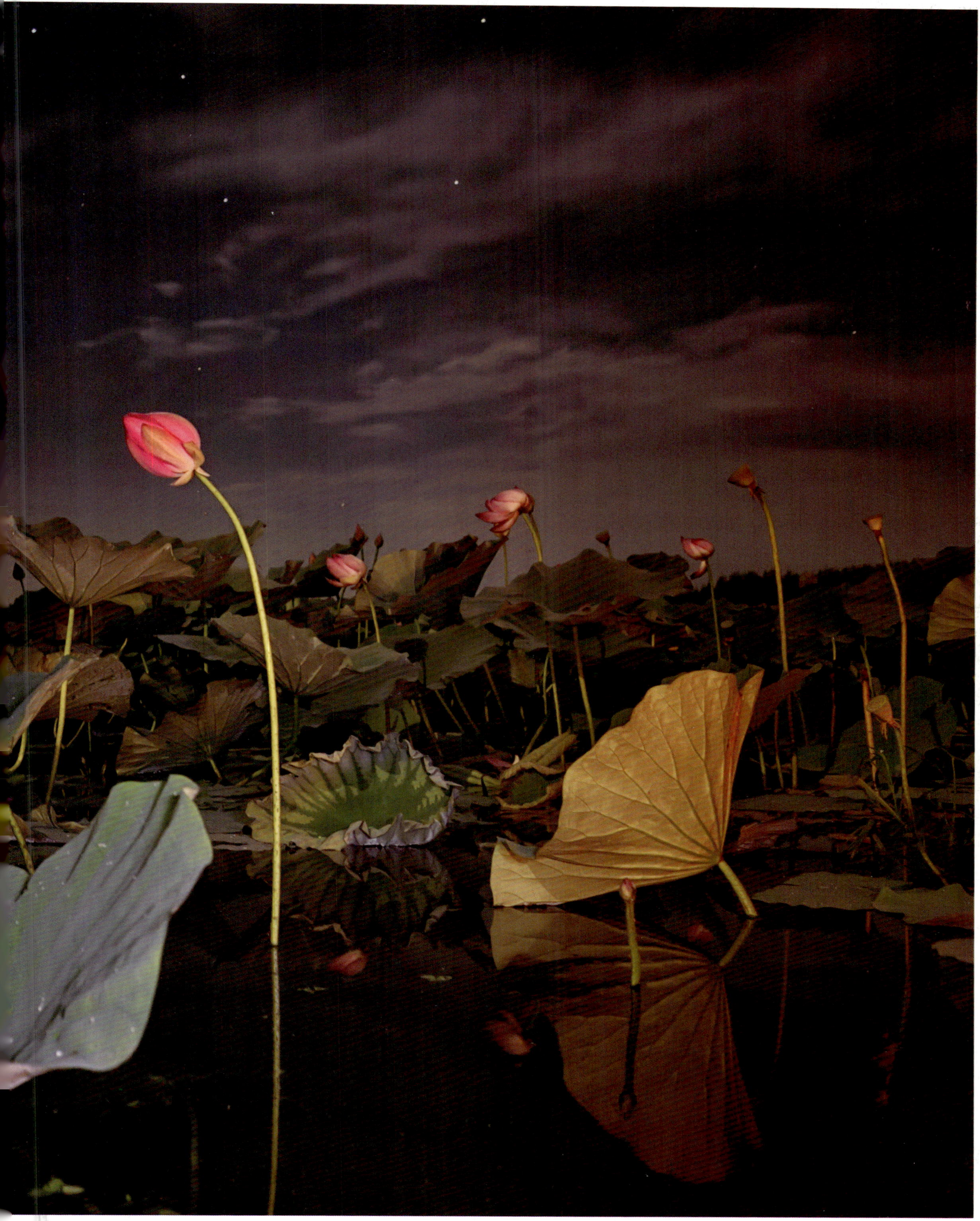

from which a lotus emerges, signifies the centre of the universe. In Lakshmi's hand, the lotus symbolises wealth, fame, fertility and water; in Parvati's hand, it symbolises detachment; while in the hands of Indra, it symbolises prosperity. The lotus also symbolises the idea of divine play where the universe and all its manifestations are described as nothing but the playfulness of supreme beings.

OM MANI PADME HUM

The Sanskrit mantra, 'Om Mani Padme Hum' means 'the Jewel in the Lotus'. Here jewel signifies the enlightenment of thought, while the lotus equals the mind. The popular translation is 'Oh jewel in the mystery of the lotus flower'; another version is 'Blessed be your speed, body and soul by the jewel of the lotus flower.' The Tibetan landscape is filled with 'mani' stones, or flat rocks that bear the mantra of compassion. Dr Pieter Verhagen, who is professor of Sanskrit and an expert on Tibet, discovered an early Tibetan grammatical treatise that goes back to the ninth century which interprets *Om Mani Padme Hum* thus: 'Because Om is the nature of the five wisdoms, it is first'. The phrase closes with the word hum or 'take to mind.' The vocative in between, 'mani', is the jewel, and the jewel-lotus is invoked in obeisance. Just as a young child fervently calls the name of its beloved mother, the practitioner fervently calls the deity Mahakarunika (Supreme Compassion), having a jewel and a lotus in his hands, calling 'O Jewel Lotus.' Mahakarunika, remembering his prior promise, comes quickly.

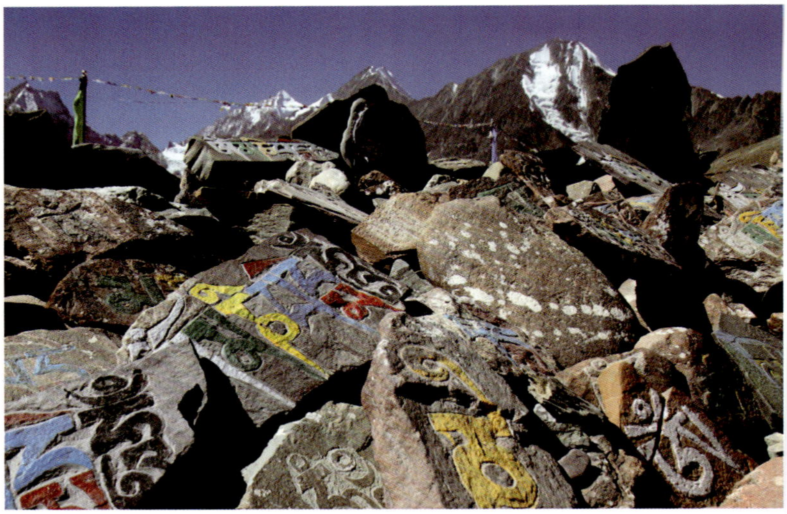

Above: Mani stones at Kunzum Pass, Himachal Pradesh

Below: Dharmachakra on Jokhang temple, Lhasa, Tibet. The eight-spoked Dharma chakra symbolises the Eightfold Path; note the inverted lotus-petalled base of the Dharmachakra

BUDDHA'S BLOSSOM

The principles of the Eightfold Path hold the lotus flower as one of the highest teachings of the Buddha. In Buddhism, the flower blooms above the muddy water, which highlights the act of rising above all desires and attachments—the key to achieve spiritual awakening. Though the flower has its root in the mud, it represents the aspiration to rise above, when it grows upward in the direction of the light. In other words, it represents the movement from the world of darkness (symbolised by the muddy

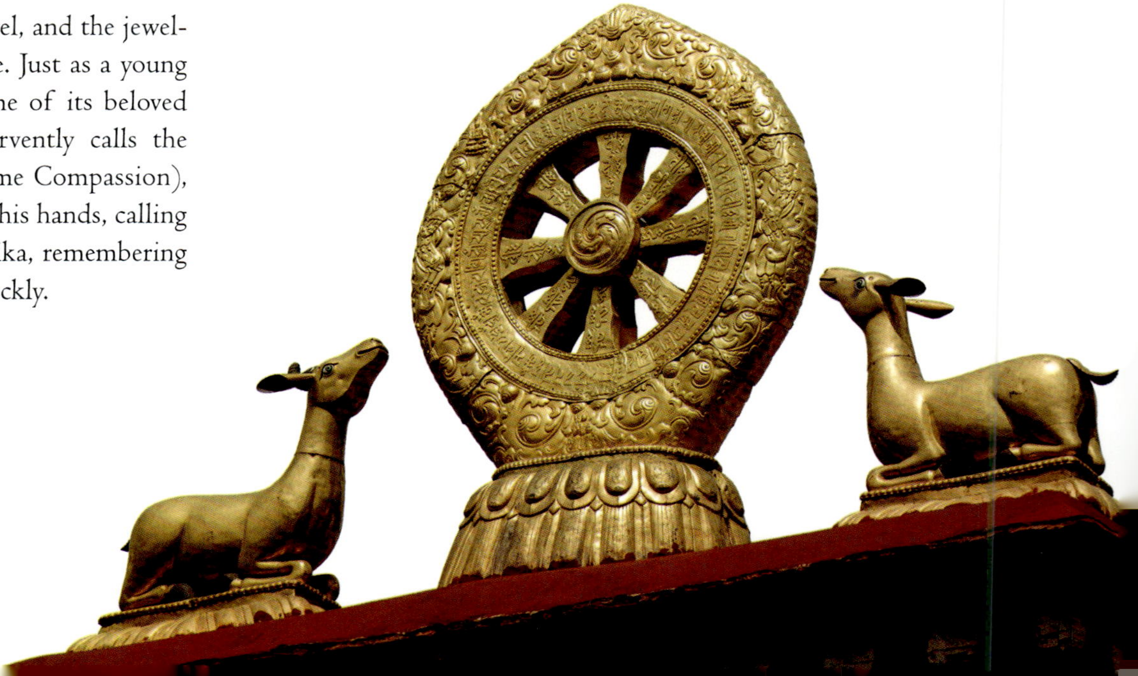

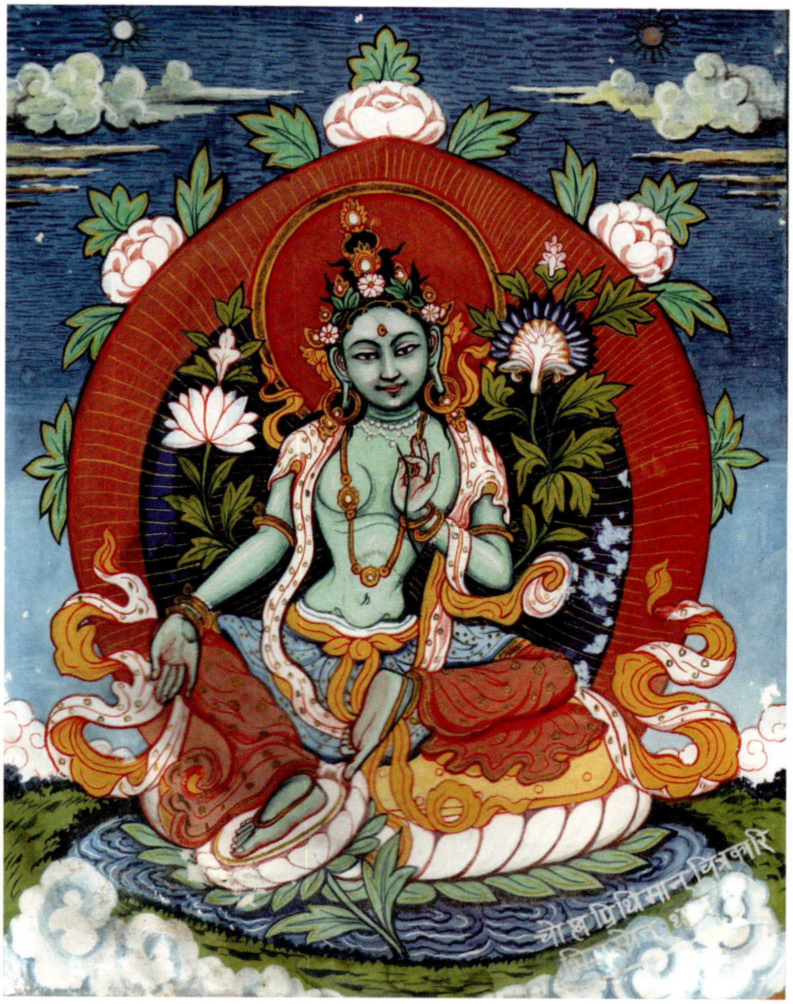

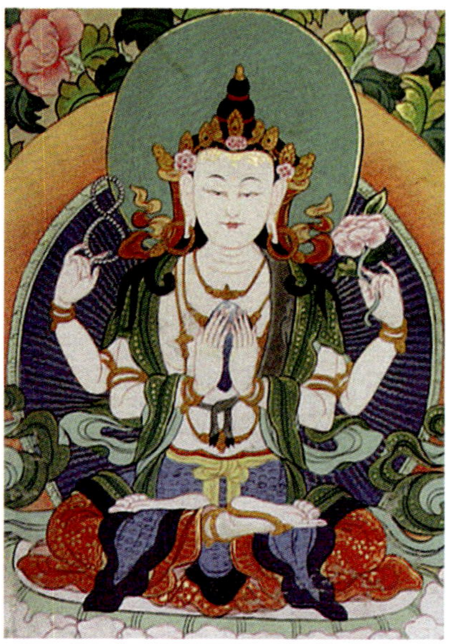

Above: 'Buddhist Goddess Green Tara seated on a lotus', painting, Prithvi Man Chitrakari, 1947

Right: Four-armed Tibetan form of Avalokitesvara, Jokhang temple, Lhasa, Tibet

pond) to the light of knowledge or wisdom.

The lotus forms one of the eight symbols of fortune as per Buddhist thought. The lotus forms a significant part of Buddhist iconography and deities are portrayed either holding lotuses or seated or standing on them (and sometimes all of the above). In fact, a famous motif in Buddhist art is of Padmapani, the lotus bearer, who is always depicted with a lotus in either or both hands.

The different colours of the lotus have spiritual symbolism in Buddhism. The pink lotus is the supreme lotus, which means the highest enlightenment and is associated with Buddha; the blue lotus is said to denote wisdom; the white lotus stands for spiritual perfection; and the red lotus suggests the heart. In the second century CE, the lotus represented the seat for the image of the Buddha. It was believed as soon as the Buddha was born, he took seven steps to announce his spiritual dominion over the entire earth; seven steps being represented in art by seven lotuses. Another tale recounts how the Buddha's mother, Maya, before giving birth to her son, dreamed of a splendid white elephant, holding a lotus in its trunk. It was also said that the night Queen Maya became pregnant with the Buddha, a lotus pushed its way from beneath the earth, its stalk snaking its way into the realm of Brahma to blossom. This could be an allegory for how the lotus symbolises spiritual growth and purity.

According to historical myth, the Buddha appeared phenomenally from the flower lotus. Some legends also state how the lotus flower appeared at every place where the Buddha placed his foot.

Avalokitesvara is perhaps the best-known figure in Tibetan Buddhism associated with the lotus flower. He is the *Bodhisattva* embodying the compassion of all the Buddhas. He is also known as *Padmapani* or Holder of the Lotus.

Tara, the female incarnation of the Buddha, is usually depicted to be so eager to help in any situation that she is not only seated on a lotus, but also has her right foot on a small lotus cushion, ready to be on the go.

SEEDS OF ETERNITY

In Hinduism, the creation myth began with the holy word Om, when the muddled expanse of water brought forth a beautiful golden lotus. From the word Om originated the three forces or trinity as we know it—Brahma (creator), Vishnu (preserver), and Shiva (destroyer). Quite significantly, the trinity was seated on golden lotuses.

The Hindu creation myth says that before time began, there was no heaven, no earth and no space between. A vast dark ocean washed upon the shores of nothingness and licked the edges of the night. A giant Cobra floated on the waters. Asleep within its endless coils lay the god Vishnu. He was watched over by a mighty serpent. Everything was so silent and peaceful that Vishnu slept undisturbed by dreams or motion. From the depths a humming sound began to emerge—Om. It grew and spread, filling the emptiness and throbbing with energy. The night ended and Vishnu awoke. As the dawn began to break, from Vishnu's navel grew a magnificent lotus flower. Lee Keith Freeman said in *Book of Sages*,

> *In the middle of the blossom sat Vishnu's servant, Brahma. He awaited the Lord's command. Vishnu commanded his servant: 'Create the world.' A wind swept the waters. Vishnu and the serpent vanished. Brahma remained in the lotus flower, floating and tossing on the sea. He lifted up his arms and calmed the wind and the ocean. Then Brahma split the lotus flower into three. He stretched one part into the heavens. He made another part into the earth and with the third part of the flower he created the sky.*[10]

In the Hindu tradition, Vishnu sleeps at the surface of the 'Primeval Waters.' A pink lotus

Sculpture of Lord Vishnu with a lotus emerging from his navel, with Lord Brahma seated upon it, My Son style of Cham sculpture, sandstone, Vietnam, 7th century

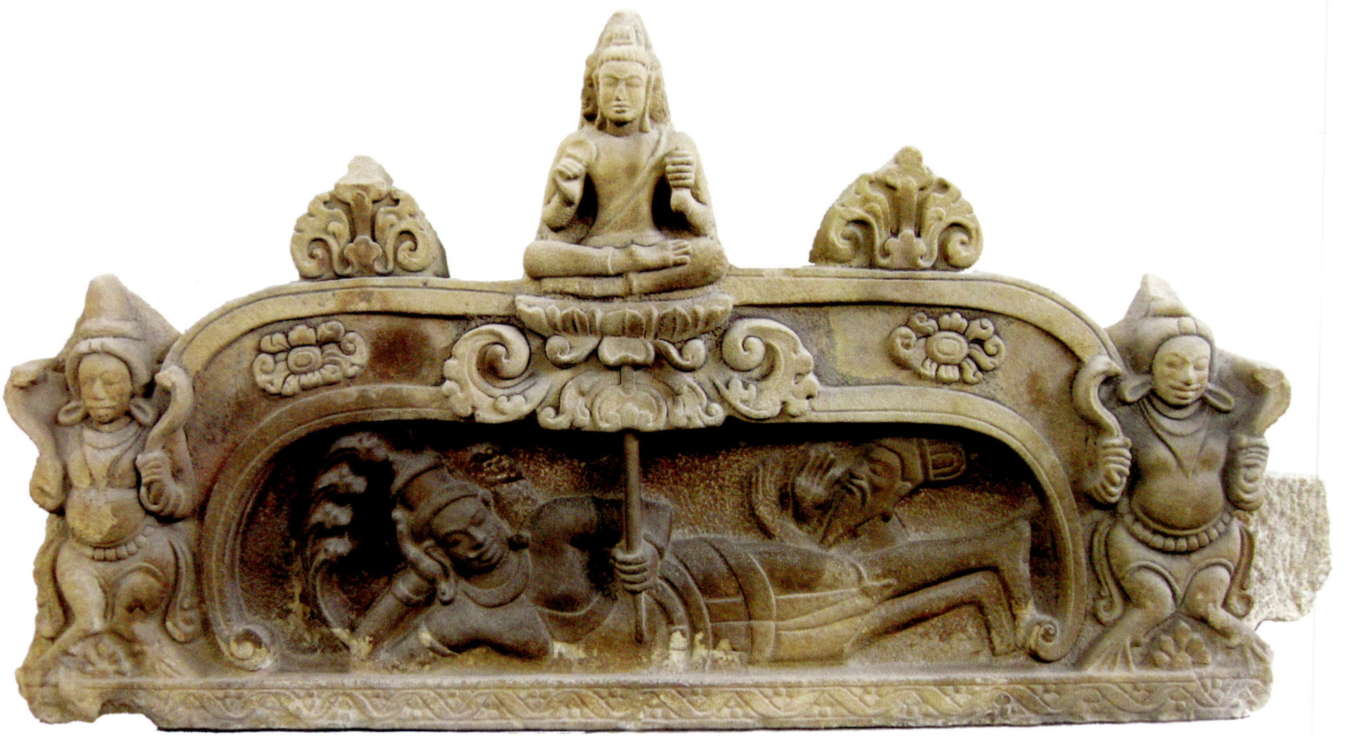

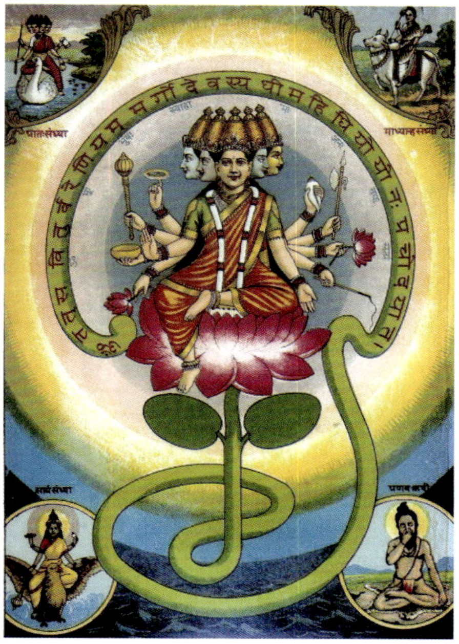

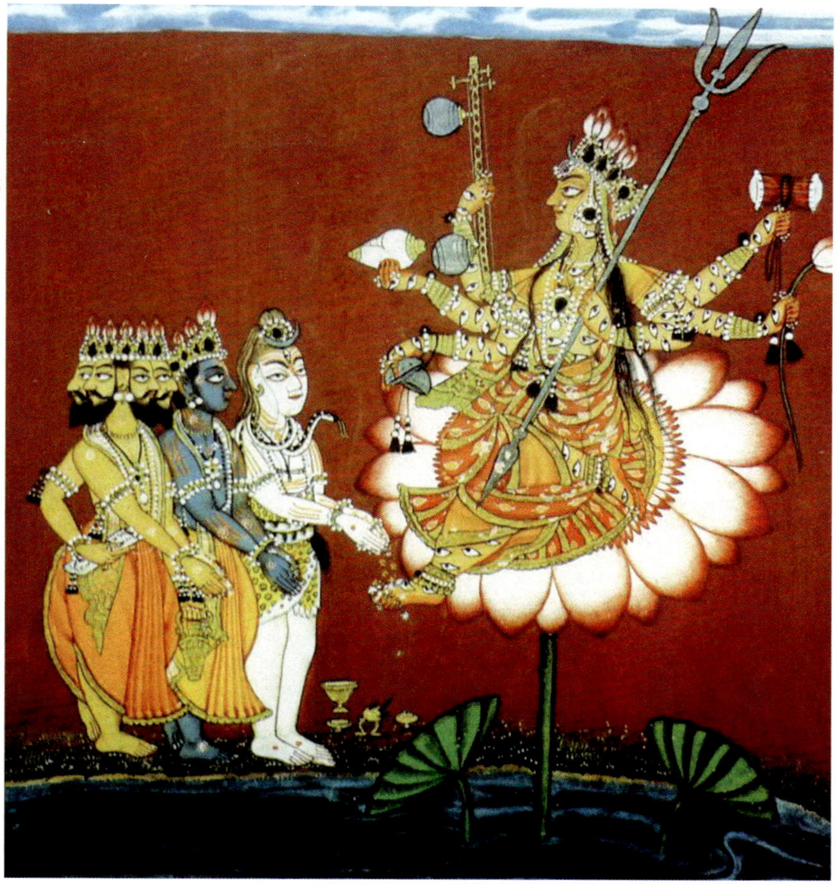

Above: 'A representation of the Gayatri Mantra', Raja Ravi Varma School, poster art

Above right: 'Indrakshi seated on a lotus flower adored by the gods', opaque watercolour on paper, Basohli, c. 1660-70

emerges from his navel. In the middle of the flower stands Brahma. The pink lotus (*Padma*) synonymous with Vishnu, is a daytime in contrast to the blue lotus (*Utpala*) particular to the night, which is in relation to Shiva. These three Gods represent the triple manifestation (*Trimurti*): Brahma the Creator, Vishnu the Preserver and Shiva the Destroyer or more accurately the Transformer. The preservation is done during the day while the transformation operates by night. Lakshmi, consort of Vishnu and lotus goddess, appears standing or sitting on a lotus, holding up lotuses in her hands and surrounded by flowers depicted as buds or corollas.

SACRED STORIES

THE LOTUS OF BRAHMA The name of the magical flower Brahma Kamal is translated to mean, 'the lotus of Brahma.' Brahma, the Creator of the Universe, is always shown to be sitting on the huge lotus and holding a lotus in one of his hands. Another name for the flower is 'Brahma Kanja', which means the one born out of the water body. Some of the schools of thought also believe that Brahma was born from the navel of the Lord Vishnu, while other historians say that he was born from a white lotus that is known to us as the 'Brahma Kamal'.

THE LIFE-GIVING LOTUS The popular story of the birth of Ganesha is also associated with the lotus flower. When Shiva attached the head of an elephant on Ganesha's body, he was bathed with water that was sprinkled from a Brahma Kamal. This is the reason the lotus is also given the status of a life-restoring flower of the Gods, and more importantly the holy Trinity.

THE GOLDEN LOTUS In the epic *Mahabharata*, when the Pandavas were in exile in the forests with Draupadi, she could not forget the insult she had to bear at the court of Kauravas. During one of the evenings in the forest she saw a beautiful lotus in the stream. She felt a spiritual joy at the sight of the blooming 'golden' lotus (as a lotus is usually pink, 'golden' can be taken for a shimmering 'white'). However the lotus faded as quickly as it had bloomed. Later, Draupadi asks her devoted husband Bhima to look for the blooming lotus, and on his quest for the flower he met Hanuman (the monkey god from the *Ramayana*).

THE DIVINE SISTERS

According to the Hindu mythology, the lotus flower is associated with Lakshmi, the goddess of wisdom and generosity. She is said to be born of the pink lotus and before the arrival of the more patriarchal Aryan society, she was of supreme import to the inhabitants of India. Lakshmi is often portrayed as sitting or standing on a completely blossomed lotus and holding lotuses in her hands that gives a sense of purity to her form. The fully blossomed lotuses indicate that she is showering her generosity upon mankind.

Goddess Saraswati is worshipped by those seeking knowledge, especially students, teachers, scholars and scientists. In popular images, Saraswati is shown with four arms, donning a white sari and seated on a white lotus. The iconography signifies that she is rooted in the supreme reality, and symbolises supreme knowledge. This is a popular prayer to Saraswati:

Yaa kundendu tushaara haaradhavalaa, Yaa shubhravastraavritha
Yaa veenavara dandamandi:akara, Yaa shwetha padmaasana
Yaa brahmaachyutha shankara prabhutibhir Devaisadaa banditha
Saa maam paatu Saraswatee Bhagavatee nishesha jaadyaapahaa.

May Goddess Saraswati, who is fair like the jasmine-colored moon, and whose pure white garland is like frosty dew drops; who is adorned in radiant white attire, on whose beautiful arm rests the veena, and whose throne is a white lotus; who is surrounded and respected by the Gods, protect me, May you fully remove my lethargy, sluggishness and ignorance.

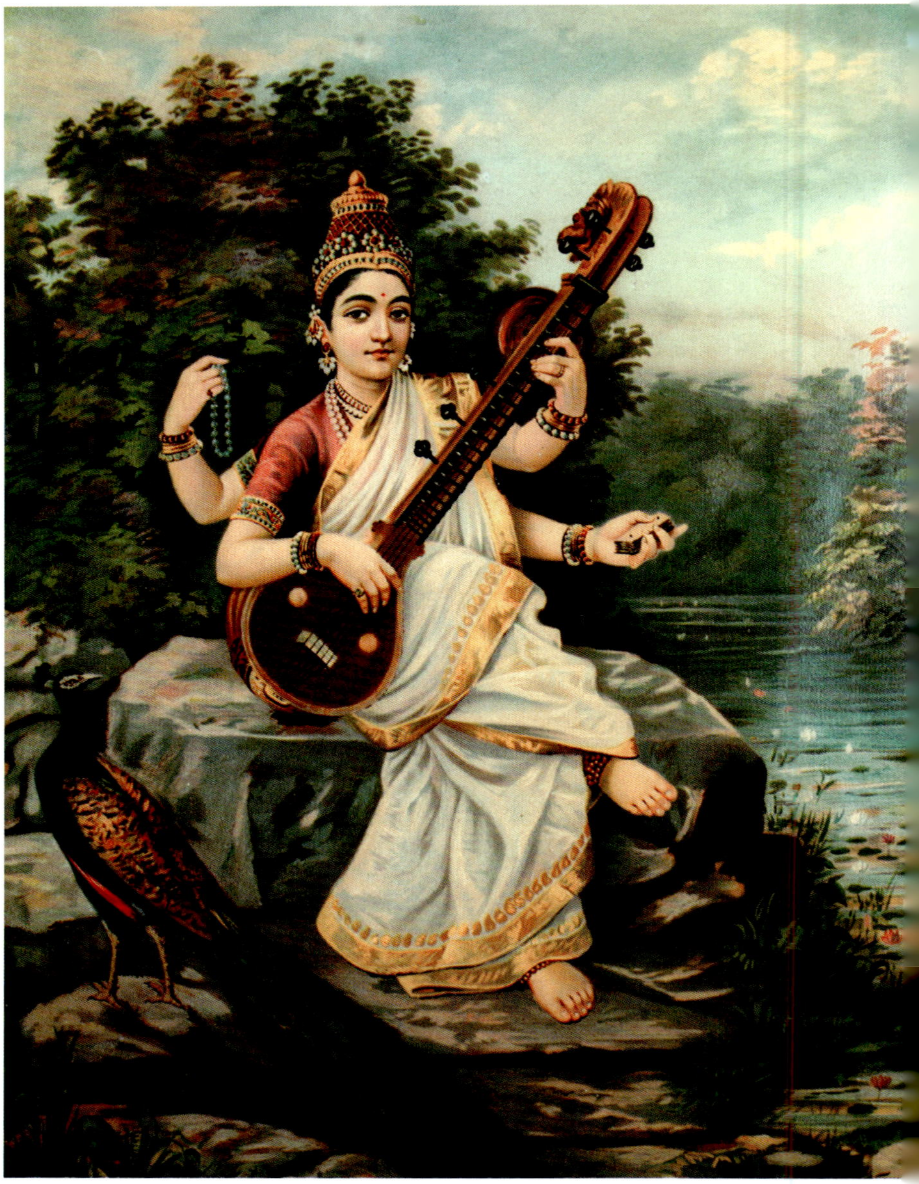

Above: 'Goddess Saraswati', Raja Ravi Varma School, lithograph, late 19th century

Facing page: Golden Lotus in Meenakshi Amman Temple, Tamil Nadu

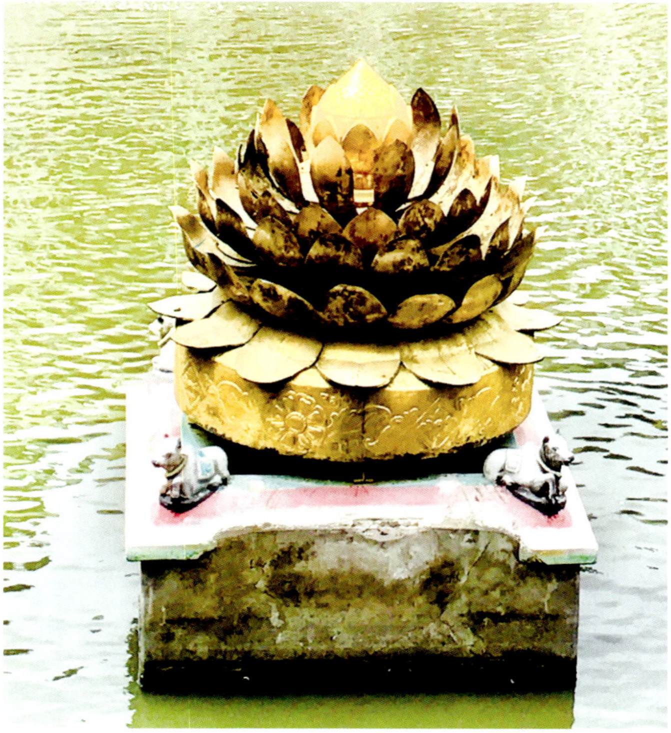

THOUSAND-PETALLED WONDER

THE LOTUS IN ARCHITECTURE AND CULTURE

> *Joyful birds of every kind seem to inflame my love as I think of my dark, lotus-eyed beloved whose face is like the moon.*
>
> —The *Ramayana*[11]

The lotus is the oldest and the most universal floral symbol in Indian art. Its important status is highlighted by the multiple interpretations of its forms. The lotus is said to indicate beauty, prosperity, wisdom, sanctity, purity and eternity. It is also a solar as well as a geographical symbol. There are many beliefs regarding the lotus through the many religions that have seeped into or emerged from India. Among them, Hindu thought seems to indicate that the spirit of the lotus flower resides within every person, symbolising a higher state, in which one rises through the mud to turn one's face to the sun—pure and joyous. It is a symbol of the Buddha and is very closely tied to Buddhist philosophy and art.

ACROSS FAITHS

The motif of the lotus on pillars and railings can be found alongside yakshis, jinas, apsaras across monuments such as the Bharhut, Sanchi, Amaravati and Bodhgaya Buddhist structures.[12] Rosette roundels then became popular, taking their cue from the religious architecture; they began to appear in palaces, forts and homes, as well. The rock-cut temples of Western India, such as the Ajanta caves, also feature lotus designs. The Jain shrines in Dilwara and Ranakpur having carved lotuses on the ceilings in the *navacaukis* (sanctum) and in the marble domes of the *rangamandapa* (performance area). Lotuses also populate the ceilings in the porches and corridors, adding a breath-taking effect as they adorn works featuring dances, celebrations and penances—the whole spectrum of human life.

On the ceiling of the corridor in the Vimalavasahi temple, in Dilwara, dating back to the 11th century, there is a remarkable work that depicts Narashima tearing the belly of the demon Hiranyakasyapa in immense fury; violence that is quite rare in the Jain tradition. A lotus flower with 16 petals is the motif within which the two are engraved in their combative embrace.

The lotus holds a reverent position in the realm of Buddhist art, symbolising enlightenment; with the white lotus indicating purity—it is said to be a symbol of the Buddha himself. Buddha attained enlightenment under a banyan tree in Bodh Gaya, part of the Mahabodi temple. Here a platform of some height called the Jewel Promenade Shrine is where the Buddha paced, deep in his meditative state. Eighteen lotuses indicating knowledge and enlightenment are carved into the platform. Stories tell that with every step he took in his trance, lotus flowers bloomed. Close to the temple is the Mucalinda tank, once dry, now a flourishing lotus pond adjacent to which Buddha resided for nearly two months after he attained enlightenment.

OVERLAPPING EMBLEMS

The thousand-petalled lotus or the *hiranya kosa* is evidenced to have been of great import to South Asia.

> *One can't say with certainty that one comprehends the meaning of these images and metaphors fully. One thing is certain, however, that those wondrous hands that created these eternal images of exquisite lotuses, some as large as 25 feet in diameter, several hundreds of years ago, must have known and experienced at some level the glory that resides in the treasure chest, hiranya kosa, of a 'thousand-petalled lotus'.*[13]

The shape of the lotus is circular, similar to the wheel, and the solar disc though each indicate different concepts. The lotus is an auspicious symbol of wisdom and purity. In Indian sculpture, the three circular motifs come together as part of a composite whole.

The heart of the devotee is another thing abstractedly indicated by the lotus. The heart of the lotus seems to represent space, the chief residence of the Gods. Its eight petals point in four directions. They can also be representative of the four points of the compass, as can be witnessed in the mandalas.

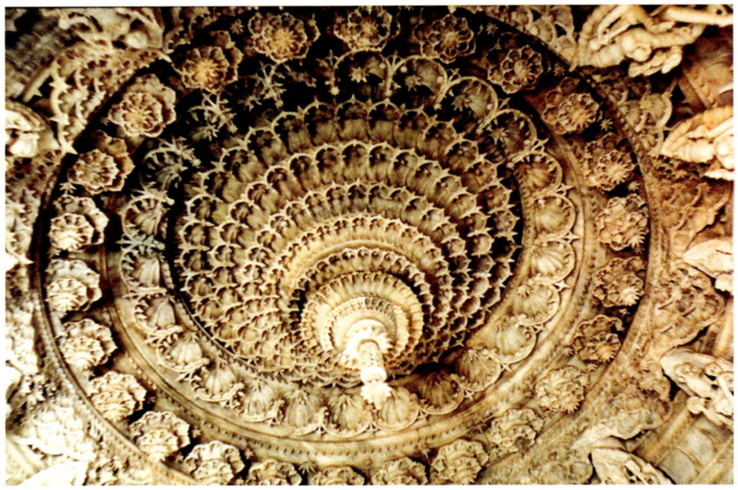

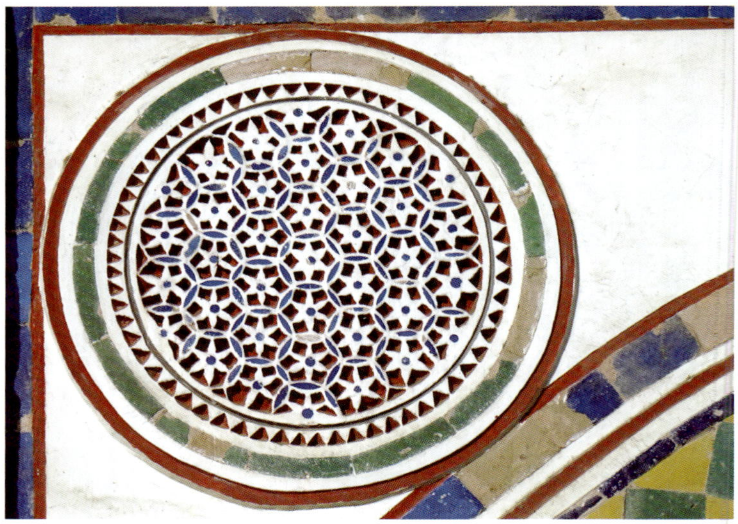

Top: Lotus motifs carved in white marble on the ceiling of the Dilwara Temples, Rajasthan, 11th–13th century

Above: Lotus-like floral motifs carved on a monument in the Red Fort, Delhi

The lotus is the centre of a number of creation myths. In Hindu mythology, the Supreme Creator Brahma, himself is said to have emerged from a lotus that extended from Vishnu's navel. It was a way for ancient Indians to indicate the created world, as F.D.K. Bosch said, 'the very symbol of life risen from the waters.'[14]

Additionally an analysis of the water on which the lotus is said to float, the elixir of life, *amrita*, indicates its potency as a symbol; a similar train of thought compares lotuses to kings and their grandeur, which they propagated everywhere as elixir. As the lotus bloomed on its stalk that held it above the water, so the gods crossed from the beyond into the now.

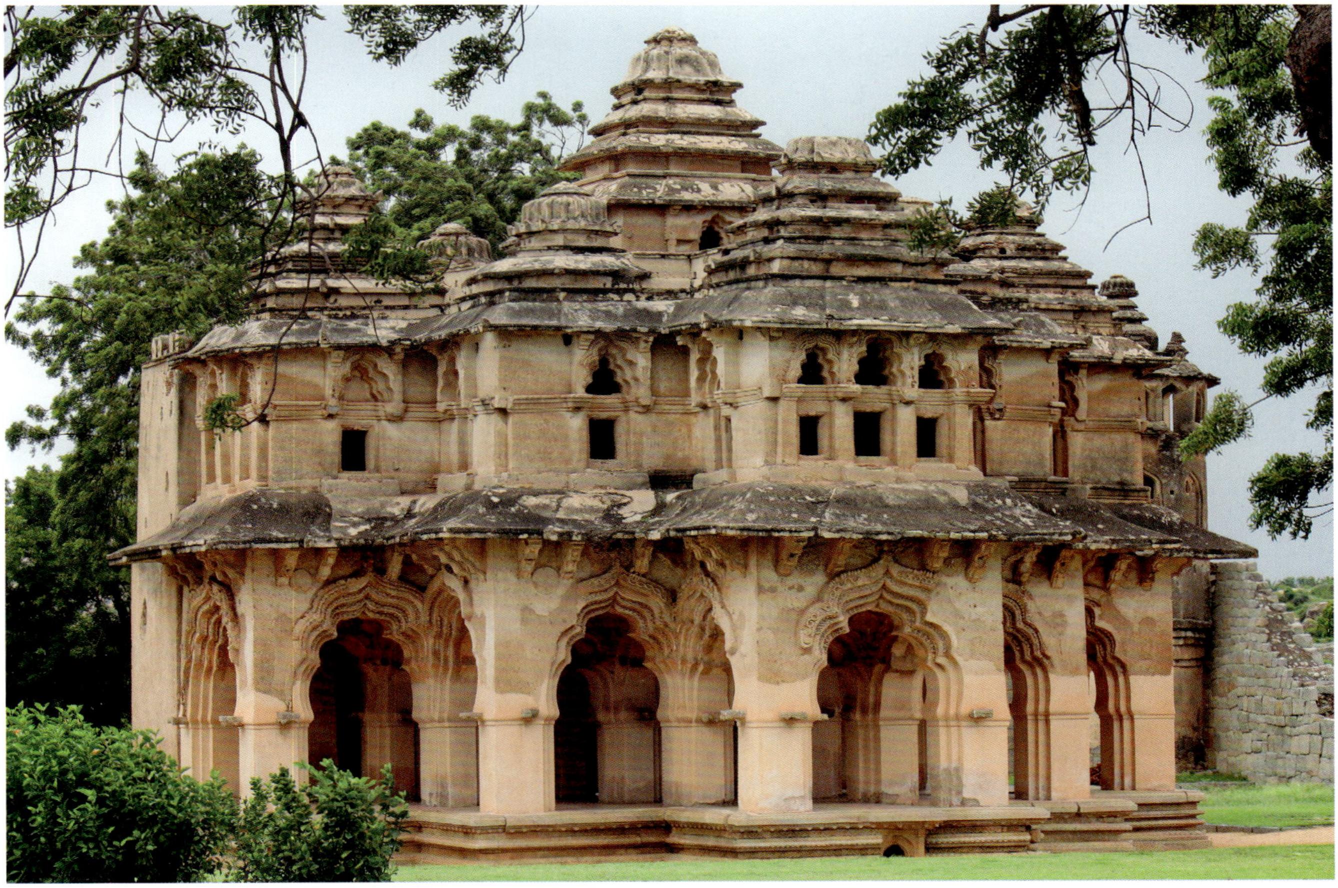

Lotus Mahal in Hampi, Karnataka

ETCHED IN STONE

The lotus made its first appearance in the ceilings of the Buddhist cave temples at Ajanta, which were created during the Gupta period. Pillars and railing medallions at important Buddhist sites at Sanchi, Amarvati and Bodh Gaya featured lotus motifs as did western Indian rock-cut temples. In early Buddhist art,[15] 'the vase of plenty' was shown with a lotus or a Bodhi tree emerging from within it, signifying spiritual growth from the reservoir of *dhamma* it symbolised. In the early phases of Gandhara Art, the lotus is used to celebrate the Goddess Saraswati. The one feature that connects it with the Buddha, is the border of conventional lotus leaves fringing the edge of the stool on which he is seated, and the larger architectural setting.

In the Jain temples at Dilwara and Ranakpur, in Rajasthan, the lotus shapes were carved in a multitude on the ceilings in the corridors, with mesmerising magnificence and subtlety. Hundreds of dancers and musicians adorn the ceilings, columns and domes of these temples, and are seen meditating or celebrating in the middle of lotus petals.

The Vijayanagar Empire was a mighty Hindu kingdom that reigned over much of South India from the 14th to the16th centuries. During this time, magnificent temples dedicated to the various Hindu gods were built in the Dravida style. The buildings are eclectic in character where the tall pyramids or *vimanas* tower over the *garbha griha*, the inner sanctuary. One example of such a two-storey monument of an uncertain function is the Lotus Mahal, which is located in Hampi, Karnataka. This structure is like the entrance pavilion of Delhi's first mosque, the Quwwat-ul-Islam, and brings out the stylistic cross-currents of South Asian and Islamic architecture of the second millennium.

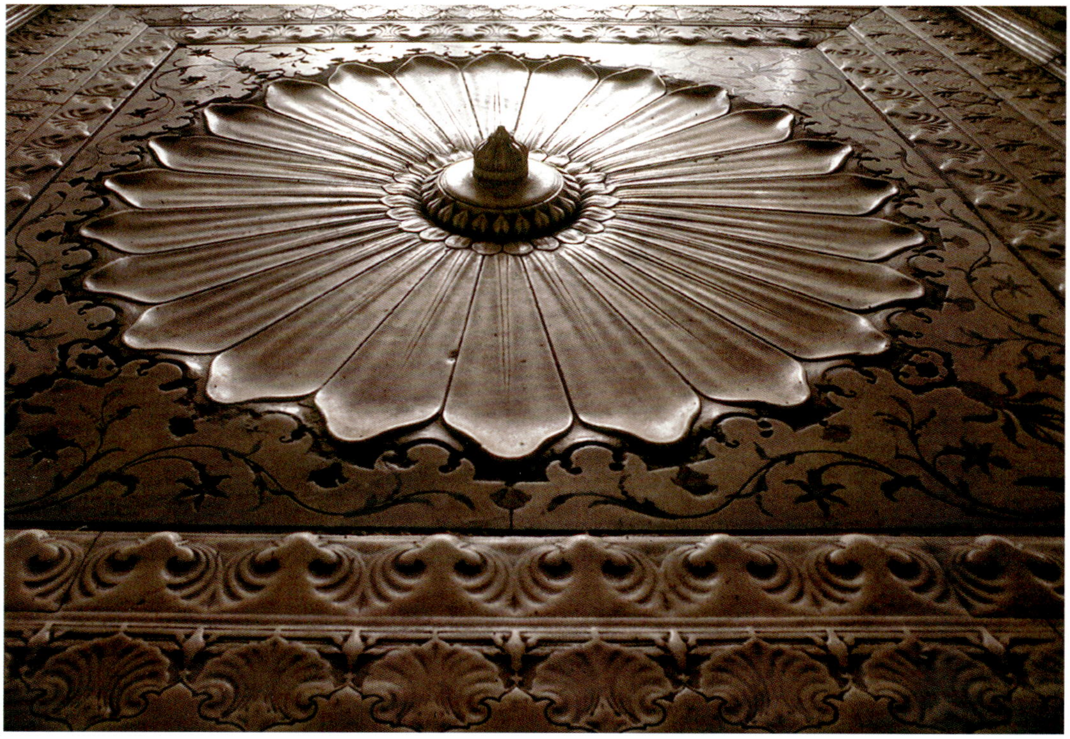

The Mughal rulers also had a fondness for lotus designs, that seeped into their architecture from the native craftspeople. In the Red Fort in Agra and Delhi we see the lotus in a marble avatar, carved into the floors of certain rooms, such as the Diwan-i-Khas (hall of private audience).

THE ARTIST'S MUSE

In Hinduism, the lotus was incorporated into the myth and belief systems, unlike Buddhism, where it would represent the Buddha himself, or the entire Buddhist universe. The native painters would express their creativity intuitively without giving a thought to the rules of formal artistic expression. In Rajput art, we come across the symbolic lotus flowers blooming in the river, although in reality it is a perennial flower that grows in the mud and water. The main significance of the lotus here is the availability of rainwater for crops; thus in traditional Indian culture it would symbolise prosperity and wealth. In Sanskrit, a lotus (*kamala*) is also the name of Lakshmi, the goddess of wealth. The Jains of Gujarat also worshipped her, where rice was depicted as if it was a flowering plant like a lotus.

In the Ajanta caves, one views an amorous couple around whom the female musicians are serenading the rains as well as the lovers. After the rain starts to pour, the couple retire to their small pavilion where a bed awaits them. Down in the garden is a lotus pond and a tree filled with birds. The ceilings at Ajanta also incorporate lotus motifs, both petals and buds, something that can also be found at the rock cut caves at Ellora, which features rosettes of the lotus blossom.

Over the centuries, the lotus continuously appears in the Indian ethos. In various miniatures, Radha and Krishna were illustrated as walking on a bank of the lotus pond and holding fresh lotus flowers in their hands. In one image they are shown entirely covered in lotuses. The shape of their eyes is also rendered in the shape of a lotus.

Above left: Lotus-shaped fountain on the floor of the Diwan-i-Khas (hall of private audience), Red Fort, Delhi

Facing page above: Lotus-like floral motifs in paintings in the Ajanta Caves, Maharashtra

Facing page below: 'Goddess Chinnamasta', Bundi School, opaque water colour on paper, c. 1775; the Goddess has a lotus crown on her severed head and a large lotus rises in the pond before her, symbolising wisdom and purity

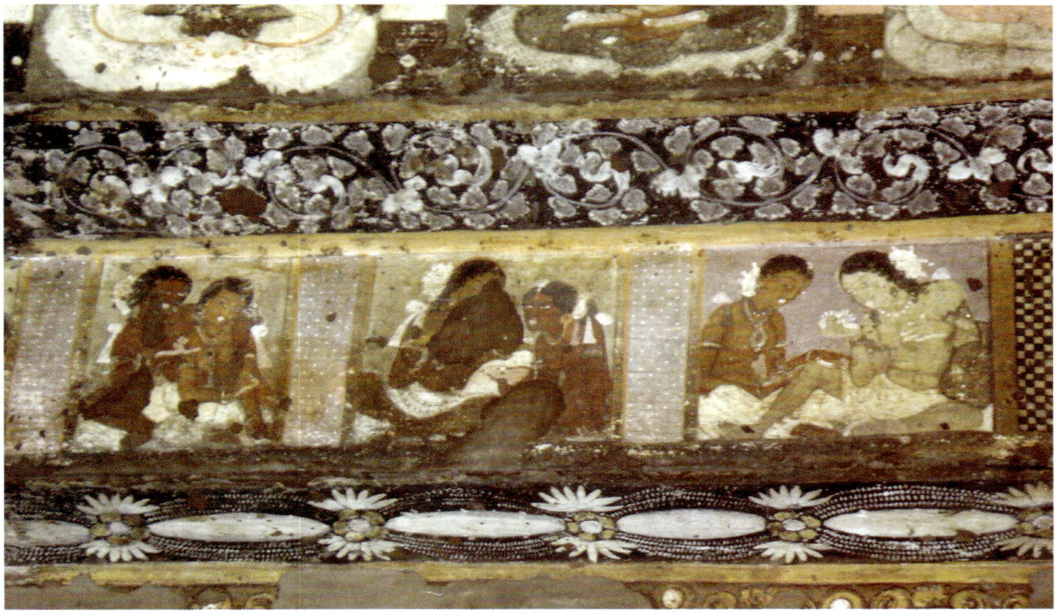

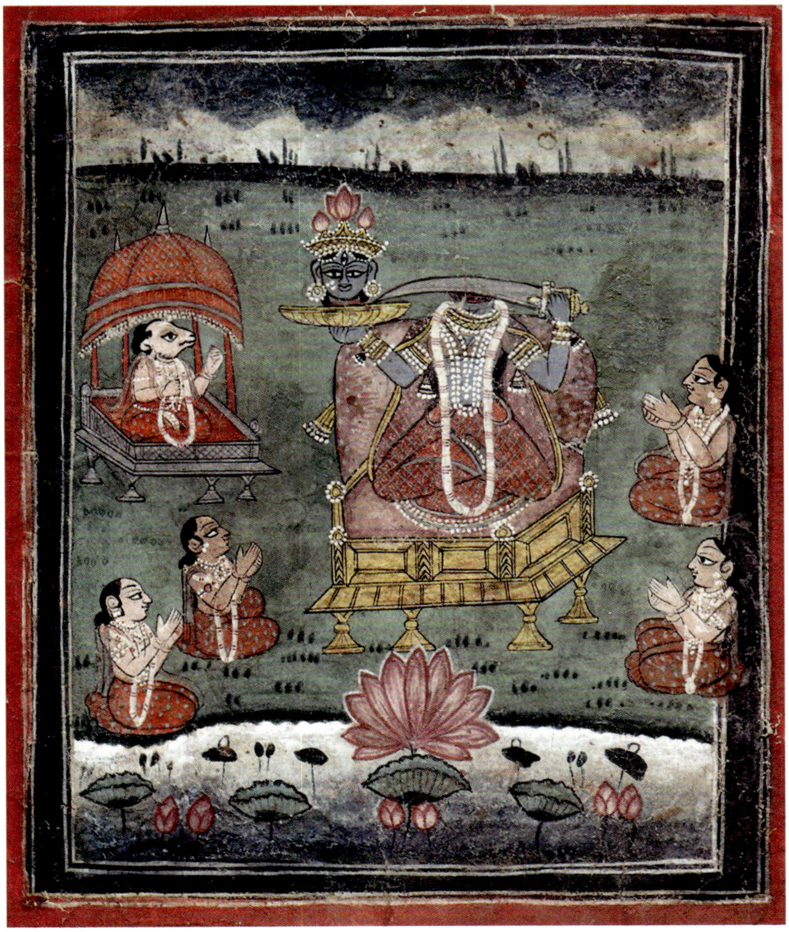

In the Bundi School of painting, the lotus pond has a significant place, as with the other aspects of the painting. They are normally placed in the foreground (occasionally in the background) surrounded by floral elements, foliage and aquatic life such as fish and ducks. This was further enhanced through the constant use of grey, silver and blue hues to in order to create the ripples.

In the manuscripts of 17th-century Nepal, we see that paper had become common at the time, even though it was still designed like the elongated palm leaf with a thread binding the manuscript together in the centre. The illustrations[16] were not painted into the manuscript, but stuck in their respective places. Vasudhara, a well-known Buddhist deity, is seen in the middle of a horseshoe-shaped scarlet body halo. She is sitting on the lotus pedestal with both legs folded on the seat, with the right knee raised and the other flat on the seat. This posture that communicates an attitude of royal facility is known as *rajalilasana*. Originally, Vasudhara was the goddess of rice, and the Buddhist Newars of Kathmandu worshipped her in the beginning of autumn at the time when the rice paddy was ready for harvest.

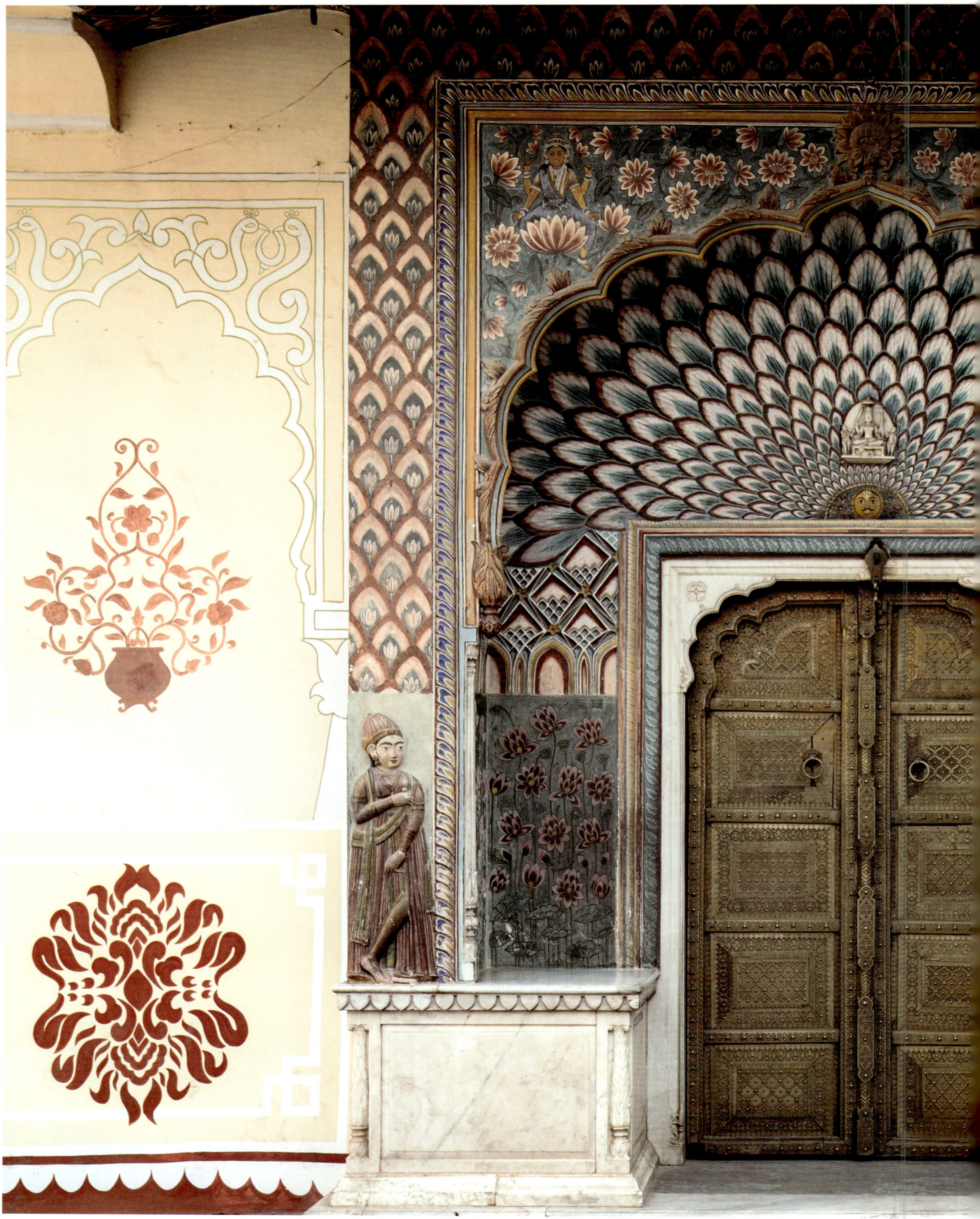

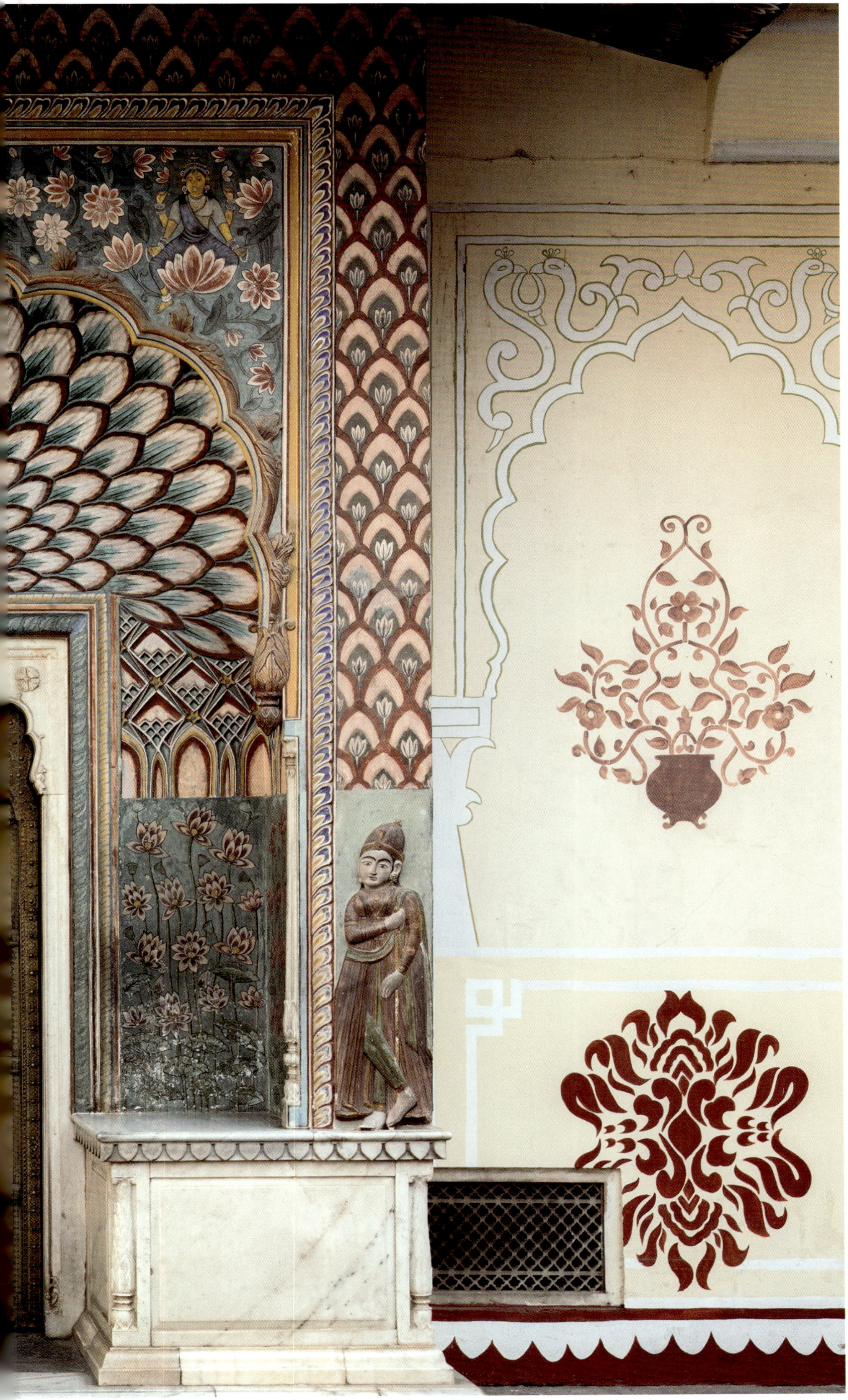

The Lotus Gate in City Palace, Jaipur, Rajasthan

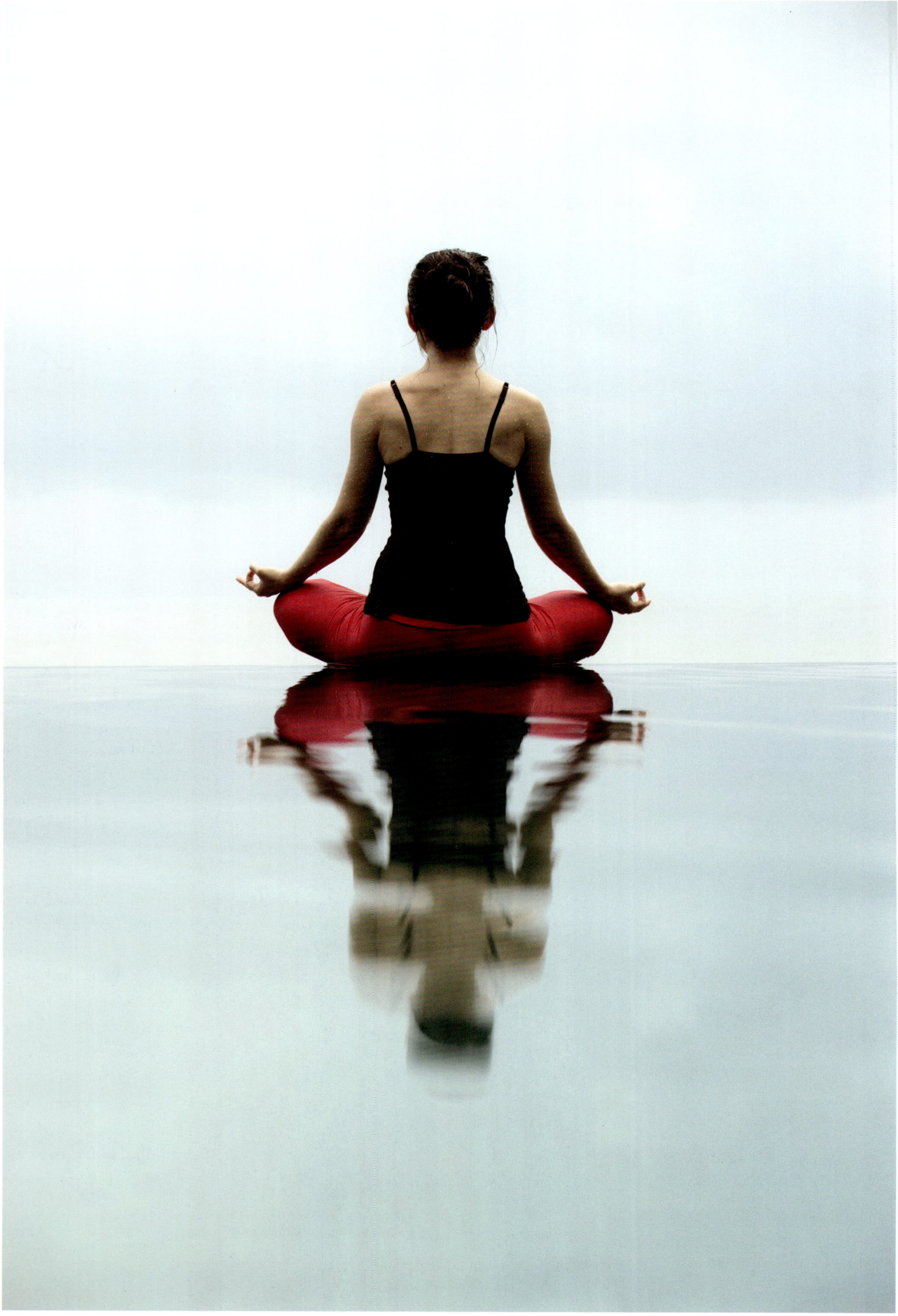

Facing page: Woman in *padmasana* pose

Right: Radha and Krishna from a Gita Govinda series, Basohli School, attributed to Manaku of Guler

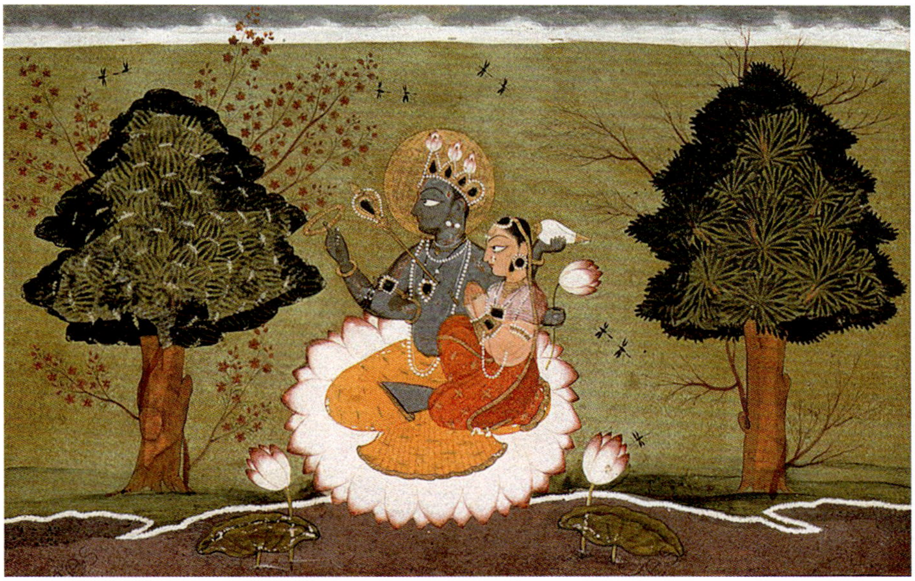

The Basohli School has many paintings depicting the gods such as Vishnu and Shiva seated on lotuses. We often see Krishna and Radha embracing each other on a lotus pedestal where the lotus is painted in red, and the background in sap green. The lotus has a double significance here. Firstly, it signifies the emphasis of the transcendent aspect and secondly, it symbolises love and union.

Ernst Waldschmidt, the German Orientalist and Indologist, on studying these miniatures, highlighted A.K. Coomaraswamy's observation that the three lotus petals that were on the crown of Krishna, showed that the custom of wearing fresh flowers in the turban was confined to the Himalayas.

MULTIPLE MEANINGS

What other can convey the ideas of beauty, wealth and fertility as swiftly and elegantly as the lotus? The spirit of eternal purity, rejuvenation, the touch of the divine, the fertility it represents makes the enduring lotus, in Hindu thought, the spirit that resides within every individual.

Womanly beauty is indicated by the lotus, as well. The lotus is a symbol of eternal beauty of spirituality and of the enlightenment. That great text holy to Hindus, the Bhagavad Gita holds that a man can only be judged to be lotus-like if he works detachedly in service of God, impervious to sin as the water on the leaf of a lotus, like the flower reaching from the mud and water to the sun. It is unsurprising that the *padmasana* or the lotus pose is the yogic pose that means that one is exerting to reach the highest level of consciousness and it can be located, with a thousand petals on the head and in the lotus chakra as well.

In modern times, the lotus had been assimilated and has become an important architectural and sculptural element. Nowhere is this more evident than at the Lotus Temple, the Baha'i House of Worship, located in the capital city of New Delhi. A contemporary architectural marvel built in the1980s, it took inspiration from the lotus blossom, honouring the bloom's sacred place in the

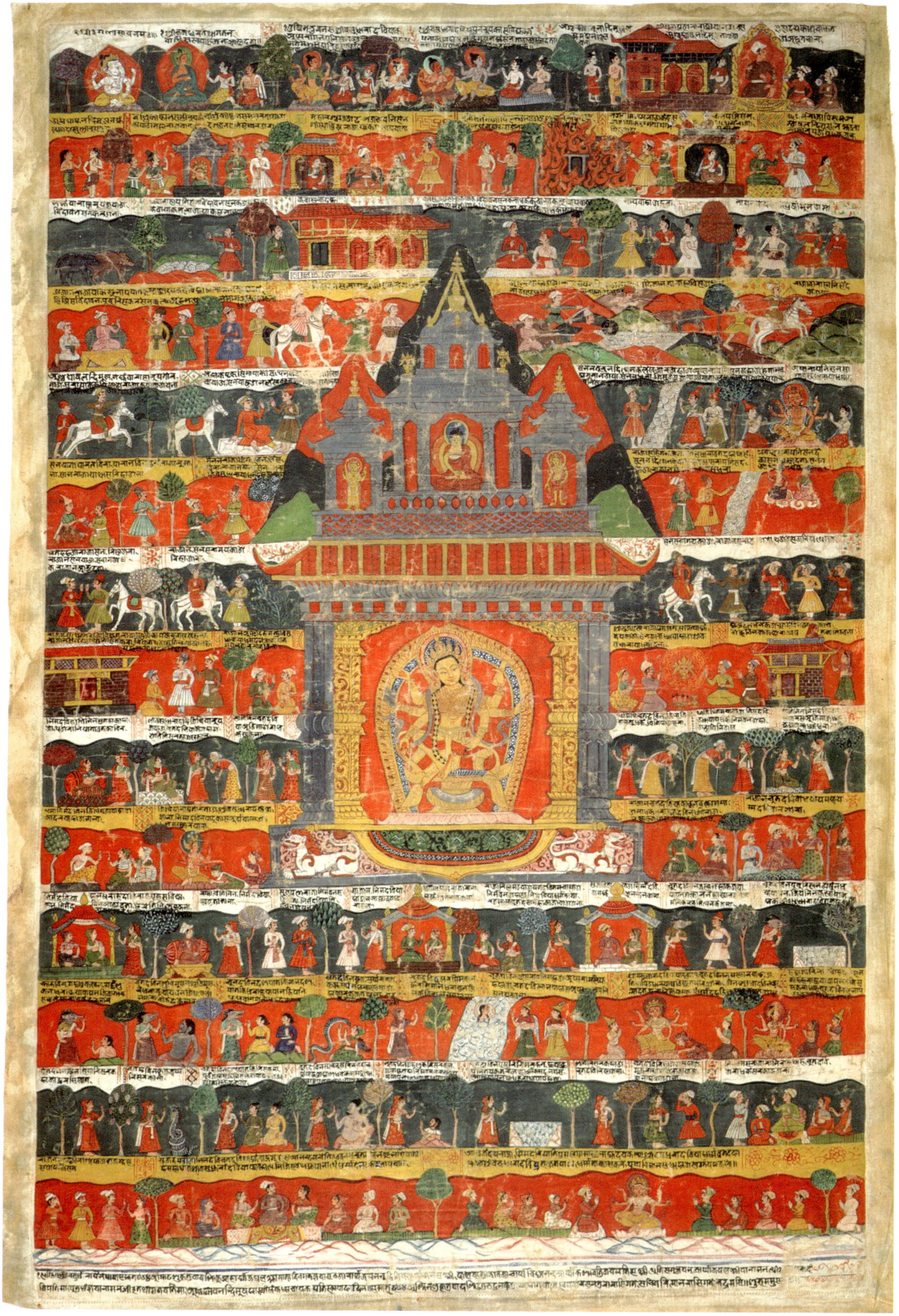

Facing page: 'Legend of Vasudhara' (Ashvaghosha Avadana), mineral pigments on cotton cloth, Nepal, 1744

realm of spirituality. The manner in which it was designed and erected makes it appear like a lotus flower floating upon the nine reflecting pools around it. The pinnacle of modern beauty and architecture, it is small wonder that its design was influenced by the flower that has inspired millions through the ages, touching all faiths and creeds.

ENDNOTES

1. Dilip J. Karnik, *Rays of Wisdom* (New York: Xlibris Corporation, 2012), p 10.
2. Bharata Muni, *Natyashastra – Volume II*, trans. Manmohan Ghosh, p. 294
3. Divya Nandyala, *Bhajanika* Lulu.com, 2013. p. 21.
4. Mahaperiyavapuranam.org. *In the Presence of Divine, Vol 2, Chapter 6, Balu Mama (Part 4)*, p. 11.
5. Sehdev Kumar, *A Thousand Petalled Lotus: Jain Temples of Rajasthan: Architecture and Iconography* (New Delhi: IGNCA and Abhinav Publications), p. 42.
6. Ibid. p. 44.
7. Ibid. p. 43.
8. Paul A. *LaViolette, Genesis of the Cosmos: The Ancient Science of Continuous Creation* (Inner Traditions: Bear & Co), 2004. p 124.
9. Jennifer Potter, *Seven Flowers And How They Shaped Our World* (London: Atlantic Books, 2014), pp. 17–19.
10. Lee Keith Freeman, *Book of Sages* (Pennsylvania: Temperance Publishing, 2000), p. 19.
11. T.R.S. Sharma, ed., *Ancient Indian Literature: An Anthology* (New Delhi: Sahitya Akademi, 2000), p. 96.
12. Sehdev Kumar, *A Thousand Petalled Lotus: Jain Temples of Rajasthan: Architecture and Iconography* (New Delhi: IGNCA and Abhinav Publications, 2001), p. 44.
13. Ibid. p. 47
14. F.D.K. Bosch, *The Golden Germ: An Introduction to Indian Symbolism* (-Gravenhage: Mouton. 1960), p.82.
15. Damien Keown and Charles S. Prebish, eds, *Encyclopedia of Buddhism* (London: Routledge, 2013), p. 112.
16. Navina Najat Haidar, and Marika Sardar, *Sultans of the South: Arts of India's Deccan Courts, 1323-1687* (New York: Metropolitan Museum of Art, 2011) p. 45.

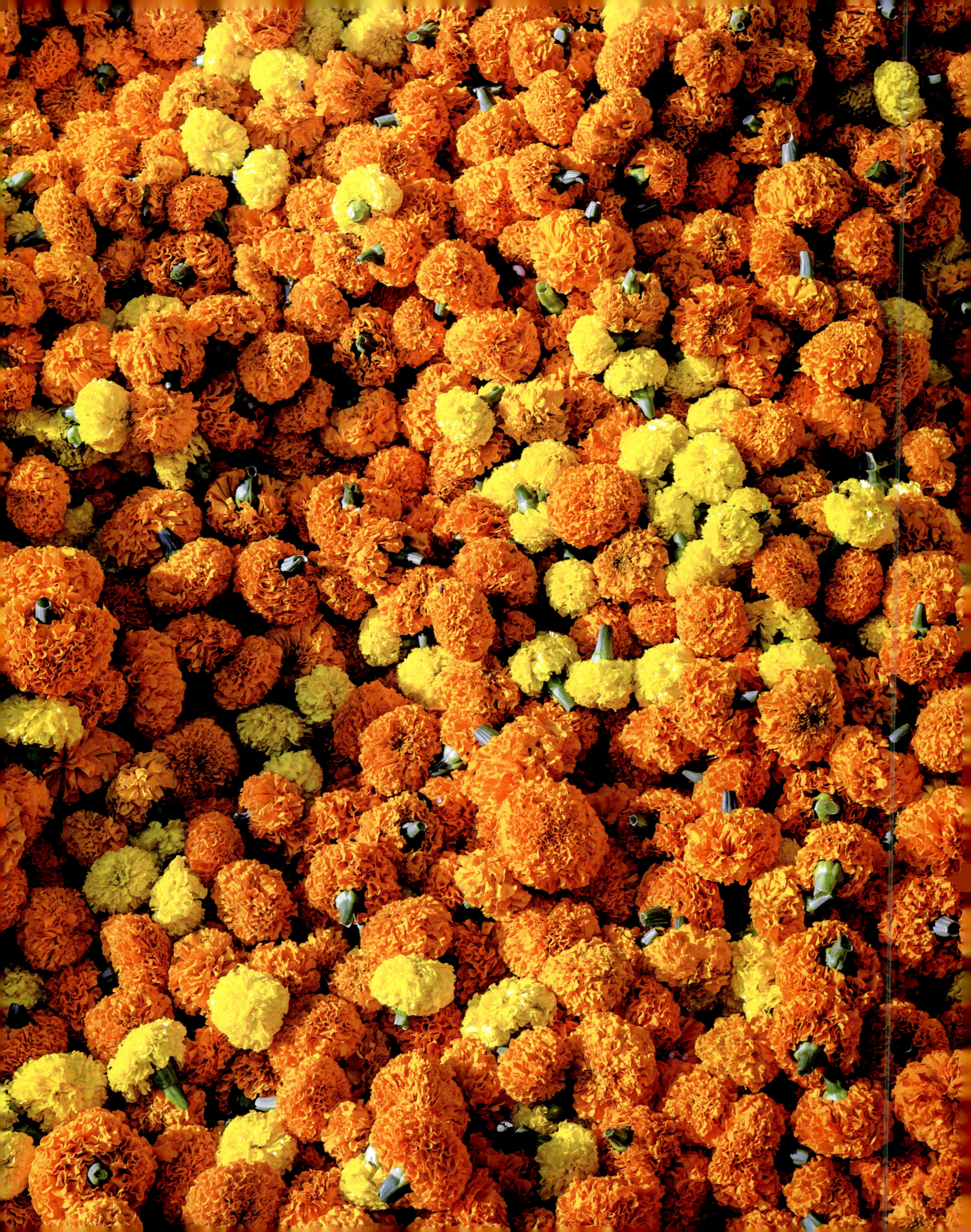

Blossoms of the Sun and Moon

First it took on a thick coat of sandalwood
then rubbed up against lotus pollen
It grazed the jasmine vine and roses
fondling how many pine flowers?
It was tinted yellow in a field of champa flowers
and embraced the banana tree along the way.
—Description of the wind by Ritikaleen poet Keshavadas[1]

Along with the emperor of Indian flowers, the lotus, and the fragrant Indian rose, the *champa* or *champaka* and the *genda* (or marigold, derived from Mary's gold) enjoy a special place in the cultural consciousness of India. Velvety in form, sensual in connotation, the *genda* is one of the hardiest of flowers. Easily grown, yet ornamental in appearance, it is also used extensively in the making of garlands for auspicious occasions and exchanges. The vivid orange colour of the marigold is eloquent at weddings in the form of garlands, shredded into petals for rangoli, floating on water displays, or as a decoration on large platters of food. Its golden hue makes every occasion, whether religious or secular, vibrant and charming. The *champa,* on the other hand, is a favourite motif in Hindu temple sculptures, as well as a recurring leitmotif in Rajput paintings and Pahari miniature paintings. Even now, *champa* flowers are used by women in the Eastern parts of India, particularly Bengal, in their coiffure, and the delicate fragrance of their creamy petals adds to their subtle charm. *Champa* flowers are also used for making aromatic oil.

Facing page: Orange and yellow marigold flowers enhance a festive atmosphere

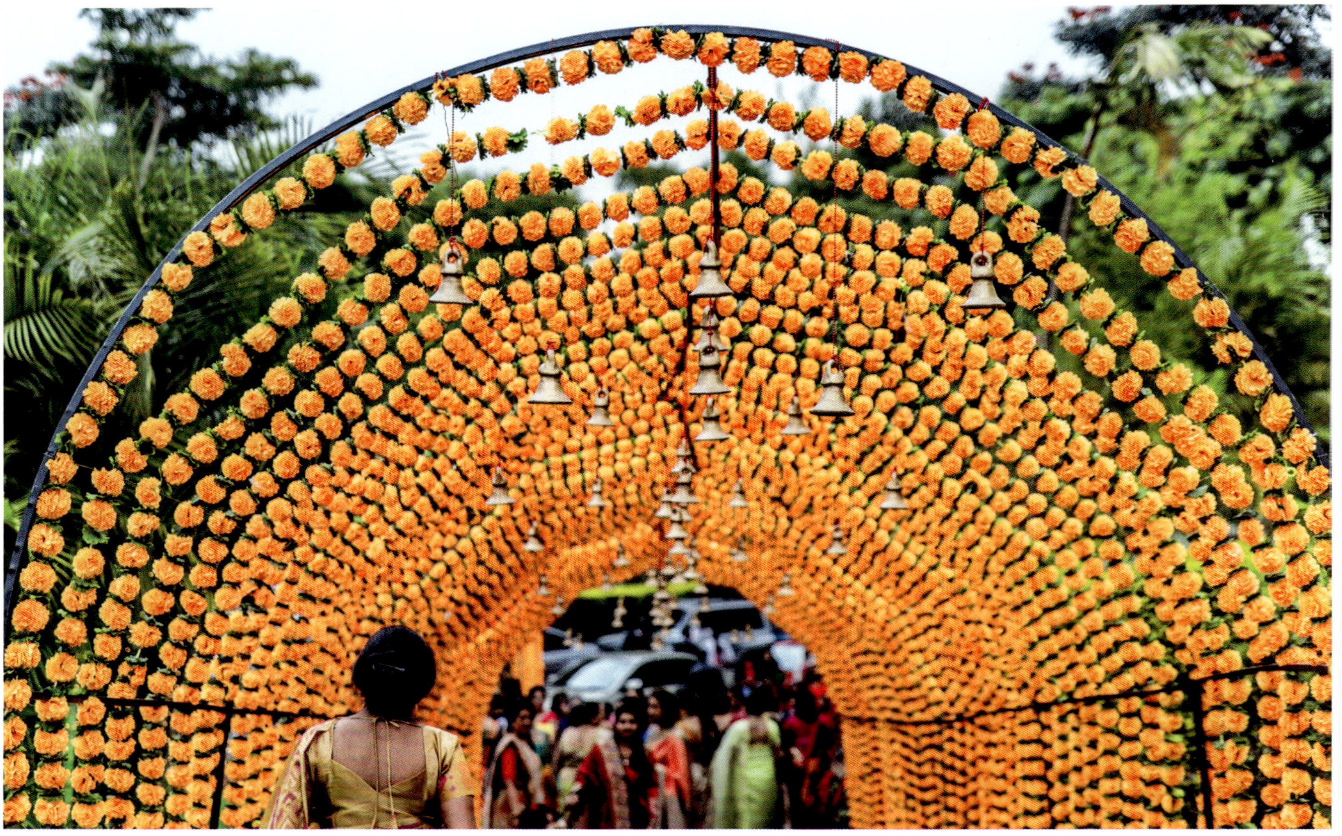

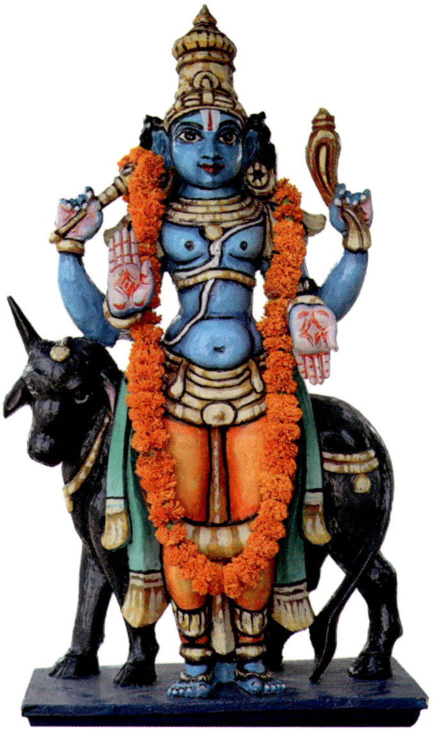

THE GOOD LUCK FLOWER

Famous as much for its medicinal goodness as for its eye catching beauty, the *genda* is known as *Calendula officinalis* to botanists. Like several plants now commonly seen across India, it was introduced to the country not long ago, in this case by Portuguese colonists. In addition, it is also used as a culinary ingredient, for adornment, divine offerings and also for making an effective pesticide. Interestingly, it can also symbolise cruelty and jealousy. *Genda* was once thought to protect against the plague and to be effective in stopping gossip. Practitioners of voodoo believe that used in combination with the proper spells, however, it is an antidote for the sharp-tongued and promotes cheery conversations.

Above: An archway decorated with orange marigold flowers welcomes guests at a traditional Indian wedding

Left: Idol of Lord Vishnu adorned with a garland of marigold flowers

Genda is easy to grow. It blooms from mid-summer throughout the winter and comes in white, yellow, orange, red and mixed colours. Its real significance is seen around Dussehra and Diwali, when it is used to make garlands to offer to the gods. Garlands of *genda* convey honour and respect, especially to Ganesh and Lakshmi, and are also used to welcome all visitors.

Known in Sanskrit as *sthulapushpa*, *genda* is used in weddings in all parts of India because of its profound symbolism. One popular belief is that when it is offered to Vishnu and Lakshmi, the ideal couple, it will bring abundant blessings and a prosperous life to the newlyweds. In addition, since it is hardy and remains fresh long after being cut, it is a popular and beautiful option for wedding decoration. Flowers, in general, form a very important part of Indian weddings because they are seen to be a symbol of creation and regeneration, and are believed to bring good luck and happiness. Flowers are considered pure, and therefore most are an acceptable offering to ancestors, spirits, and gods. The significance of *genda* is found in a combination of the vibrant colour and the general symbolism in religious offerings.

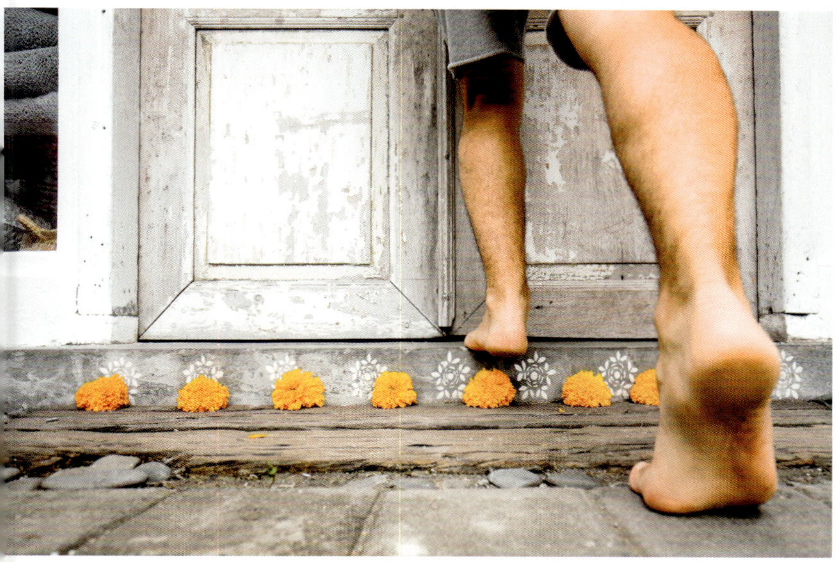

Placing marigold flowers at the entrance of a house is considered auspicious in Asia

THE COLOUR YELLOW

Colour, which is a large part of the Indian consciousness, has become synonymous with the expression of faith and belief. That's why the vibrant yellow and orange *genda* bejewels nearly every celebration, be it a wedding, the worship of a god or the welcoming of a visitor. In Hinduism, the colour of Vishnu is yellow, a symbol of purity, victory and chastity, as well as sensuality. As a sacred colour, yellow is considered symbolic and is used in different religious and social ceremonies. Unmarried girls wear yellow clothes, while some tribes believe that yellow has powers to keep evil spirits away. The spiritual guru ties a yellow thread around the wrists of the disciples as a mark of protection. It represents fire and since impurities are burnt by fire, it symbolises purity. Saffron represents religious moderation and it is a quest for light. As an auspicious and sacred colour, it signifies sacrifice, renunciation and courage. Yellow and saffron were originally associated with the sun, and as part of the solar symbolism, the colours were absorbed into the mythology of Hinduism. In Buddhism, too, saffron-yellow has great symbolic significance and monks and priests are seen with robes of the same colour.

Genda also symbolises passion and creativity, and given its many uses, it's easy to see why. This 'Herb of the Sun' can be used to dye cloth such as silk, so it is a natural alternative to the chemical dyes that pollute rivers and streams. As awareness increases about the pollution of the Ganges River caused by chemical dyes, there has been a move towards using natural alternatives. Research from Banaras Hindu University led to developing natural colours from plants, fruits and flowers, including the *genda*, acacia, madder and pomegranate. During Holi, the festival of colours, plant dyes have traditionally been used and are seeing a revival today as an alternative to synthetic dyes that can harm the skin. The dyes that are prepared from *genda* have proven to be far safer to use.

Genda is used extensively in rangoli, to decorate the floor, mainly at the entrance of house. It also makes an excellent garnish for food, adding a welcome splash of colour and a spicy tang to salads and other summer dishes. The petals are sometimes cooked with rice to give it colour. *Genda* petals also look great when floating on the top of drinks. And finally, because its musky, pungent scent repels animals and insects, *genda* is a favourite among gardeners as a natural pesticide.

IN RELIGION AND RITUAL

A day before the festival of Makar Sankranti in Odisha, myriad *genda* are seen adorning Lord Jagannath and his two companion deities, Lord Balabhadra and Devi Subhadra. This annual adornment is known as Nabanka Basha and the flowers add to the golden hue, making the ritual vibrant and highly attractive for devotees.

In the Bengali tradition, during the ritual worship of Saraswati the idol of the goddess is dressed in white to indicate purity. Of the flowers used in the puja ceremony, however, the most significant is the red *palash* (*Butea monosperma*; flame of the forest) and the yellow *genda*. The family members bathe early and dress in yellow attire before assembling in front of the goddess with offerings of flowers, fruits, sweets and books to be blessed. A special *havan pooja* is performed by the priest with incense, special wood and a diya lit in front of the *prasad*. A handful of flowers, particularly *genda* and flame of the forest, are given to each devotee to offer to the goddess. This *pushpanjali* (offering of flowers) is offered while Sanskrit verses are chanted.

The belief system in Hinduism works around the concept of reciprocity, where a devotee's life is enhanced by the gifts that are bestowed upon the deity. Many Hindu ceremonies are expressions of thankfulness through the symbolic offering of gifts that are usually in the form of flowers and food. The offering varies depending upon the financial status of the devotee and the geographical location. Those in drier environments may give sweets made of wheat and a small quantity of sugar. The devotees in northern India offer garlands of *genda* and roses, while those in the south offer flowers such as hibiscus and jasmine. During the *pooja* rituals, it is believed that the deity symbolically partakes of the food and infuses sacred energy into the flowers and the food, transforming them with vibrant divine power.

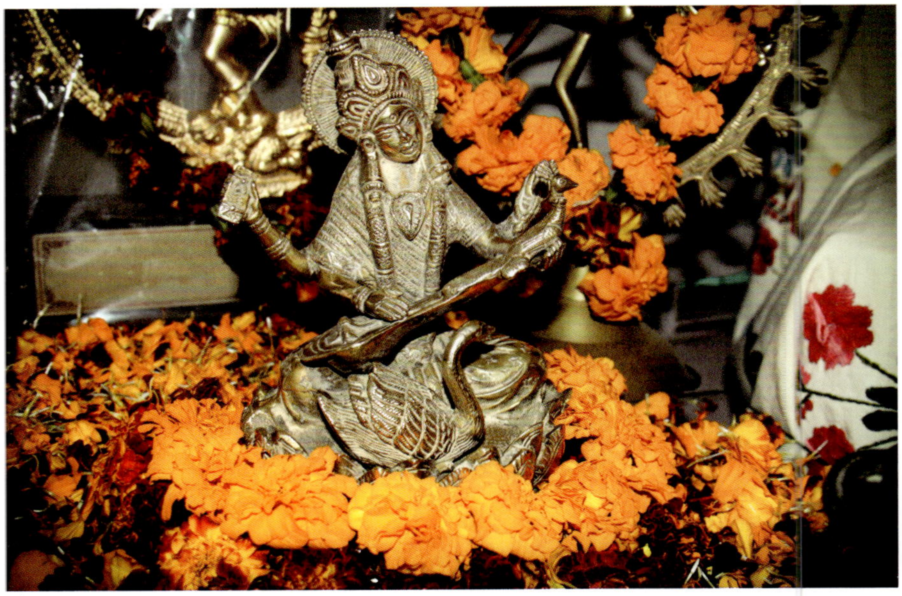

Above: Goddess Saraswati surrounded by floral offerings

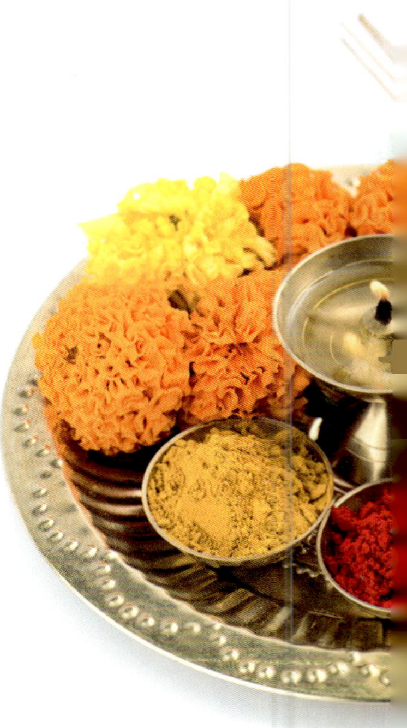

Below: Traditional *pooja thali*

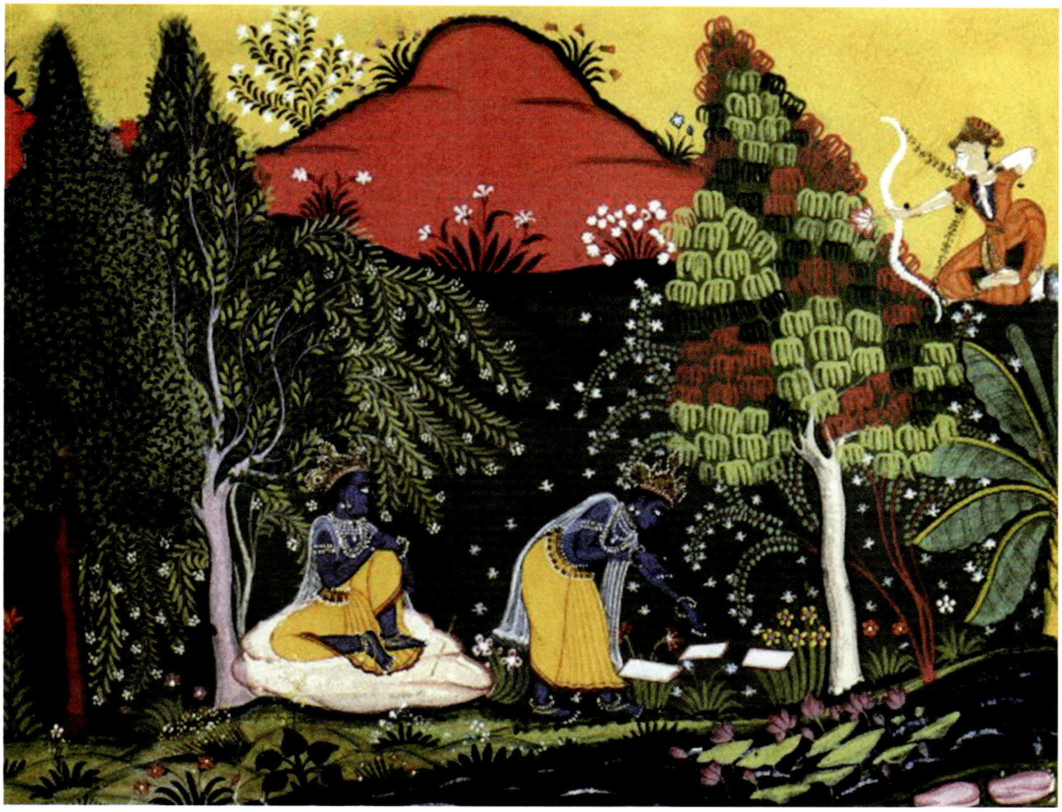

Above: 'Cupid Disturbs Krishna's Penance', Page from a *Gita Govinda* series, opaque watercolour and gold on paper, mid-17th century

Following pages: Golden champa or sonchampa flowers on a bed of yellow marigold flowers

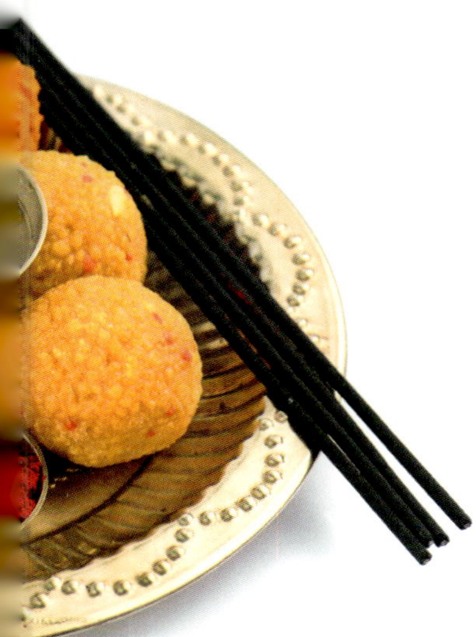

IN THE WRITTEN WORD

Mention must be made of the Sanskrit poetic tradition in which the *genda* or marigold made many appearances. A very well-known example is the Kashmiri poet Bilhana's *Chaurapanchasika*. This love poem can be translated to mean 'Black Marigolds' and speaks of the poet's own love for the king's daughter, written in her honour in Sanskrit. In medieval India's pinnacle of poetic achievement, the *Gita Govinda*, marigold is the colour of the dark Lord Krishna's attire as he waits for his lover, Radha. It is used to evoke and excite the *rasa* of *shringara*, or the essence of erotic love in the mind of the audience.

Urdu poetry has many echoes of the Persian flora and fauna and mentions trees that are found in Kashmir. However, we see the poet Mohammad Akmal Nazir includes many of the Indian flowers such as *motiya*, *chameli*, *genda* and birds and animals such as parrot (*tota*), myna, baya weaver bird, and squirrel (*gilehri*) in his poetry.

Turning to colonial times, India was an inspiration to many British writers and poets. Rudyard Kipling had found in the subcontinent a vast repository of ideas for novels such as *The Jungle Book*. In his writing, nature is a recurring theme, with the marigold featuring in many of his poems, for example *The Last Department*, a sceptical lament about administration in India, as well as in *Our Fathers of Old* and the *Undertaker's Horse*, which is ostensibly set in India.

In recent times, Kamala Das (later known as Kamala Surayya), epitomises contemporary poetic practices. Her works are replete with sexual urgency and passion, sensitive portrayals of the human mind and the feminine spirit. She employs the marigold as a symbol for the nurturing aspect of nature. So it has come to be that the marigold or *genda* has inspired many writers and poems through the ages and still continues to do so.

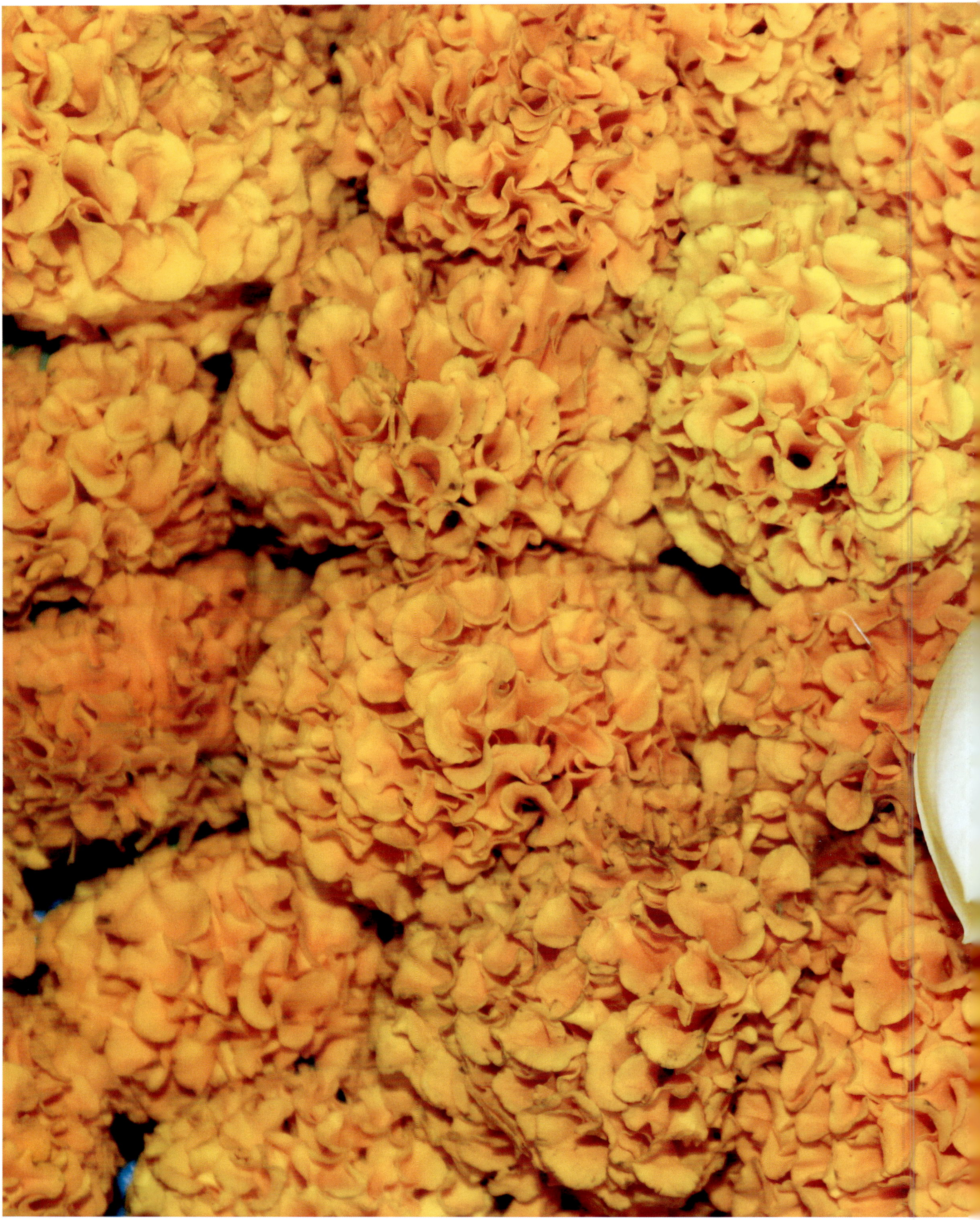

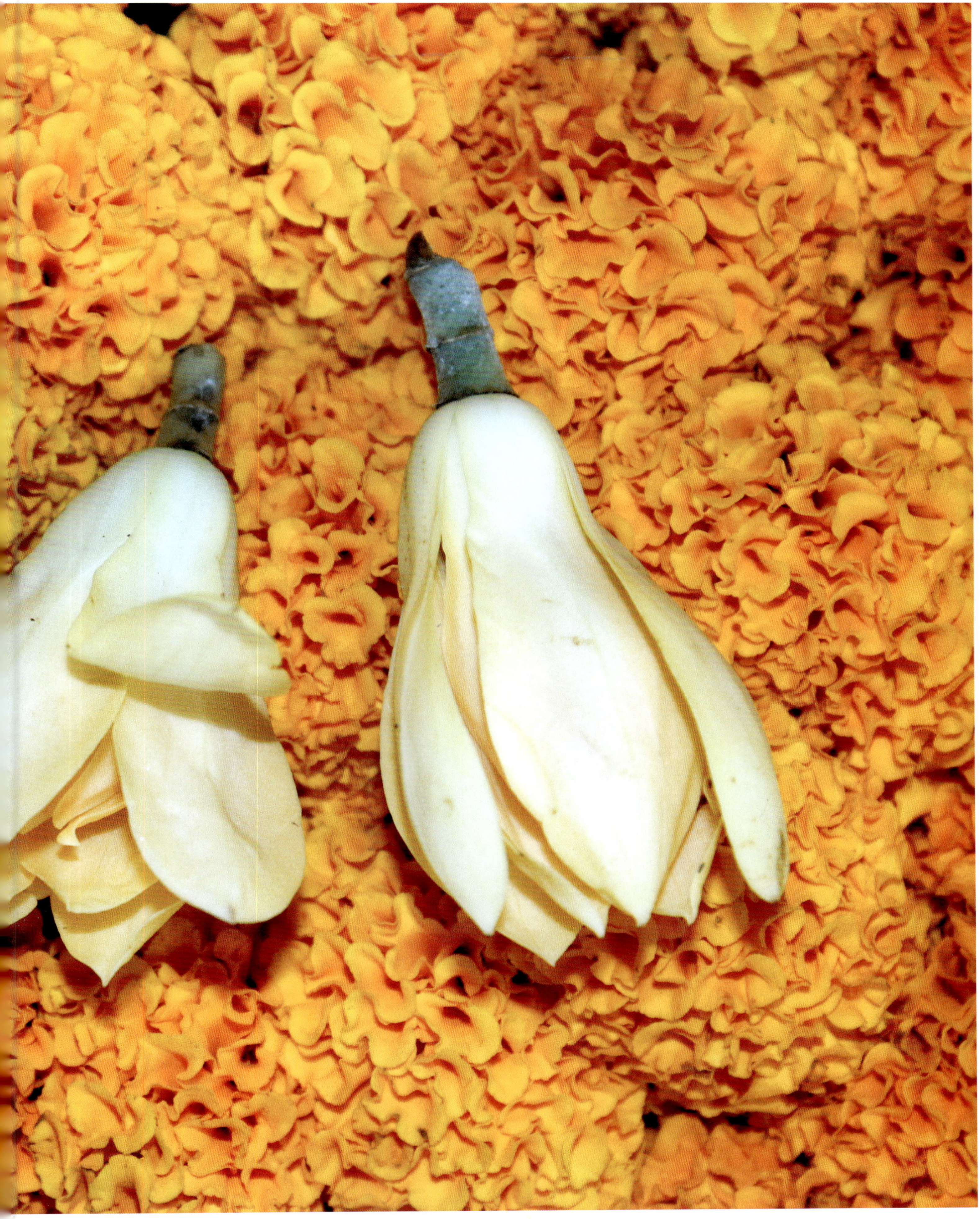

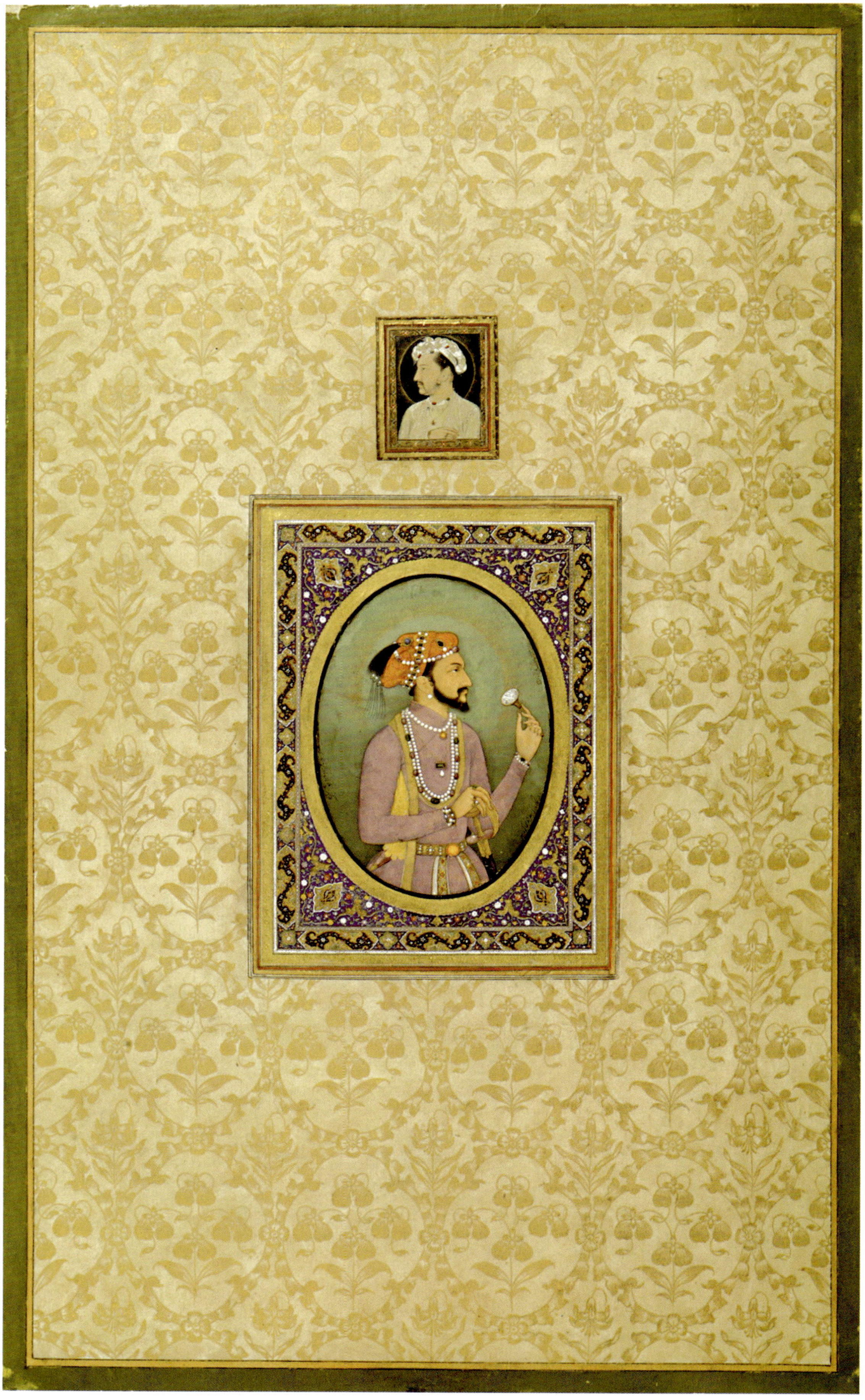

Facing page: 'Portraits of Jahangir and Shah Jahan', Abu'l Hasan, 1628, coll. Aga Khan Museum, Geneva; note the floral border and background of the paintings

TRIBAL ASSOCIATIONS

In the Munda tribe in Jharkhand, the youths usually establish friendships in the name of flowers. The name is not taken into account, instead it is the flower that is a reference to a friendship that is maintained throughout life. The friendship is normally with a person of the same sex and the friend is treated by the parents like their own son or daughter. In the friendship between those of the opposite gender, the following *Jadur* song shows how bees, as well as boys, are being drawn towards the girl decorated with flowers:

> *Genda phule jhaka-jumair goi*
> You are decked with genda flowers, O girl
>
> *Champa phule malak dair*
> Bunch of champa flowers is hanging down
>
> *Ham to alo tore love goi*
> Attracted by you we did come
>
> *Ham kaese birasal jab*
> How shall we forget you[2]

The flowers have a ritual importance for the tribals who collect them from the forest. If no flower is available in a season, it is substituted with twigs and leaves, thus they do not always form part of the offering to the Munda god.

Among the Central Indian tribe known as the Gonds, legend connects the origins of *genda* with the god of the Gonds, Gondmuli. According to legend, the God abducted the wife of another god, and in the ensuing battle, the other god emerged victorious after cutting off the head of Gondmuli. The wife wept for Gondmuli as she was dragged back by her earlier husband. She dropped her hairpin and a *genda* flower sprouted from the place it landed.

We see the bright orange and red colours of the flower everywhere in daily life in India. They are made into garlands and offered to honoured invitees and deities. The flowers stud cowdung balls that are used to decorate rice-powder drawings. In Eastern Indian folk art, the *genda* are dried and crushed to a powder form, to produce a yellow colour that is used to decorate village homes. Weaving *genda* garlands that are used for prayers and devotion and laying these at the feet of the elders while touching their feet is a further traditional sign of respect.

CHAMPA—A GIFT FROM HEAVEN

According to legend, the Buddha was born under the glorious golden-flowered boughs of a tall *champa* tree. Very often, *champa* or *champak* trees dot the landscape of Indian miniature paintings, particularly the Rajput and Pahari paintings. In miniature paintings, the *champa* tree is quite the favourite of divine beings and mortals and also a personal favourite of traditional artists as a tool of embellishment. The other name for the flower—frangipani—comes from the French *frangipanier,* meaning coagulated milk.

Champak (*Michelia champaca*) has been considered one of the three most beautiful and sacred of India's many flowering trees, along with *bakul* (*mimusops elengi*) and *parijat*. The ancient Rishis proclaimed that the gods brought these trees to earth as gifts from the heavens. Because of their many contributions to traditional societies, the trees were blessed with religious symbolism and divine qualities. Across the Indian subcontinent, the three trees are entwined in the lifestyle, culture and ancestral memories of the local people.

White *champa* came to India as a garden ornamental and is a popular plant particularly in the south of the country, the West Indies being its original home. In earlier days, the Bangaloo owners of India often used white *champa* to improve the exquisiteness of the garden.

The most fascinating part of the tree is its flowers which are not impressive but have an extremely aromatic smell. The fragrance has made *champa* flowers very popular in India from time immemorial. They are used in religious offerings in various parts of India. *Champa* was the Mughal Emperor Jahangir's favourite flower. He described it in the *Jahangirnama* saying that it had a fragrance that was increasingly sweet, but while its shape was like a saffron blossom it was 'yellow inclining to white. The tree is very symmetrical and large, full of branches and is shady. When in flower, one tree will perfume a garden.' Jahangir also said, 'From the point of view of herbs and fragrant flowers, India is preferable to anywhere else in the inhabited part of the world.'[3]

REJUVENATING VIGOUR

Champa flowers are popular in India for several purposes. They are primarily used for worship at temples and in homes and are also worn in the hair by women as ornaments and for their natural perfume. The flowers are floated in bowls of water to spread their fragrance around a room and are used to decorate bridal beds.

The *champa* tree offers various natural remedies that have been utilised for centuries, such as treating sore muscles, epilepsy and the common cold. All parts of the plant have effective medicinal properties and are used to treat ailments including gastritis, fever, constipation, bronchitis, cardiac

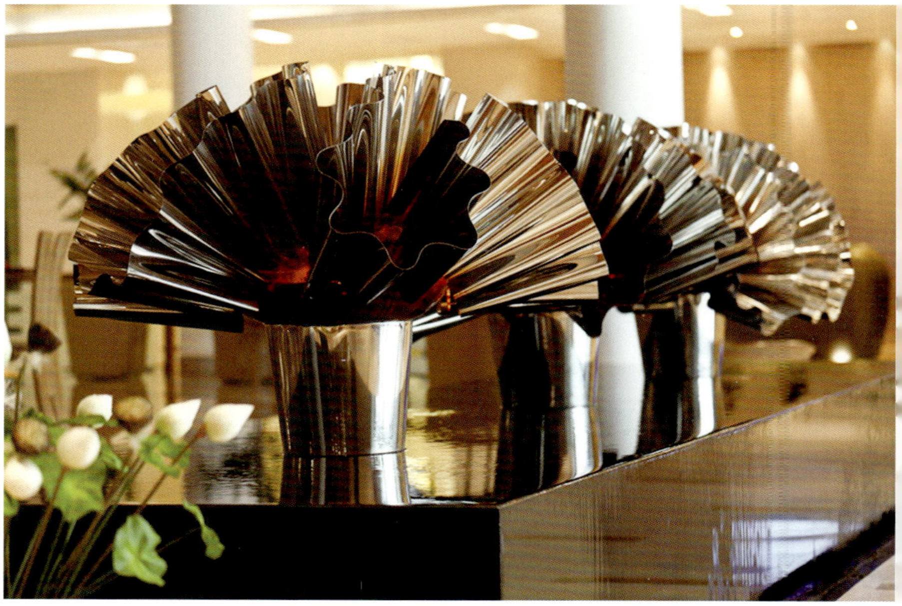

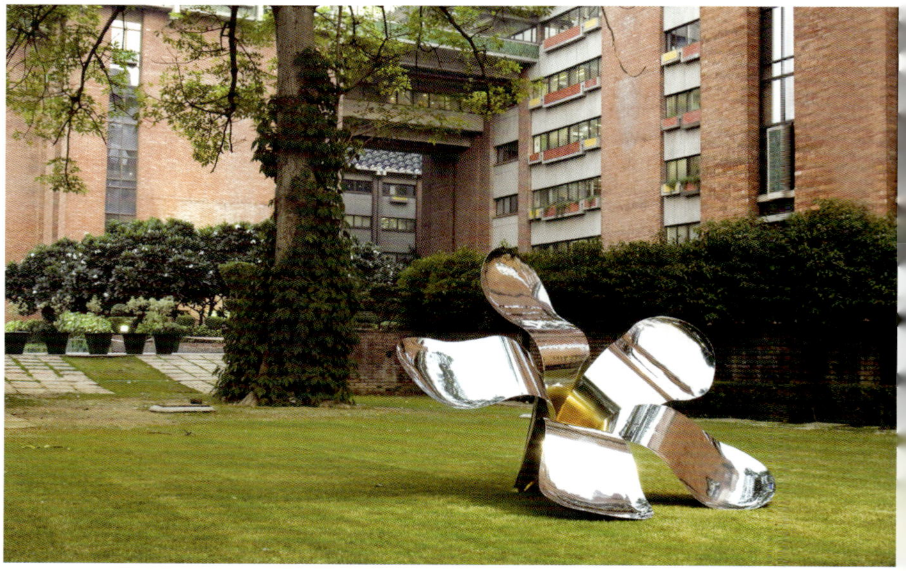

Top: Marigolds, stainless steel sculpture, Alex Davis studio

Above: Champa flower, stainless steel sculpture, Alex Davis studio

Facing page: 'Portrait of Jahangir holding a flower', ink, opaque watercolour, and gold on paper, ca 1604

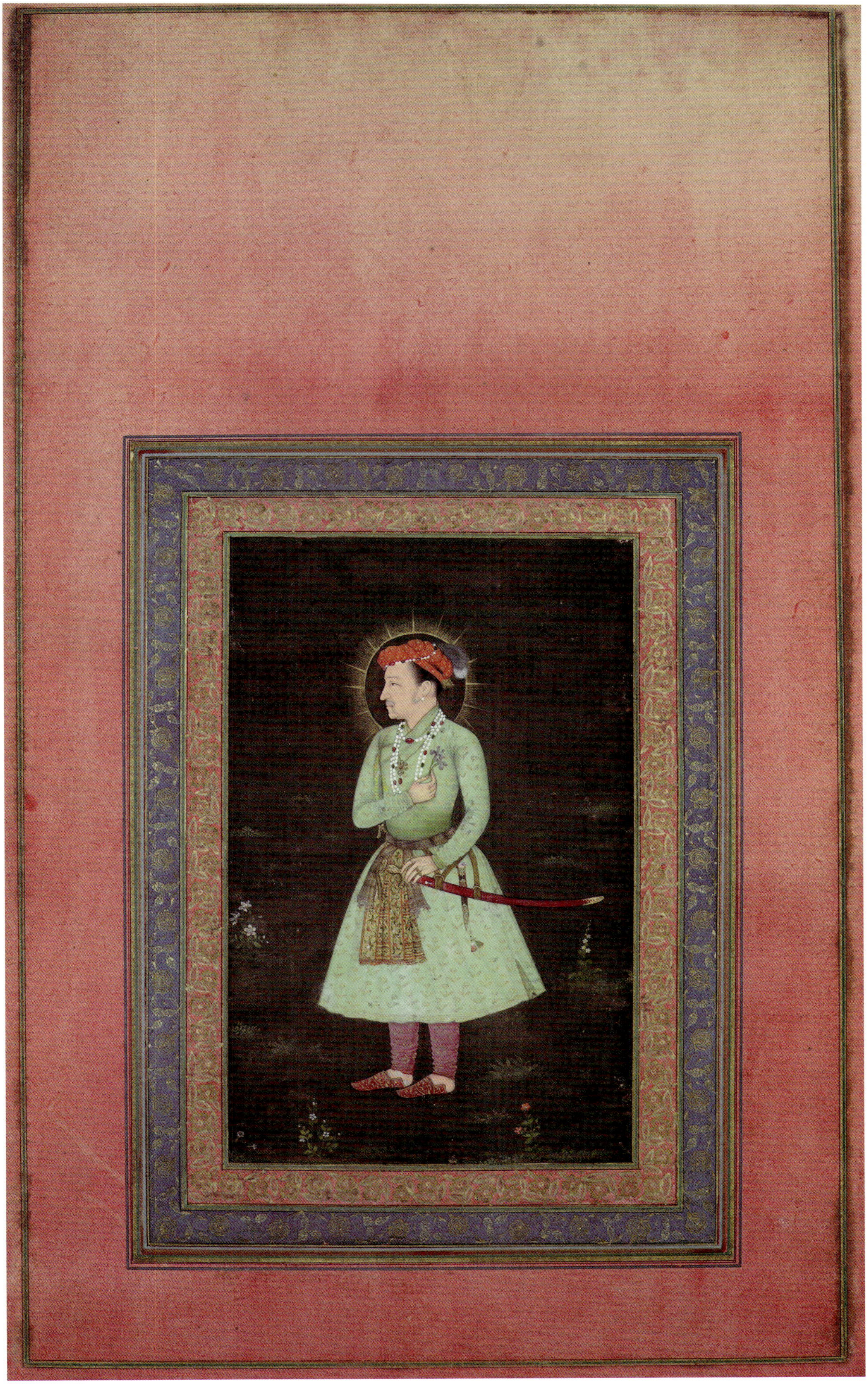

debility, ophthalmic disorders and urinary problems. But more than any cure for illness, it is the fragrance of the flower that has made it so beloved and prominent.

In traditional Ayurvedic research, the profound effect of aromas on the human consciousness has been documented. *Charaka Samhita*, the early text of Ayurveda, tells us that sweet fragrance brings out a youthful vigour and gives a long life; it is rejuvenating and boosts sensual enjoyment. Along with physical and mental recreation, it heightens alertness, enhances memory and restoration of immunity.

In perfumes, we can see the *champa* flower condensed into a base of sandalwood oil to create natural fragrances or *attar*. These floral essences, with their distinctly cooling nature, have been beneficial to lower the high temperature of both body and mind. A popular flower among women, due to its perfume-like qualities, *champa* is also used in hair oils and acts as a scalp coolant. The oil is highly valued in the international perfumery market and is used to make the very expensive perfume 'Joy'. However, the fragrance of the white-flowered *champa* does not last very long. *Champa* flowers are used as a base for other colours when dyeing silk and cotton fabrics. *Champa* leaves can be purified to make scented water and the wood of the *Champa* tree is used to build furniture, houses and carriages.

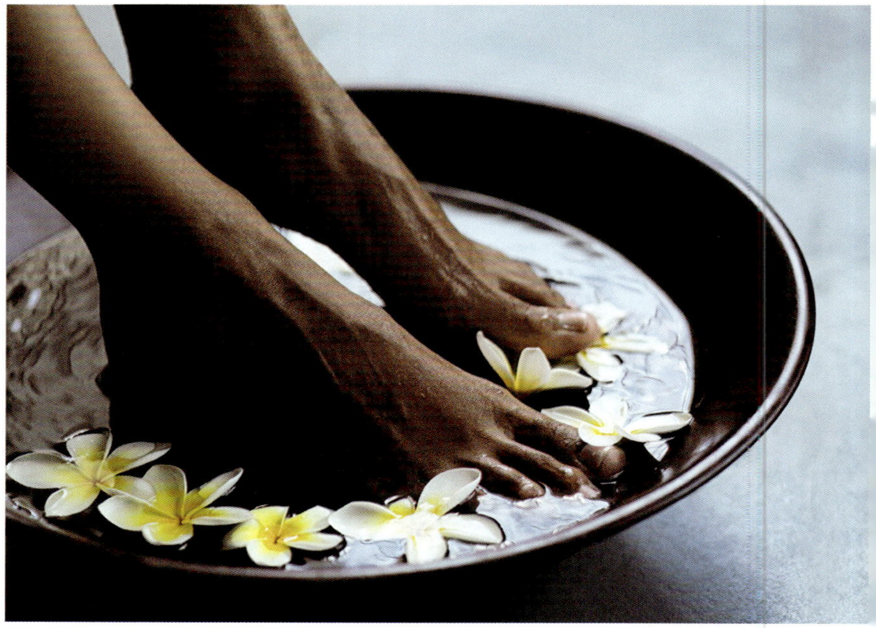

Above: A foot spa with fragrant *champa* flowers

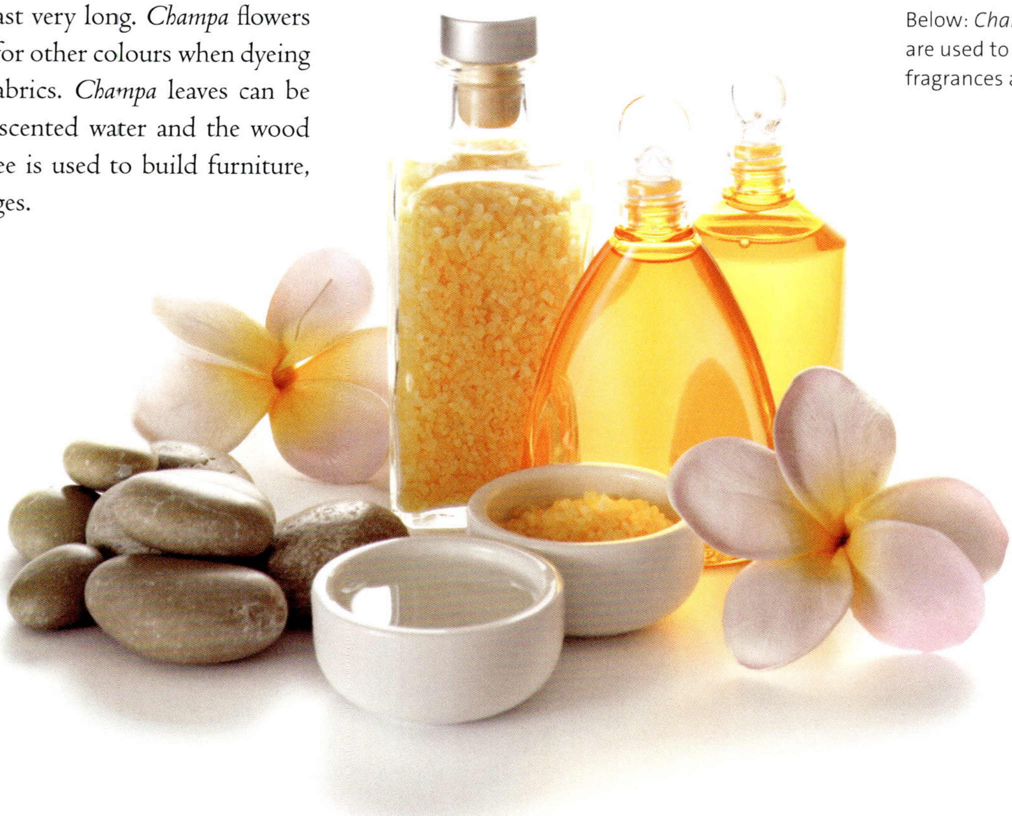

Below: *Champa* flowers are used to produce fragrances and oils

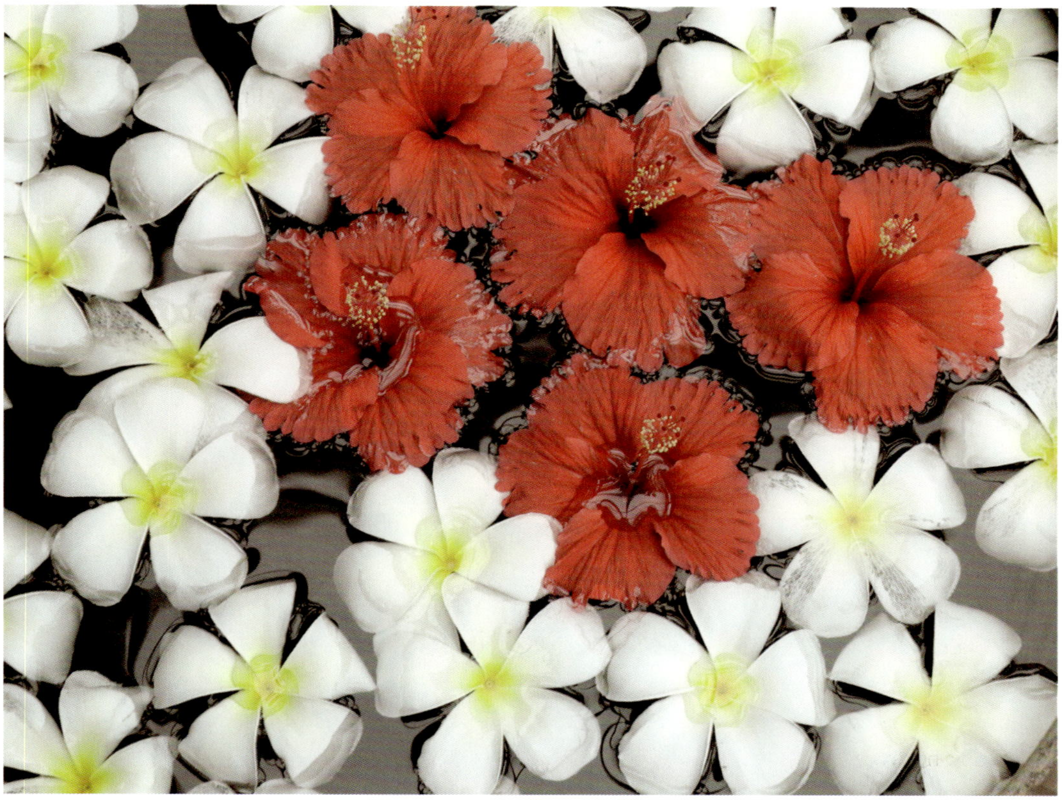

Red hibiscus and white *champa* make for a beauteous sight

SACRED SYMBOLISM

Associated with various deities and planted around the grounds of the ashrams and temples, it is easy to understand why the sages would point to the heavens as the source of these flowers. The Rishis maintained that heaven is wherever a beautiful tree, such as the *champa*, would blossom; a green universe that would split open, showing the way to a better world surrounded by living solar systems of a spiritually uplifting fragrance.

The yellow *champa* is one of the favourite trees grown in Hindu gardens; the exquisite flowers are used for prayers, especially ones offered to Lord Krishna. The glorious fragrant golden *champa* is often referred in the tribal folk tales and synonymous with goddess Lakshmi. In Orissa, some people believe Lakshmi loves the flower so she resides in it. In one folk tale, a daughter of a minister, through her devotion to the goddess Lakshmi, changed into a flowering gold *champa* tree and increased her family's wealth.

In Buddhism, Hinduism and Islam, the tree represents immortality due to its extraordinary ability to put forth leaves and blossoms even after it has been uprooted. The tree is often planted near temples and monasteries, as well as Muslim tombs, so the fresh flowers fall upon them. The tree in flower is a preferred motif in temple structures and images of the Buddha are carved from its wood.

In addition to being sacred to Krishna, *champa* is one of the five flower-darts of Kama Deva, the god of love. The tree is considered sacred to Kama Deva, so cutting it down brings bad luck.

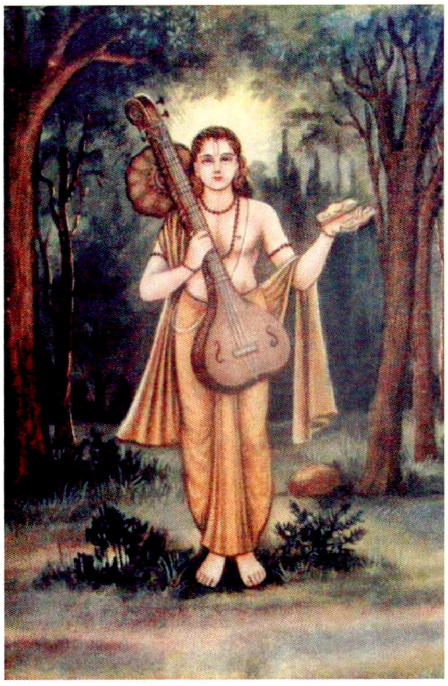

WHY SHIVA ISN'T OFFERED THE CHAMPA FLOWER

According to legend, a beautiful champa tree full of flowers stood near Shiva's temple in Gokarna. The sage Narada saw the tree on his way to the temple and admired it. He saw a Brahman one day who came to pluck the flowers from the same tree. When the Brahman saw Narada, he was reluctant to continue. Narada asked the Brahman where he was from and why he had stopped plucking the flowers. The Brahman replied that he was on his way to a village, but he continued to stand there to draw pleasure from the fragrance of the champa tree.

Narada went to the temple. Upon his return, he again saw the Brahman plucking the flowers; this time, he was hiding the flowers in a basket. Narada approached the tree and asked it whether anyone plucked its flowers. The tree answered, 'No.' But Narada shared his doubts, and went back to the temple and found that the Shiv lingam was concealed under the champa flowers. Narada spoke to a man meditating there who had offered the flowers to Shiva. The man told Narada that an evil Brahman came every day to shower the Shiv lingam with champa flowers. Shiva was pleased with this worship and blessed him. Thus the Brahman later turned powerful in the king's court and beleaguered the poor people.

Narada asked Shiva why he was helping an evil man. Shiva replied that he could not refute a devotee who worshipped him with flowers like champa. Narada went back to the tree and cursed it for not telling him the truth. He declared that champa flowers would never be recognised in Shiva worship. He also cursed the evil Brahman to be born as a demon that would only attain moksha when Lord Ram killed him.

The mother goddess Lalitambika wears *champa,* along with *ashoka* and *punnaaga* (magnolia) in her hair. In Hindu culture, the flower also signifies loyalty. Hindu women would put a flower in their hair on their wedding days to show their loyalty to their husbands.

In Indian folklore, the plants provide shelter to ghosts, demons and vampires. This is not surprising, considering how the leafless branches look in winter. *Champa* takes on a decidedly architectural look when all of its leaves and flowers vanish and the bare branches, reminiscent of winter in the north, strike a pose against the sky.

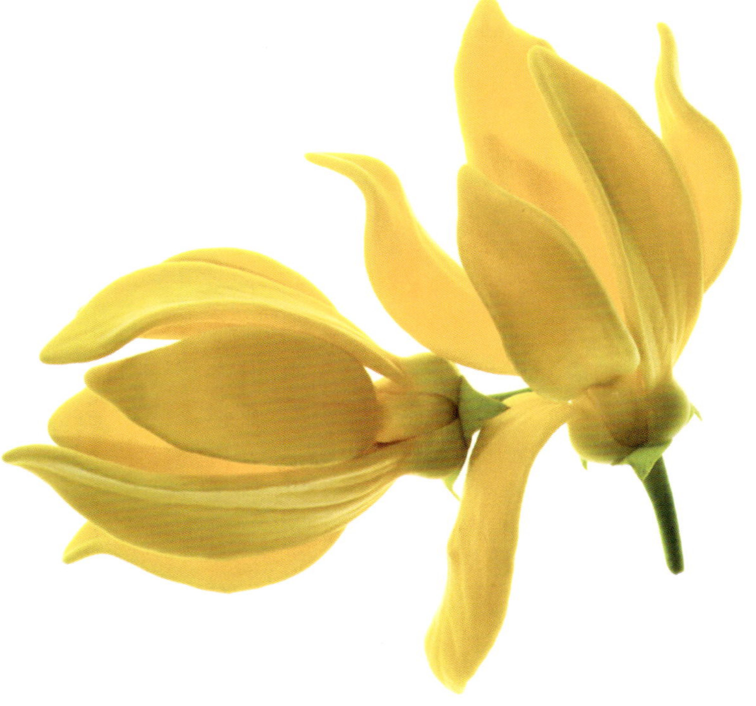

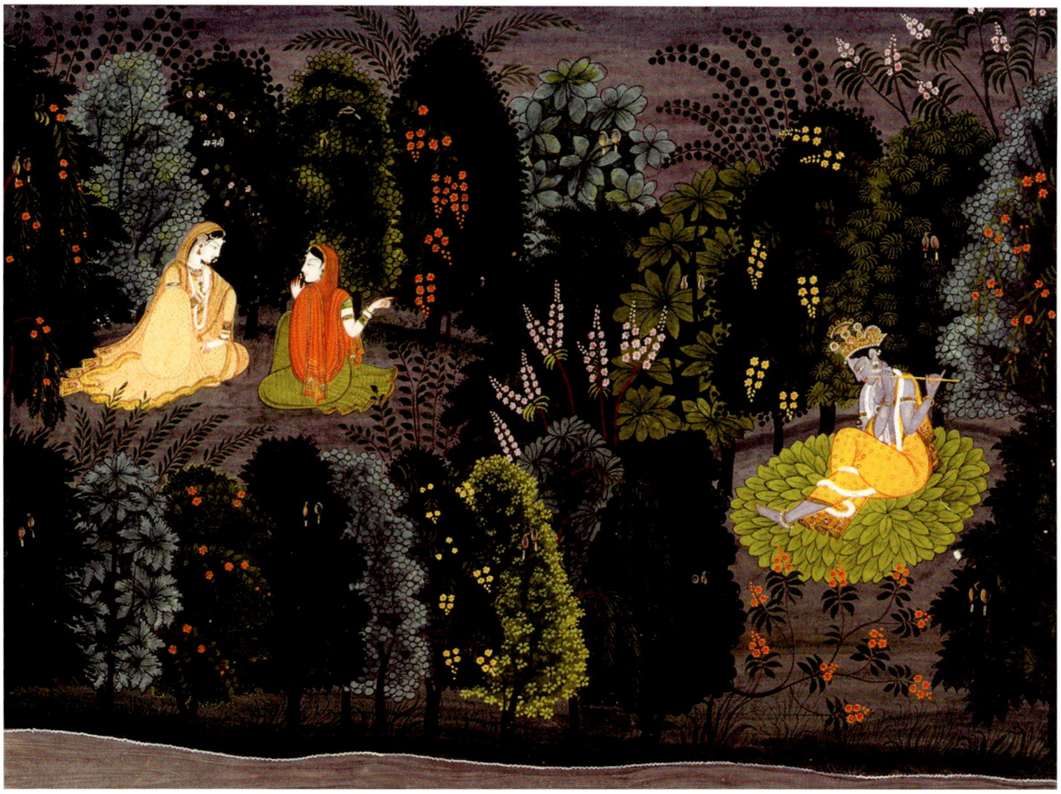

Above: 'Companion persuading Radha as Krishna plays the flute', *Gita Govinda*, Kangra, Himachal Pradesh, c. 1825

Facing page: Yellow champak flowers

LEITMOTIF OF LOVE

Tehi pura basata bharata binu raga
camcarika jimi campaka baga
rama bilasu rama anuragi
tajata bamana jimi jana barabhagi

Yet in that city dwelt Bharata free from all attachment, like a bee in a garden of *champa* flowers. (Though the *champa* bears a very sweet-scented flower, it is said that bees do not visit it.) The blessed souls who are devoted to Rama discard like vomit all Lakshmi's delights (sensual delights of wealth.)[4]

In the *Rasikapriya* of Keshavadas, there are various instances, persons and things that are illustrated further by giving examples from episodes of the love of Krishna and Radha. When describing Padmini, one of the classifications of women, Keshavadas says 'this daughter of Vrisabhanu (Radha) is like the bud of a champa flower'.

In the *Ramayana*, when Rama learns from the bird Jatayu that Sita has been abducted, he approaches a tree in his sorrow, and asks, 'Oh! Kind Compassion tree! Pray tell me where to seek for Sita? You are of pious nature, so I ask you. Why are you silent? If you say why I don't ask other trees, then hear why I don't do so: I have not asked the *champak* flower because he is an enemy of the supplicant, and he does not allow the nectar begging black bees to come near…[5]'

The *champa* is also an integral feature of medieval poetry. The Bhakti poets have incorporated it into their works as a device to denote beauty and love. The plethora of works illustrating Radha and Krishna in their passionate devotion to each other in the verdant abundance of Vrindavan, also feature the *champa* in blossom. This probably takes inspiration from classical Sanskrit poetry where the flower is a traditional leitmotif of love.

In most cases the artist endeavours to showcase the literary devices (similes, metaphors and other figures of speech) through a pictorial imagery that can translate the same into a visual language as closely as possible. Thus the Gita-Govinda remains an enduring subject of miniature paintings. In paintings from the 17th century Mewari *Gita Govinda,* trees, including the *champa,* are used to frame certain parts of the work, dividing the characters from each other, almost deliberately, as if signalling Krishna's significance as the object of worship of the *gopis* and other attendants. In this way, the trees almost take on a function of their own, seemingly performing this role with intent and purpose. In some paintings Kama, the god of love, and his consort Rati, hide themselves among the trees. Kama points his floral love arrows—of which one is a *champa* flower—at the divine coupling of Radha and Krishna.

The champa has made its presence felt in Rajput miniatures. The two schools where you see this flower depicted most often are the Rajput and the Pahari School. It stands not only for the demonstration of the ardour and natural loveliness inherent in the *nayak/nayika* relationship, but also as a visual metaphor for the *shringara rasa* or the erotic essence of a work of art.

BENGAL'S FAVOURITE

Rabindranath Tagore immortalised the blossom in one of his poems called *The Champa Flower*:

> 'Supposing I became a champa flower, just for fun, and grew on a branch high up that tree, and shook in the wind with laughter and danced upon the newly budded leaves, would you know me, mother?'
> 'You would call, "Child, where are you?" And I should laugh to myself and keep quite quiet. *I should slyly open my petals and watch you at your work.*'[6]

In Pranab Roy Choudhury's *Best Loved Folk Tales of India,* there is a famous Bengali story called 'Seven *Champa* Brothers and Sister Parul'. This

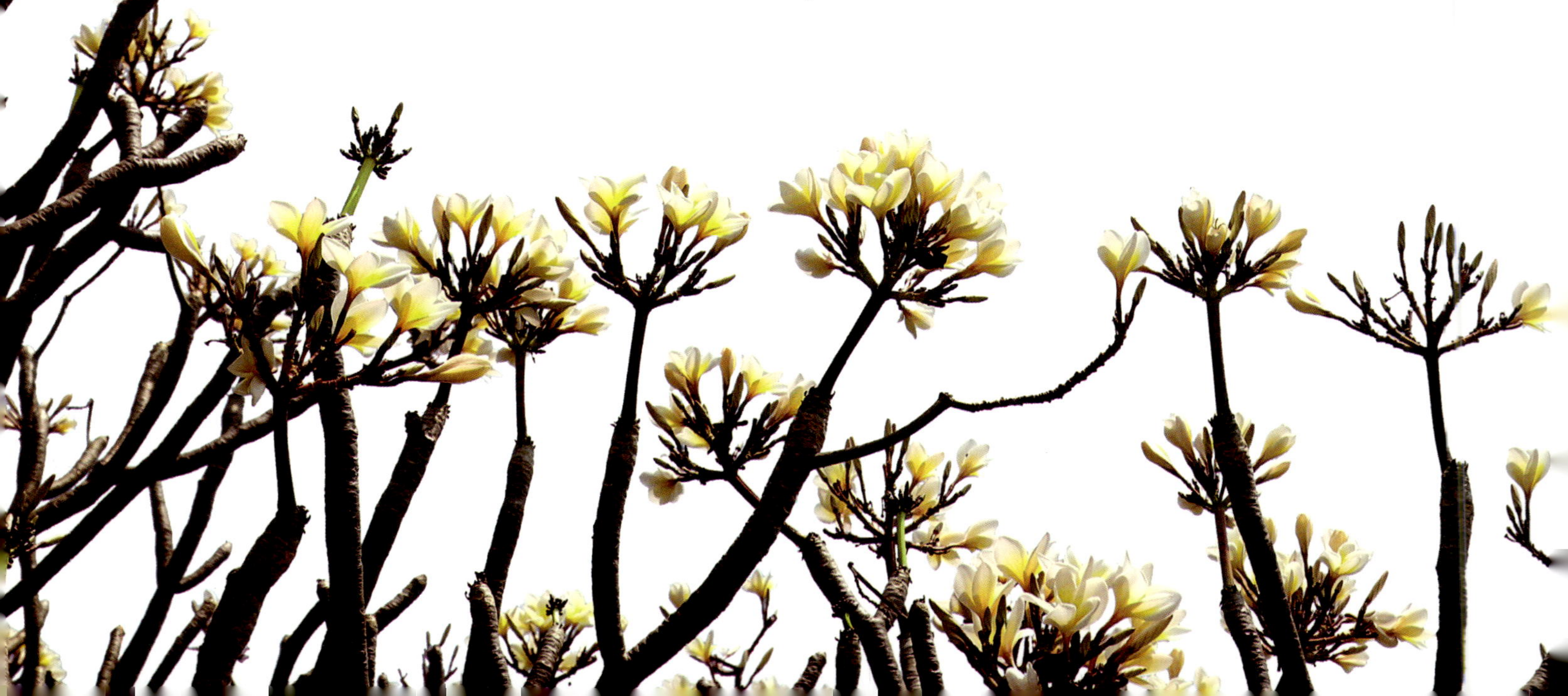

Champa flowers in full bloom

story of jealousy captures the emotions and beauty of the Bengali landscape. It tells the tale of a king who had seven wives, the first six of whom had an annoying air of superiority. The youngest queen, though, was sweet natured and a favourite in the royal household. The king loved her deeply and spent more time with her than the others. When she became pregnant, the six older queens became jealous because they had not borne any children. Between them, they conspired against their young rival who gave birth to seven handsome sons and one beautiful daughter. The elder queens killed the babies and buried them in the nearby garbage heap. Then they substituted the newborns with seven puppies and a kitten. The youngest queen was shocked to hear that she had given birth to animals. The king rushed to the harem and was shown the little animals. Aghast at the spectacle, he was told that the youngest queen was a witch. She was banished from the kingdom.

At the garbage heap, where the newborn babies were buried, seven *champa* trees and one *parul* creeper grew. The gardener working there saw the trees bloom with sweet-scented flowers. The flowers resisted anyone plucking them and asked for the banished queen who was now a beggar. When the queen arrived, she plucked the flowers and a boy emerged from each *champa* and a girl from the *parul*. The children were reunited with their mother. When the king discovered the truth, he ordered that the jealous queens be killed and lived happily ever after with his youngest wife and her children.

There is a story that the famous 19th century mystic, Sri Ramakrishna, plucked a *champa* blossom while taking a stroll under the fragrant flowering trees at the Kali temple in Kolkata. While looking at it in wonder and admiring the beauty of the flower, he saw the white sap that was coming out from the flower, just as blood would flow from a wound. He realised that he had just killed a living being and from then on, he instructed his followers to never pick flowers. If the flowers fall naturally to the ground, they could be considered as offerings from the trees, he said.

Thus, we see that the Indian cultural tradition is soaked in the significance of *genda* and *champa* flowers.

ENDNOTES

1. Allison Busch, *Poetry of Kings: The Classical Hindi Literature of Mughal India* (New York: Oxford University Press, 2011), p. 77.
2. Sem Topno, *Musical Culture of the Munda Tribe* (New Delhi: Concept Publishing Company, 2004), p. 313.
3. Eugenia W. Herbert, *Flora's Empire: British Gardens in India* (Philadelphia: University of Pennsylvania Press, 2012) p. 108.
4. Tulasidasa, *Shri Ramacharitamanasa* (New Delhi: Motilal Banarsidass Publishers, 1989) p. 458.
5. K.P. Bahadur, *Rasikapriya of Keshavadasa* (New Delhi: Motilal Banarsidass Publishers, 1990), p. xc.
6. Rabindranath Tagore, *The English Writings of Rabindranath Tagore, Volume 1: Poems* Introduction by Mohit K. Ray (Atlantic Publishers & Distributors, 2007), p. 118.

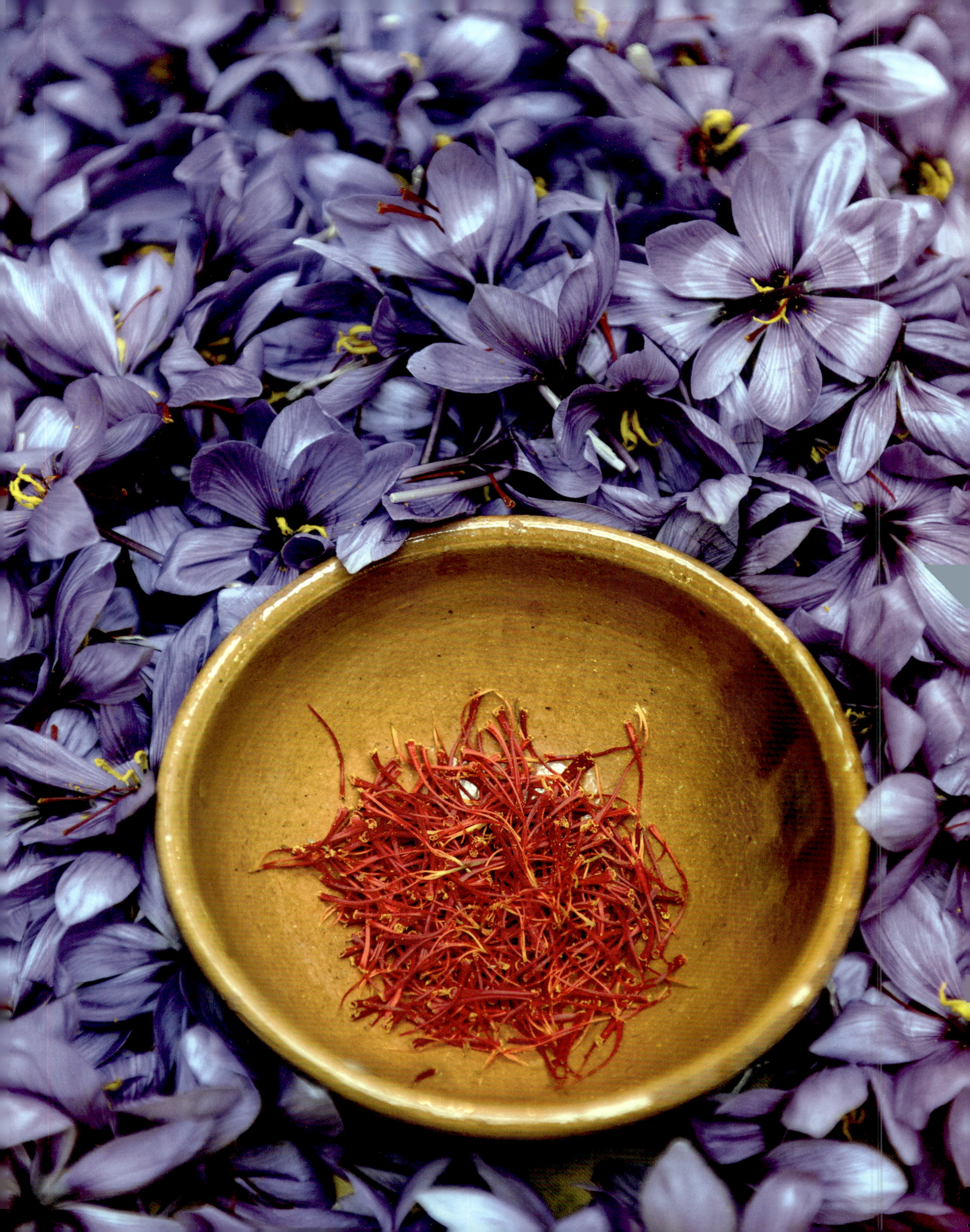

Delectable Blooms

Flowers always make people better, happier,
and more helpful,
they are sunshine, food and medicine to the soul.
—Luther Burbank

Flowers have always been a part of the Indian kitchen cupboard—as *kewra* essence (extracted from the pandanus plant) used to enhance the taste of *kheer* and Mughlai chicken curry, *zafran* (saffron) which is the prized possession of every gourmet cook, and rose water—often used in north Indian cuisine. Many cultures use flowers in their traditional cooking—squash blossoms in Italian food and rose petals in Indian food are just two examples. Adding flowers to your food is an easy way to add colour and flavour to certain dishes. Their range—from spicy to herbaceous and floral to fragrant—is surprisingly huge.

The *desi gulab* or the bright pink Indian rose is probably the most widely-used flower in the Indian kitchen. *Gulab ki kheer* or the rose-petal rice pudding—a gift from the Mughals to the Hindu kitchen—as well as *gulab ki kulfi* or rose ice cream, are all rich examples of the use of this flower in Indian food.

Some of the most refreshing Indian coolers use *kewra* essence that is made by distilling pandanus flowers. The pandanus is a perennial plant which grows wild in Asian countries. It is commonly referred to as screw pine. *Kewra* is widely used in Indian cuisine—from flavouring biryani and kebabs, to adding aroma to desserts and sherbets.

Facing page: Saffron pistils form a colourful contrast to lavender saffron flowers

Zafran or *kesar*, consists of deep orange stigmas of the *Crocus sativus* flower grown in Kashmir and costs as much as gold. It is added as a flavouring agent to the best of *shrikhand*, a yoghurt-based delicacy from Maharashtra, to the *kesar rabri* or saffron-flavoured Indian cream, to that very delicate foam as light as a whisper called *daulat ka chaat* made by few surviving *halwai*s in the traditional towns of Kanpur and Varanasi.

Floral essences in Indian cuisine became popular with what we call 'Mughlai' food—the cuisine that evolved in the courts of the Mughal emperors—a lovely example of the plurality of Indian culture. The love of flowers as an integral part of food, shared by both the indigenous people of India and the later Central Asian settlers, began with the Mughal emperor Babur, who made India his home in 1526 CE.

Using flowers in cooking can been traced back to the Roman Empire, when wine was flavoured with roses and violets, and earlier to the Ancient Greeks, who used flowers and herbs for medicinal purposes. The Greek physician Dioscorides in the first century CE invented a herbal remedy made from lavender flowers. Grandmothers in Crete drink an old-style mix of thyme, balsamic vinegar, and honey to stave off sore throats.

VILLAGE KITCHENS

The sub-Himalayan region of north India, comprising Himachal Pradesh, Uttarakhand and Garhwal, includes herbs and flowers as part of its indigenous cuisine. Due to extreme and long winters, vegetables throughout the year are far and few between. Wild edible plants thus become an important source of nutrition for people living in hilly or remote areas. In the Kangra District of Himachal Pradesh, for example, traditional dishes are made from the leaves of plants such as basil, fig, buckwheat and watercress. Nearby, in Uttarakhand, people use the flowers of the rhododendron tree which they call *baransh* in the local dialect to make squash and syrup. Korial, a local flower which grows in March, is consumed both by people and cattle. The stem of the flower is given to the cattle and the soft buds are made into a vegetable.

The cuisine of both Garhwal and Kumaon is highly nutritious, primarily based on green leaves and local cereals like *madua* (Finger millet) or buckwheat flower and *jhingoria* (rice made of the seed of a local plant). The stinging nettle is also used to make a particular dry green vegetable known as *shishunkey ka saag*.

In verdant areas, such as the Valley of Flowers, it's not surprising that flowers feature prominently in local cuisine. But it's not just in the northern mountainous regions that wild edible plants are so useful. If you venture into the

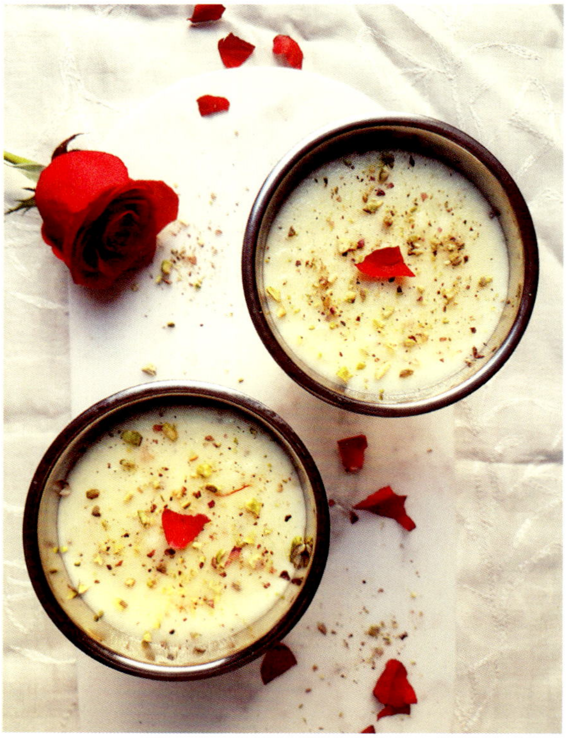

Phirni—saffron and pistachio flavoured rice pudding

Facing page, right: Kachnar ki kali is eaten in Garhwal, Uttarakhand

Facing page, below: Peacock flower (*Caesalpinia pulcherrima*) is an edible flower villages and forests of north Karnataka, you'll find people using the flowers of the *agasatibak* (swamp pea), *gulutura* (peacock flower), *tarwar* (tanner's cassia) and the *gulmohar* (flame tree) in their food. In fact, the popularity of these plants has spread and people living in the cities are also buying edible flowers sold by villagers. This is an encouraging trend because there is no accurate documentation of the vast folk knowledge on wild edible plants. And this knowledge is rapidly disappearing as villagers migrate to cities and lose touch with their traditional way of life.

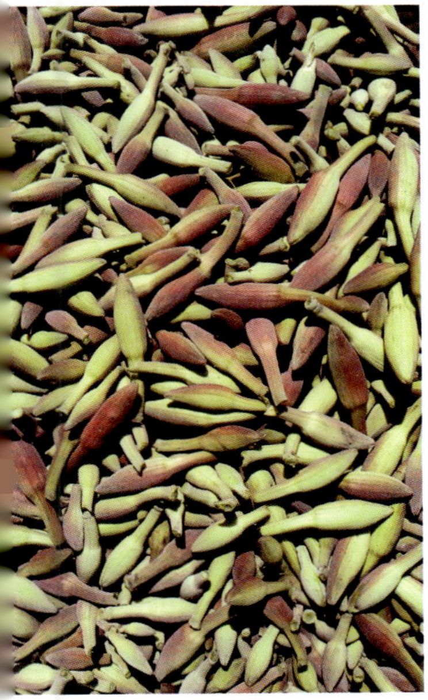

Wild edible plants, with their high vitamin and mineral content (potassium, calcium, phosphorous, magnesium, iron, etc) are essential for combating malnutrition in harsh environments. In addition, certain wild plants help in the regeneration of barren soil and are immune to many diseases. Unfortunately, the popularity of these wild plants has recently decreased. We must act quickly to preserve the diversity of wild edible plants and reap the benefits of their immense nutritional and medicinal wealth.

Traditional recipes are also in danger of dying out, thanks to the lure of fast food, so it's time to revive this unique cuisine. These recipes use not only the fruits or leaves of the plants, but also the buds, flowers, stems and roots. Here is a simple suggestion to get you started on the path to healthy, organic and nutritious food:

TIPS

Once you have identified which flowers are safe to eat, there are some tips on how to collect them. Edible flowers should be picked when they are fully open and in the cool of the day, after the dew has faded. If you are harvesting flowers from areas aside from your garden, sample several flowers before harvesting. Like any other produce, the taste of flowers grown in different locations can vary, depending on the soil type, fertilisation used and eco-friendly conditions. The time of year can also affect the taste of the flowers.

KARALEN

Ingredients

100 gms kachnar buds
A splash of mustard oil
¼ teaspoon coriander seeds
¼ teaspoon cumin seeds
¼ teaspoon turmeric powder
1 red chilli
1 green chilli
A pinch of asafoetida
Salt to taste

Method

Boil the kachnar buds for 10–15 minutes until they are soft. Heat 1 teaspoon of mustard oil in a pan. Add the coriander seeds, cumin seeds, red and green chillies, asafoetida and turmeric powder to the dish. Sauté for a few seconds until brown and add the boiled kachnar buds. Add salt to taste. Continue to cook over a low flame for 5 minutes. This dish is great with hot, buttered chapattis.

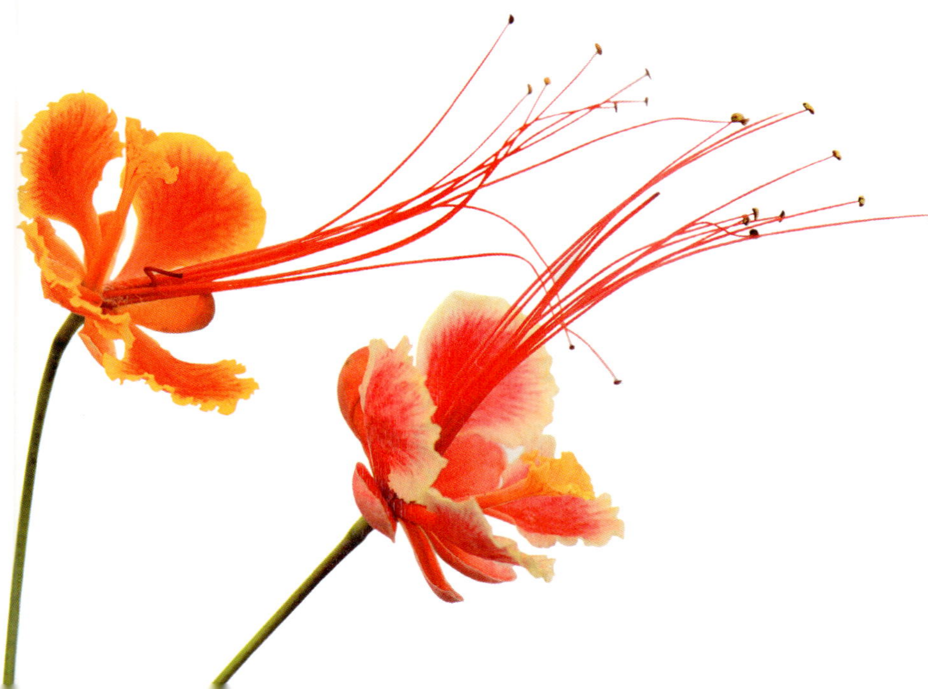

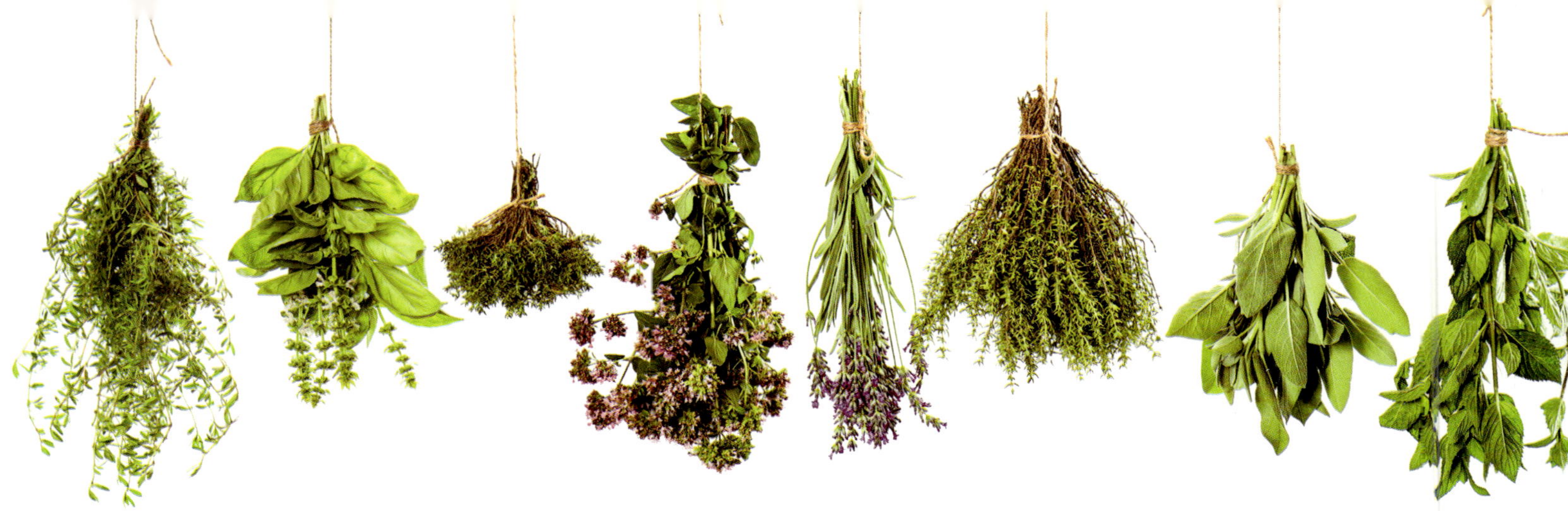

Bundles of fresh herbs—basil, sage, dill, thyme, mint, marjoram and lavender

MODERN-DAY MORSELS

These days many chefs are reviving the culinary art of using flower petals and blossoms to add both flavour and beauty to their menus. Growing edible flowers and using vegetable and culinary herb blossoms enlarges the pleasure of cooking from the garden. They add a delicious touch to everyday cooking and elegance to special-occasion dishes. It is not unusual to see flower petals used in salads, teas, as garnish for desserts, and they stimulate innovative uses as well—you can roll spicy flowers (like chive blossoms) into handmade pasta dough, combine fragrant ones into homemade ice cream, pickle flower buds to make ersatz capers, or use them to make a floral syrup for use in cocktails.

Edible flowers may be preserved for future use with methods like drying, freezing or steeping in oil. They can be used in drinks, jellies, salads, soups, syrups and main dishes. Flower-flavoured oils and vinegars are made by steeping edible flower petals in these liquids. Candied flowers are crystallised using egg white and sugar.

There are several benefits to cooking with flowers. Most are rich in nectar and pollen, and some, like rose hips, are very high in vitamin C. Marigolds and nasturtiums also contain vitamin C, while dandelions have vitamins A and C. As an added bonus, flowers are almost calorie-free. The blossoms of dill, arugula, thyme, fennel, oregano, basil and garlic chives are all attractive and savoury when scattered over salads, soups or open-faced sandwiches. Allowing your herbs flowers is noble gardening practice as well; most blossoms entice beneficial insects and pollinators.

So you may be an expert at making rose sherbet and lotus stem chips (*kamal kadki* or *bhee kachri*). And you have been using saffron in your sweets for years and have cooked with onion, drumstick and banana flowers. Well, consider branching out to include garden flowers and/or imported delights such as zucchini in your kitchen. The blossoms of your culinary herb plants can be treasures in the kitchen. There are several types of common garden flowers that are both edible and palatable. Some of the more popular edible flowers include hibiscus, nasturtium, sage, marigold, chive flowers, mint, rose and violet.

THE ROSE IN THE KITCHEN

The rose is one of the most delightful flowers to use as a secret or main ingredient in the kitchen. With its vibrant blooms, velvety petals and primeval perfume, the rose exceeds any floral status and springs from the realm of the divine. Roses are the most beloved of all cultivated flowers. Apart from its decorative qualities, coupled with aroma and medicinal uses, the rose is one of the most popular edible

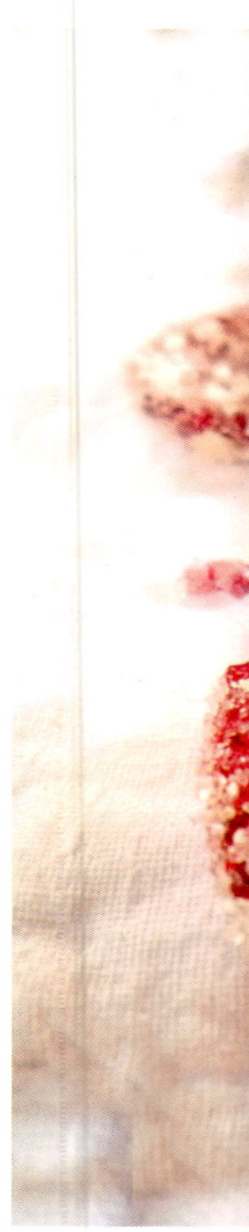

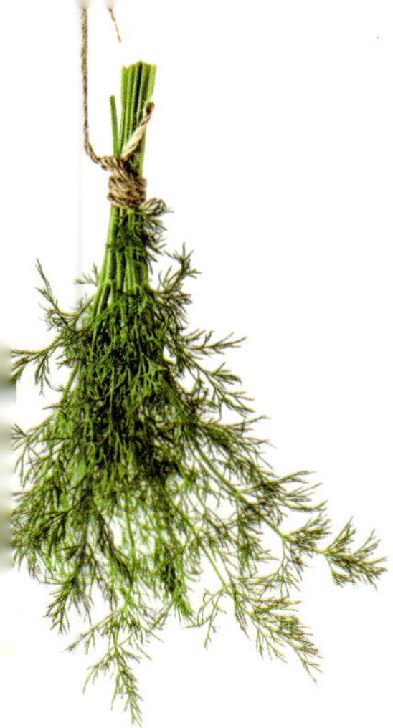

CHOOSE WITH CARE

There are some precautions you must take before you begin using flowers in the kitchen. The most important rule to follow is to choose only varieties that you're completely sure are edible While cooking, it is wise to use flowers from your own garden or those grown organically. You must also be aware that while some flowers are not toxic, they might cause irritation or allergy in some people. If anyone suffers from allergies or asthma, eating pollen might trigger a reaction. Also, keep in mind that while a flower might be edible, it might not be delicious. For example, not all varieties of marigold are tasty. There are several dangerous plants, so please be careful before you start cooking with flowers. Certain flowers are outright poisonous, such as azaleas, calla lilies, castor beans, crocus, daffodils, datura/angel's trumpet, foxglove, larkspur, lily of the valley, lupine and sweet pea.

Sugar-candied rose petals

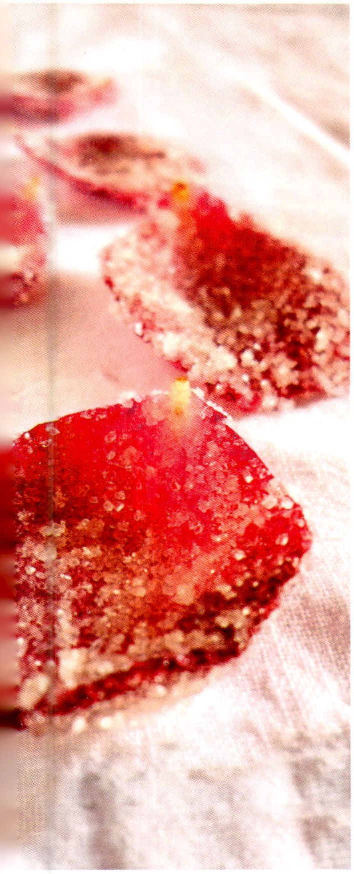

flowers. The rose is considered as a cooling stomach settler and a digestive aid, as well as an antioxidant, tonic, mild laxative; and is good for memory and eyesight.

Roses have added their rich scented perfumes to confections, casseroles and beverages in Europe, Central Asia and Asia for centuries. The aroma and flavour of roses vary, ranging from faint to richly sweet. All roses are comestible, with true rose taste and they are usually sweet, with differing undertones of apple, cinnamon, orange or lemon. Older diversities have more flavour than fresher hybrids. It is believed the stronger the fragrance, the stronger the flavour. It is important to smell and taste roses before using them, to check which are the most appealing to your palate.

It is important to find how the roses were grown, and never to use florist-grown roses that may have been treated with chemicals. To use in the kitchen, rinse them well, then shake off the excess water or spin them gently in a salad spinner. This way, rose petals can be used to make jelly and jam, vinegar, flavoured honey, syrup and tea, or simply as a garnish for cakes, salads and desserts.

CANDIED ROSE PETALS

Ingredients

15 Rose petals
2 Egg white
1 tablespoon Granulated sugar

Method

Rose petals must be dry and clean. Dip or brush both sides in whipped white of egg, and then coat both sides of the petals with granulated sugar. Lay carefully on waxed paper and allow to dry thoroughly before packing in boxes. To hasten drying, turn the petals once. They will be preserved for a year.

ROSE HONEY JELLY

Ingredients

3 cups strained honey
½ bottle Fruit pectin
1 teaspoon extract of roses

Method

Mix three cups of strained honey with one cup of hot water and bring to a quick boil. Add 1/2 bottle of fruit pectin. Bring to a full, rolling boil and remove from the flame. Skim the mixture, add one teaspoon of extract of roses, stir thoroughly and pour quickly into sterilised jelly glasses and seal. This makes five glasses of jelly.

ROSE HONEY

Ingredients

1 lb. / 0.45 kg of trimmed rose petals
1 pint / 0.56 litre of water
2 lbs. / 0.45 kg of strained honey

Method

Cut the white base of the sweet-scented rose petals. Crush the trimmed petals with a wooden masher. Boil in water for 15 minutes. Strain and add the honey. Boil it down till it becomes a thick syrup. Pour into scalded glasses and seal. Add a washed rose leaf and cover with a thin layer of paraffin.

ROSE ICE CUBES

Ingredients

2 cups water
1/2 cup of sugar
1/2 cup of finely-minced rose leaves
3 lemons

Method

Boil 2 cups of water and 1/2 cup of sugar for five minutes. Pour over 1/2 cup of finely-minced rose leaves. Add the juice of 3 lemons and strain. Freeze them in an ice tray and use the cubes in fruit drinks and herb tea.

ROSE WATER AND ESSENTIAL OIL

Ingredients

Rose petals
Ice

Method

Place a clean brick in a large pot. Fill the space around the brick with rose petals and pour just enough water to cover the petals. Place a small glass dish on top of the brick. Cover the pot and place it on a burner. Place a stainless steel bowl with ice on top of the pot. Simmer about 3 hours, depending on the number of petals used, and replenish the ice as needed. The ice will condense the steam within the pot, causing it to drip into the glass dish.

The water in the glass dish is the rose water. There will be a layer of oil on top of the rose water—this is the essential oil, which can be separated from the water and used in potpourri, soap or lotion.

GULKAND SHAKE

Gulkand is an all-time favourite with kids

Ingredients

3 cups ultra-toned milk
4 tablespoons of Gulkand (*see suggestion below*)
2 tablespoons of rose syrup
1 cup crushed ice
2 to 3 peppercorns
For the garnish: a few chopped rose petals

Facing page: Gulkand is prepared by soaking rose petals in sugar

Below: Frozen rose ice cubes

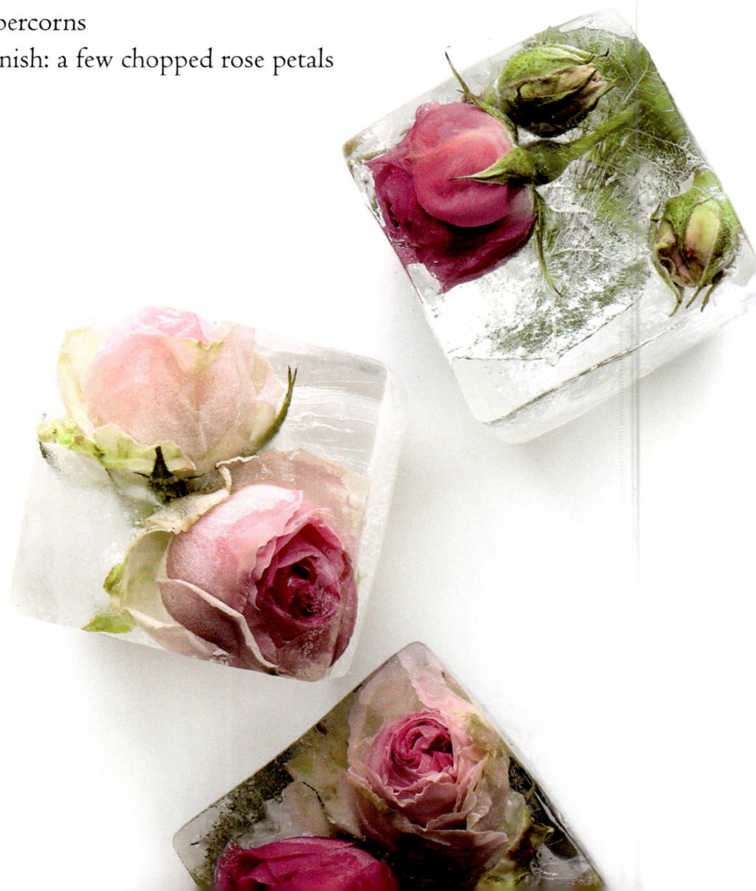

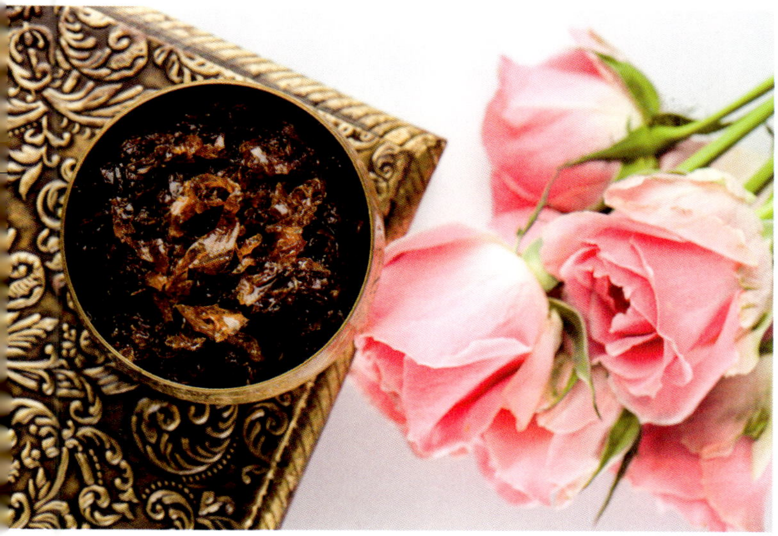

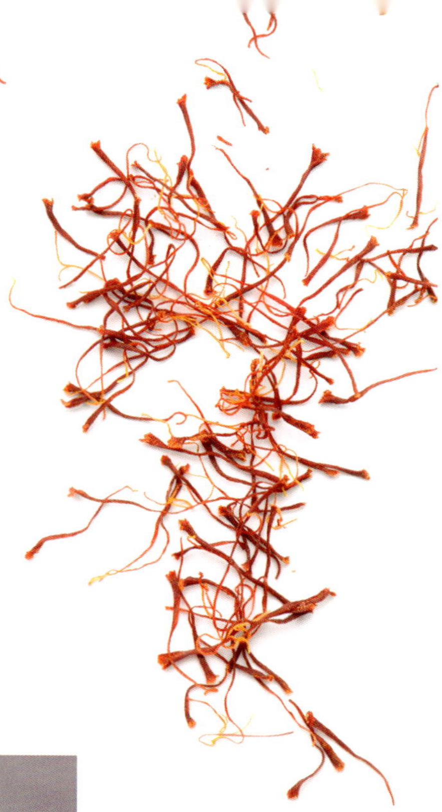

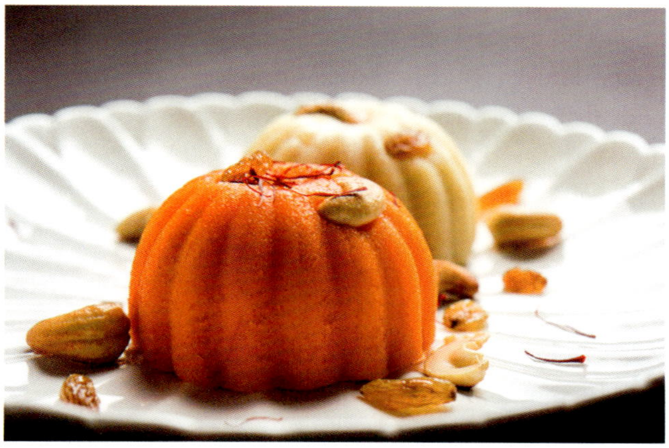

Top right: saffron strands

Above: Rawa kesari

Method

Combine all the ingredients together and blend in a mixer till smooth and frothy. Pour an equal quantity of the drink into four individual glasses and serve immediately, garnished with rose petals.

Handy Tip: Gulkand is made by soaking rose petals in sugar. It is a coolant and tastes great when used in ice creams, milkshakes and other desserts. Store in an airtight container or in the refrigerator.

ROSE PETAL TEA

Rose petals are available at specialist tea stores and if you have the time, you can also dry your own rose petals. It is also thought to have a cooling effect on the body and reduce fevers along with associated rashes. A mild diuretic, rose petal tea is recommended as a digestive tonic to restore balance in the gut.

SAFFRON, THE MAGICAL HERB

Frequently called the magical herb, saffron has a wealth of colouring uses in addition to its culinary values. In pre-modern India, upper class citizens would dye their clothes with saffron because of its warm, natural fragrance. It was soon discovered that these clothes could also last for hundreds of years because the saffron protected the cloth from insect damage.

When it comes to cooking with saffron, the options are endless. You could blend it into tea, as has been done for centuries in royal Arab kitchens. Or you could add a pinch, with a little honey or sugar, to a glass of milk. If you want something special for dessert, especially during the festive holidays, here's a recipe:

RAVA KESARI

Ingredients

6–7 cardamoms
½ teaspoon saffron threads
½ cup ghee
I cup semolina
10–12 cashew nuts
I cup sugar
I cup water

Method

Crush the cardamoms to a semi-fine powder using a mortar and pestle. Set aside.

Next, crush the saffron threads with the mortar and pestle.

In a separate pan, bring the sugar and the water to a boil. Shake in the powdered cardamom and saffron and reduce the heat to a simmer.

In a pan over a low flame, add the ghee. Once the ghee is heated, gradually add the semolina, while continuing to stir. Throw in the cashews, broken in half, and keep stirring for 7-8 minutes. When the semolina grains change colour and the cashews become a light golden hue, increase the flame under the pan of sugar and water. Once it begins to bubble, add the hot sugar and water mixture to the semolina.

Continue stirring the whole mixture as it begins to thicken and pulls away from the edge of the pan. Cover the pan with a tightly fitting lid or place a heavy weight on top of the lid.

Turn off the flame and allow the contents of the pan to steam for 5–6 minutes. Remove the lid and serve the *rava kesari*.

KEWRA SYRUP

Think of kewra essence as the vanilla of Indian desserts! If you feel like having a nice sherbet but want something besides rose, then try kewra.

Ingredients

2¼ cups of water
4 cups of sugar
¼ cup of fresh lime juice
I¼ teaspoon of kewra essence

Method

Mix water with sugar in a heavy bottom pan. Heat till all the sugar is dissolved.

Add lime juice and bring to a boil. Turn off heat and let the mixture cool.

Stir in kewra essence and put in a bottle.

To serve, mix one part kewra syrup and 3 parts water. Serve over ice.

CONFETTI OF COLOUR

Every kitchen garden should include showy and flavourful nasturtiums. They come in a wide range of saucy colours, and both the leaves and flowers are delicious, with a peppery flavour. Chop the petals and add them to rice, potatoes, pasta or green salads. Mix the petals into cheese or sandwich spreads. Nasturtium flowers blend well with seafood; add them to crab or shrimp salads and use them to garnish grilled or poached salmon. The flowers also can be steeped in mild rice or cider vinegar to add a bit of spice. You can also pickle the flower buds as a substitute for capers. Different varieties offer leaf colours including dark green, blue-green and even variegated green and white, all with pretty lily-pad-shaped leaves.

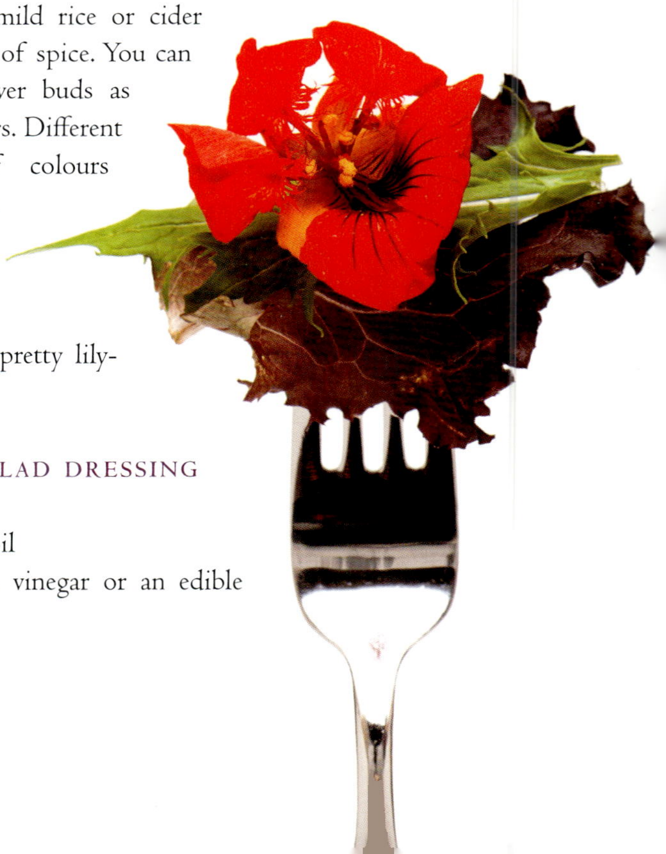

Red nasturtium flower salad

NASTURTIUM SALAD DRESSING

Ingredients

¾ cup of vegetable oil
¼ cup of red wine vinegar or an edible flower vinegar

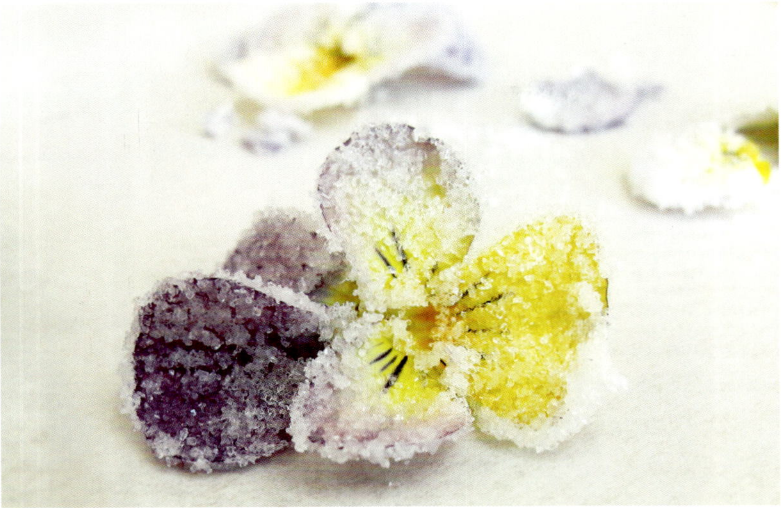

Candied pansy

3-4 garlic cloves, minced
½ cup nasturtium petals
¼ cup snipped chives and florets
salt and pepper to taste

Method
Whisk together the oil, vinegar and garlic cloves. Add the petals and chives. Gently blend and add salt and pepper. Serve over salad greens (iceberg lettuce and rocket, for example), garnished with nasturtium flowers.

SUN-KISSED MARIGOLDS

Marigolds add a bright touch of colour to vegetables, pastas, rice dishes and garden salads. The petals offer a delicate flavour and colour to both egg and cheese dishes. Marigolds adds a hint of saffron flavour to rice dishes and breads. You can also make delicious marigold vinegar by taking two cups of sun-dried petals and adding one teaspoon each of sea salt and sugar and some white wine vinegar. Seal the bottle and let it steep for two weeks under a shade. When ready, use it freely in salads and dressings. The vinegar will last for about six months.

PANSIES AND VIOLETS

Both these flowers have a slightly sweet taste, although the flavour varies among the different varieties. They are gorgeous in salads and as a garnish, and can be added to stuffing for fish and poultry. You could also try freezing them in ice cubes for a welcome burst of colour to add to your drinks. The crystallised petals can be used to decorate cakes and candies. Here's a simple way to crystallise the flowers:

CRYSTALLISED PANSIES AND VIOLETS

Ingredients
15 pansies and violets
2 egg white
1 tablespoon granulated sugar

Method
Select clean, dry flowers. Keep a length of the stem to help you hold the flower—you can cut this off later.

Whisk 1 egg white until stiff.

With a small brush, paint the flower petals with the egg white mixture. Make sure to cover the petals completely.

Gently sprinkle the painted petals with finely ground, granulated sugar (superfine sugar or bartender's sugar).

When the flowers are completely sugared, place them on a rack covered with parchment paper. Move the flowers once in a while so that they don't stick.

Place the flowers on a baking sheet and put into an oven at low heat—approximately 65° C—and keep the oven door open a little.

It will take 10 to 36 hours for the flowers to dry. Once they are dry, place them in a sealed, dry container. Some varieties will last for two days; others will keep for several months.

LUSCIOUS LAVENDER

Lavender buds taste herbal and sweet and are great for flavouring macaroons, mousses, biscuits, scones, shortbread and sugar. If you want something that will add a kick to salad dressing, mix a handful of dried lavender flowers and slightly bruised mint leaves to a jam jar. Add good olive oil and let it infuse for a week. Strain the oil and transfer it to a bottle. For a final flourish, add a fresh lavender stalk. This delicious oil can be used on salads or eaten plain with bread and balsamic vinegar.

CHEEKY CHIVE BLOSSOMS

These spiky lavender-pink blossoms shining above the grassy green leaves, can be sprinkled over omelettes or scrambled eggs, potatoes, rice or cooked vegetables. Toss chopped onion flowers with boiled baby potatoes. Add it in a dressing made from mustard, lemon juice, salt, pepper and a pinch of red chillies.

CHIVE BLOSSOM BUTTER

Ingredients

2 tablespoons of chive flowers
½ cup of unsalted butter
I tablespoon of chive leaves
I teaspoon of salt

Method

Chop up the chive flowers either roughly or finely. In a small bowl, mash up the unsalted butter until it is fairly soft. Slowly add the chive flowers and the chive leaves. Mix in one teaspoon of salt. Cover and keep in the fridge. Use the butter to season cooked vegetables, such as peas or carrots.

VEGETABLE DELIGHTS

The flowers of various vegetables provide a singular flavour. Vegetable flowers of broccoli in salads or stir-fry are delicious—they add a lightly spicy, broccoli taste. Radish flowers, meanwhile, are more pungent. Squash and pumpkin flowers can be used as a garnish and can also be stuffed, fried or sautéed. Zucchini flowers or squash blossoms are delicate buds. Picked when slightly opened, they can be fried like *bhajiyas*, reminiscent of what our grandmothers used to make. They can be stuffed with paneer for a delicious change of pace. Squash blossoms can be baked in a pie or provide an interesting ingredient in pesto—simply replace half of the basil in the pesto recipe with the squash blossoms. Roast a handful of pine nuts, walnuts or almonds in a skillet. After they've cooled, pulverise them in a food processor with basil, squash blossoms, a handful full of garlic cloves, grated parmesan cheese and salt and pepper to taste. Gradually drizzle in olive oil as you mix the ingredients.

FLOWERY INFUSIONS

There is a whole host of flavours to choose from with regard to floral infusions. Flowers used in herbals teas or tisanes can be soothing for a number of ailments, while also serving the dual purpose of pleasing the taste buds. The delicate flavours of the dried flowers from rhododendron to rose to chamomile combined with the relaxing properties of the infusion makes them a terrific herbal cure for problems such as upset stomachs, insomnia, congestion, stress and more.

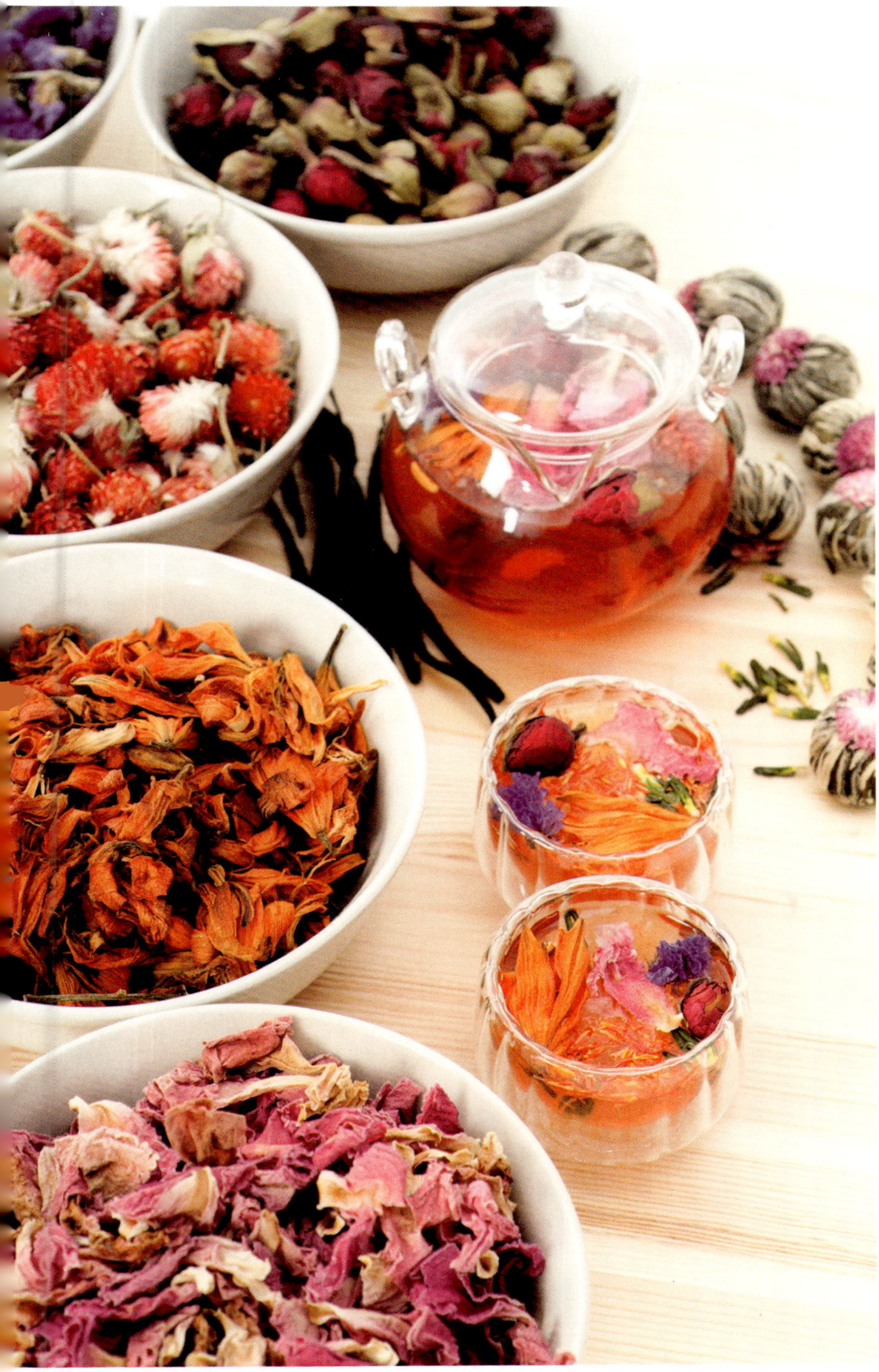

Herbal tea infusion with dry flower ingredients

And finally, if none of the recipes have inspired you, then consider drinking your flowers! You can make a heavenly elderberry, lavender, honeysuckle or dandelion or citrus flower sparkling wine by adding the dried petals. There are plenty of sparkling wines that will benefit from this addition. If you prefer martinis, then mix a vodka martini and add hibiscus extract or syrup for a memorable taste. Flowers can, thus, do much more than gladden the eyes, they can truly tickle your taste buds and delight your senses.

Glossary of Indian Flowers and Terms

Ancient Indian literature is replete with the names of various flowers. Many of these flowers are referred to by names that are not in common use today. Let us take a closer look at some of these flowers and get to know their present-day names. The glossary also includes terms from Indian languages that are mentioned in the book.

A

Aak: Calotropis or milkweed

Aarama: flowers grown in a garden

Achkan: long coat worn over fitted trousers

Adytum: innermost shrine of a temple that was open only to priests

Akasha: cosmos, space, ether

Akhlaat: in Unani medicine, a humour in the body

Akshat: rice

Akundor arca: Crown flower (*Calotropis gigantea*)

Alasa: relaxed and indolent

Alpana: traditional decorative art form of Bengal; patterns made by drawing designs on the floor with a mixture of rice flour and water

Amrita: elixir of immortality

Amsha: munificent

Aparajitha: Butterfly pea *(Clitoria ternatea),* also called *Shankpushpi* and *Vishnukanta*

Apsara: in Hindu and Buddhist mythology, a female spirit of the clouds and waters

Arali: *Nerium indicum* or *Nerium oleander*

Aranya: forest

Archa: worship

Arjuna: *Terminalia arjuna*

Artha: prosperity

Arusa: Malabar nut (*Justicia adhatoda*)

Ashoka: *Sita Ashoka* (*Saraca asoca*)

Atasi: *Flax (Linum usitatissimum)*

Attar / Itr: natural perfume extracted from flowers

B

Bahuvriti: in Sanskrit, a compound noun that specifies a characteristic or quality; e.g. sabre-tooth is neither a sabre or a tooth, but a feline with sabre-like teeth

Bajubandh: armlet or armband

Bakul: Spanish Cherry, *Maulsari* (*Mimusops elengi*)

Bandhook: Midday Flower, Scarlet Mallow (*Pentapetes phoenicea*)

Baraat: procession of the family and friends of a groom to the bride's home on their wedding day

Baransh: rhododendron

Bavari / Baoli: step-well

Baya: weaver bird

Bhajiya: a type of pakora; chopped vegetables mixed in a spiced batter and deep-fried

Bhakti: religious devotion; the love felt by the worshipper for a personal god

Bheda: distinction

Bilva: *bael* or wood apple

Brahma manduki: Indian Pennywort, Coinwort

Bulbul: a commonly seen passerine songbird

Buta: floral motif in stitching and embroidery

C

Ceplokan: traditional batik design based on circular rosettes or stars

Chabari: bamboo basket

Chakra: disc; war discus held by Vishnu and Krishna; spinning wheel; energy centre in the body

Chahar bagh or *Charbagh:* symmetrical garden designed in a four-part plan
Chameli: jasmine
Champa: Plumeria
Champaka: Champak (*Magnolia champaca*)
Chandrika: Sanskrit for *Sarpagandha*
Chautha: prayer service for the departed
Chemparathy / Jabakusumam: Hibiscus (*Hibiscus rosasinensis*)
Chethi, Thechi or *Thetti:* Ixora, Jungle Geranium (*Ixora coccinea*)
Chikankari: white thread embroidery done on white muslin or mulmul cloth
Chinar: oriental plane tree (*Platanus orientalis*)
Churidar: fitted trouser-like garment worn with a long shirt-like *kurta* or *kameez*

D

Darpana: looking into a mirror
Datura: white thorn apple or angel's trumpet
Deep or *Dip:* light, flame
Desi gulab: Indian rose
Devi: goddess
Devta: god
Dhamma / Dharma: righteous living
Dharmashastra: Theology
Dhavani: piece of cloth draped around the upper body of a women as part of a half-sari
Dhol: drum
Dhoop: incense
Dhoti: traditional men's garment; rectangular piece of unstitched cloth, wrapped around the waist and legs, then knotted at the waist
Dosha: in Ayurveda, one of the three bodily humours
Durva: Couch grass (*Elymus repens*)
Dussehra: Hindu festival celebrating the victory of good over evil

G

Gada: war mace held by Vishnu
Gajra: chain of flowers
Gandharava Vivah: celestial marriage
Gandhies: experts in making itrs
Gandhkarika: women who prepared itr or fragrances
Ganga-jamuni: the mutually participatory co-existence of Hindu and Muslim culture of northern India
Garba griha: inner sanctum of a temple
Genda / Chendumalli: Marigold (*Tagetes*)
Ghoonghat: veil
Gilehri: squirrel
Gopi: female cow herder
Gudahal: red hibiscus
Gul or *gulab:* rose (flower)
Gulab jal: rosewater
Gulab ki kheer: rose-flavoured rice pudding
Gulbadan: one whose body is rose-like
Gulistan: garden
Guna: quality
Gunja / Gunj: Crab's eye (*Abrus precatorius*)

H

Haat: market
Haldi: turmeric
Halwai: sweet-maker
Hansas: birds in textiles or other decorative arts
Harsingar: parijat
Havan: fire ceremony
Henna: flowering plant; also, the dye prepared from it used to decorate women's hands and feet on festive occasions
Hiranyagarbha: womb of the universe
Hiranya kosa: treasure chest
Holi: Hindu festival of colours celebrated annually in spring

I

Itihasa: history
Itr: see *Attar*

J

Jaimala: wedding garland
Jal: water
Jalebi: Indian sweet, fried batter spiral immersed in sugar syrup
Jamanthi / chandramallika: Chrysanthemum (*Chrysanthemum indicum*)
Jamdani: fine muslin textile of Bengal
Jatakas: stories of the various incarnations of the Buddha
Jhingoria: barnyard millet
Jina: genie

K

Kadai: thick, circular, deep pot used in Indian cuisine
Kadamba: Kadam (*Neolamarckia cadamba*)
Kala bashimb: Gandarusa or water willow
Kala siris: Ceylon rosewood
Kalamkari: ancient form of textile painting of southern India
Kalhara: white lotus
Kalka: paisley or mango motif used in embroidery

Kama: desire
Kamadeva: Hindu god of human love or desire
Kamal: Lotus (*Nelumbo nucifera*)
Kanakambaram: Firecracker flower (*Crossandra infundibuliformis*)
Kantha: traditional embroidery of West Bengal and Bangladesh; originally used to join layers of old saris to make quilts
Karir: *Capparis decidua*; small tree or shrub that produces red flowers and berries
Karnikar: Kanak Champa (*Pterospermum acerifolium*)
Kashida: Kashmiri embroidery
Kasuti: folk embroidery from Karnataka
Kasturi: musk
Kavya: heavily embellished, idealised Sanskrit poetry
Kembang: floral motifs in Batik textiles
Kesar: saffron
Kesar rabri: saffron cream
Kesri: flavoured with saffron
Ketaki: Kewda (*Pandanus odoratissimus*)
Ketaki-bandha: adorns herself with ketaki flowers
Kewda / Kewra: padanus or screw pine flower
Khagi Leihao: frangipani / plumeria
Khadir: acacia tree
Khanjar: personal dagger of the Mughal period
Khapwah: double-edged dagger of the Mughal period
Kila: nail, peg
Kimkhab: cloth of gold
Kimsuka: *Butea monosperma*
Kolam: decorative design of rice flour and other colourful powders, made at the entrance of houses and in courtyards in South India, to welcome deities and guests
Kovidar: Purple orchid tree (*Baukinia purpurea*)
Kumkum: vermilion
Kumud: White Water Lily (*Nymphaea lotus var. pubescence*)
Kund: Star jasmine (*Jasminum multiflorum/pubescens*)
Kunkuma: Sanskrit for saffron
Kurava: Asiatic Tarenna or bottle-flower (*Tarenna asiatica*)

L

Lung-lungan: Batik designs depicting creepers and vines

M

Maan: soul
Madhavi Lata: *Hiptage benghalensis*
Madua: buckwheat flower
Maina: mynah bird
Malati: Clove scented echites (*Aganosma dichotoma*)
Mallika: jasmine
Mallipoo or *Mallipu:* jasmine
Mallipoo jade: hair ornament used in southern India consisting of jasmine buds sewn on a piece of cardboard
Mandala: sacred drawing
Mandapam: pillared, outdoor hall or pavilion for public rituals
Mandaram: Buddhist Bauhinia or White Orchid Tree (*Bauhinia alba*)
Mandorla: almond-shaped aureole of light surrounding the entire figure of a holy person in religious art
Manjari: tulsi or basil flower
Manshila: Realgar (an arsenic sulphide mineral, As_4S_4)
Maulsari: Spanish cherry
Maya: illusion, unreality
Mehndi: see *Henna*
Mogra/Mulla: Jasmine (*Jasminum grandiflorum*)
Mohalla: neighbourhood subdivision
Moksha: spiritual liberation or release
Motiya: jasmine
Mozham: measurement from the tip of the finger to the elbow
Muggu: see *Kolam*
Mukkutti: Life plant or *Biophytum sensitivum*, similar to the Touch Me Not; called *Viparitalajjalu* in Sanskrit
Mukula: bud

N

Nadi: channel
Nag kesar: evergreen tree with fragrant white flowers
Nagare ki jor: musical drums
Nagsura: *Mesua ferrea* or Indian rose chestnut; hardwood tree with large, fragrant white flowers; the new leaves, which come out in March, are crimson and make the tree resemble a flame
Naivedya / Naivedyam: offering to deity; in pooja, an offering of food
Nalini: lotus
Nandvanam: temple flower garden
Nandyaarvattam: Crape jasmine or pinwheel flower (*Tabeamaemontana diverticata*)
Nartaki: a dancer
Nargis: narcissus, poet's daffodil
Natak: drama
Nayak or *nayaka:* hero, leader
Nayika: heroine
Neel Kamal: Blue water lily (*Nymphaea nouchali/stellate*)
Nili nargandi: gandarusa or water willow

P

Paavadai: long skirt worn by young girls between puberty and marriage in South India; this becomes a two-piece sari or half-sari when worn with a blouse and a *dhavani* draped around the upper body

Padma: lotus

Padmasana: lotus position in Hatha Yoga

Palasa or *palash:* flame of the forest, parrot tree

Pancha Bhoota: the five elements of Nature—air, water, fire, earth and ether

Panch Karma: five Ayurvedic therapies to detoxify the body

Panch panche: a folk game played with five pebbles

Panchopachaar Pooja: ritual of worshipping the deity with five items

Pandan or *Pandanus:* screw pine

Pankha: hand fan

Parang: batik pattern

Parichaya: worship

Parijata: Harsingar, shefali (*Nyctanthes arbor-tristis*)

Parul: five-petalled flower known for its beauty

Parva: book

Patal/Paral: Fragrant padri tree (*Stereospermum chelonoides*)

Patala: a low stool for sitting

Peelu: *Salvadora oleoides*; small, bushy evergreen tree

Phoolon ka Gehna: Kashmiri wedding ritual where groom's family sends flower jewellery and tinsel to the bride

Phool Waalon ki Sair: 19th-century festival of flower vendors held annually in Delhi

Pichakam or *Pichi:* Spanish jasmine

Pisang Bali: traditional Batik design with a motif of stylised banana leaves and flowers

Pooja or *Puja:* prayer ceremony

Prajapati: Supreme Being said to be the lord of the firmament

Prakriti: nature

Prasad: religious offering of food in Hinduism and Sikhism that is consumed by devotees

Punnaaga: Sultan Champa (*Calophyllum inophyllum*)

Purohit: family priest; pandit

Purusharthas: four goals of life—dharma, artha, kama, moksha

Pushp: flowers

Pushpa varsha: celestial flower shower

Pushpanjali: offering of flowers to Hindu gods

Q

Qawwali: Sufi devotional music

R

Raat ki raani: parijat or night-flowering jasmine (*Nyctanthes arbor-tristis*)

Rajamalli: Peacock flower (*Caesalpinia Pulcherrima*)

Rajasa or *rajo:* middling or mediocre

Rangoli: decorative design usually made of coloured powder on floors and in courtyards to welcome deities and guests

Ranjan: delight

Ras-leela or *rasleela:* dance that is part of the traditional story of Krishna in scriptures and literature

Rishi: a seer or sage

Ritikaleen: Hindi poetry that emphasises love (especially eroticism and the feminine form), beauty, fate and duty

S

Shalmali: Silk-cotton tree (*Bombax ceiba*)

Salwar kameez: traditional outfit of loose trousers and a long shirt or tunic

Samskara: rite of passage in Hinduism, Jainism and Buddhism

Samudra Manthan: churning of the milky ocean

Sanjeevani: magical herb that is believed to cure any illness (*Selaginella bryopteris*)

Saparya: to serve attentively, worship

Sarga: canto

Sarpagandha: Black snakeroot (*Rauwolfia serpentina*)

Sattharar: garland-maker for temple deities

Sattva / sattvika: superior

Swayamvar sabha: ancient Indian practice of a girl choosing a husband from a list of suitors

Sembagen: stylised fruit and flower design in Batik textiles

Shakti: primordial cosmic energy

Shami: *Prosopis cineraria*; flowering tree in the pea family

Shanka: conch or trumpet held by Vishnu

Sharbat: sherbet

Shehnai: musical instrument similar to the oboe

Shisha: mirror, glass

Shiuli: parijat, night-flowering jasmine

Shivling: representation of the Hindu deity Shiva

Shloka or *sloka:* verse line from Vedic Anustubh (poetic meter)

Shraddhanjali: tribute

Shrikhand: yoghurt-based delicacy from Maharashtra

Shringara: adornment

Sindhu: a variety of seedless mango

Sinhas: animal depictions in textiles or other decorative arts

Sirisha: *Kala siris* or Ceylon rosewood

Soof: folk embroidery of western India

Sthulapushpa: the name for marigold or *genda* in Sanskrit
Sujini: embroidery technique, originating in Bihar, used to recycle worn-out or damaged clothing

T

Tamala: Indian bay leaf
Tamasa or *tamo:* inferior
Tan: body
Tatpurusha: class of compound words; the first noun modifies the second word, e.g. sheepskin, motorboat, shoemaker
Thiruvilayadal purine: the sacred sports of Shiva
Tilak: coloured spot or mark worn by Hindus on the forehead
Tinduka: Gaub or Malabar Ebony tree
Trimurti: trinity of Hindu deities—Brahma, Vishnu and Shiva
Torana: gateway in Hindu and Buddhist architecture
Tota: parrot
Tulasi or *tulsi:* Holy basil (*Ocimum tenuiflorum*)

U

Ubtan: facial mask
Udajati: Blue justicia (*Ecbolium linneanum*)
Upachara: ritual in worship
Upasna: worship
Upvan: orchard or grove
Utpala: blue lotus or water lily

V

Vadamalli: Bachelor's Buttons (*Gomphrena globosa*)
Vajra: thunderbolt
Vajradanti: Porcupine flower (*Barleria prionitis*)
Van / vana: forest
Varivayasya: worship
Varjita: daily temple worship
Vatika: orchard; plantation
Veena: plucked string instrument used in Indian classical music
Veni: hair plait or braid
Vidyadevi: Goddess of knowledge
Vidyadhara: supernatural being in Hindu mythology that has magical powers and lives in the Himalayas
Vikrita: flowers that are bought
Vimana: adytum of a temple
Vinyasa: pensive
Vishellakarani: Hophead (*Barleria lupulina*)
Vishnu: Hindu deity who is the preserver of all living beings
Vivaha: marriage

Y

Yaksa: demigod
Yaksi: demigoddess
YuthikaJuhi: *Jasminum auriculatum*

Z

Zafran: Saffron (*Crocus sativus*)

Bibliography

All-India Fine Arts and Crafts Society. *Roopa-Lekha.* New Delhi: All-India Fine Arts and Crafts Society, 1987.

Amirthalingam, M., and Nanditha Krishna. *Sacred Plants of India.* New Delhi: Penguin Books Limited, 2014.

Bahadur, K.P. *Rasikapriya of Keshavadasa.* New Delhi: Motilal Banarsidass Publishers, 1990.

Balakrishnan, Usha R. *Alamkara: The Beauty of Ornament.* New Delhi: National Museum, 2014.

Batth, Amarjeet Singh. *Indian Garden Flowers: Home Gardener's Guide.* New Delhi: Prakash Books, 2007.

Beck, Brenda E.F., Peter J. Claus, Praphulladatta Goswami, and Jawaharlal Handoo, ed., *Folktales of India.* New Delhi: Motilal Banarsidass, 1989.

Bier, Carol, and Mary M. Dusenbury. *Flowers, Dragons & Pine Trees: Asian Textiles in the Spencer Museum of Art.* Manchester: Hudson Hills Press, 2004.

Bosch, F.D.K. *The Golden Germ: An Introduction to Indian Symbolism.* ('s-Gravenhage: Mouton, 1960).

Brown, Percy and George Watts. *Arts and Crafts of India: A Descriptive Study.* New Delhi: Cosmo Publications, 1979.

Buck, William. *Ramayana.* New Delhi: Motilal Banarsidass Publishers, 2nd edition. 2000.

Busch, Allison. *Poetry of Kings: The Classical Hindi Literature of Mughal India.* New York: Oxford University Press, 2011.

Campbell, Joseph. *The Mythic Image.* Princeton: Princeton University Press, 1974.

———. *The Power of Myth.* New York: Anchor Books, 1991.

Chaturvedi, Makhanlal. *Modern Indian Literature: An Anthology – Volume One.* Translated by Shaileshwar Sati Prasad. New Delhi: Sahitya Akademi, 1994.

Coomaraswamy, Ananda K. *Essays on National Idealism.* New Delhi: Munshiram Manoharlal Publishers, 1981.

Debroy, Bibek. *The Mahabharata – Volume 10.* New Delhi: Penguin India, 2015.

Dhody, Chandan Lal. *The Adhyatma Ramayana: Concise English Version.* New Delhi: M.D. Publications Pvt. Ltd., 1995.

Faiz, Faiz Ahmed. *100 Poems by Faiz Ahmed Faiz, 1911-1984.* New Delhi: Abhinav Publications, 2002.

Freeman, Lee Keith. *Book of Sages.* Pennsylvania, Temperance Publishing, 2000.

George, K.M., ed. *Modern Indian Literature: An Anthology – Volume One, Surveys and Poems.* New Delhi: Sahitya Akademi, 1994.

Gopalakrishnan, Paravur. *The Gitagovinda as the Song Celestial.* New Delhi: Sanbun Publishers, 2008.

Gupta, Shakti M. *Plant Myths and Traditions in India.* New Delhi: Munshiram Manoharlal Publishers, 2001.

Haidar, Navina Najat, and Marika Sardar, eds. *Sultans of the South: Arts of India's Deccan Courts,* 1323–1687. New York City: Metropolitan Museum of Art, 2011.

Hatanaka, Kokyo. *Textile Arts of India,* Kokyo Hatanaka Collection. San Francisco: Chronicle Books. 1996.

Heine, Heinrich. *Great Thoughts from Master Minds, Volume 2.* London: A. W. Hall, 1884.

Herbert, Eugenia W. *Flora's Empire: British Gardens in India.* Philadelphia: University of Pennsylvania Press, 2012.

Hutton, Frankie. *Rose Lore: Essays in Semiotics and Cultural History.* Lanham, Maryland: Lexington Books, 1955.

Jayadeva. *Love Song of the Dark Lord, Jayadeva's Gitagovinda.* Translated by Barbara Stoler Miller. New York: Columbia University Press, 1997.

Jayaram, Suresh. *Whatever He Touched, He Adorned.* Bangalore: Goethe Institute, 2011.

Kalidasa. *Abhijñānaśākuntalam.* Edited by C. R. Devadhar. New Delhi: Motilal Banarsidass Publishers, 1996.

Karnik, Dilip J. *Rays of Wisdom.* New York: Xlibris Corporation, 2012.

Keown, Damien and Charles S. Prebish, eds. *Encyclopedia of Buddhism.* London: Routledge. 2013.

Kumar, Sehdev. *A Thousand Petalled Lotus: Jain Temples of Rajasthan: Architecture and Iconography.* New Delhi: IGNCA and Abhinav Publications, 2001.

Kundalia, Nidhi Dugar. *The Lost Generation: Chronicling India's Dying*

Professions. New Delhi: Random House India, 2015.

LaViolette, Paul A. *Genesis of the Cosmos: The Ancient Science of Continuous Creation*. Rochester, Vermont: Inner Traditions / Bear & Co, 2004.

Manilal, K.S. *Van Rheede's Hortus Malabaricus: Malabar garden with annotations and modern botanical nomenclature, Volumes 1 to 12*. Trivandrum: University of Kerala, 2003.

McHugh, James. *Sandalwood and Carrion: Smell in Indian Religion and Culture*. New York: Oxford University Press. 2012.

Moor, Edward. *The Hindu Pantheon*. London: J Johnson, 1810.

Moore, Thomas. *The Poetical Works of Sir Thomas Moore*. Philadelphia: J. Crissy, 1835.

Mother, The. *The Spiritual Significance of Flowers, Part I and II*. Pondicherry: Sri Aurobindo Ashram Publications Department, 2000.

Moynihan, Elizabeth. 'The Lotus Garden Palace of Zahir Al-Din Muhammad Babur,' in *Muqarnas Volume V: An Annual on Islamic Art and Architecture*. Edited by Oleg Grabar. Leiden: E.J. Brill, 1988.

Mulchandani, Sandhya. *The Five Arrows of Kama: The Art of Love, Sex and Desire*. New Delhi: Penguin Books India, 2010.

Muni, Bharata. *Natyashastra – Volume I*. Translated by Manmohan Ghosh. Calcutta: Asiatic Society, 1951.

Muni, Bharata. *Natyashastra – Volume II*. Translated by Manmohan Ghosh. Calcutta: Asiatic Society, 1961.

Nandyala, Divya. *Bhajanika*. Lulu.com, 2013.

Pattanaik, Devdutt. *The Goddess in India: The Five Faces of the Eternal Feminine*. New Delhi: Simon & Schuster India, 2000.

Paul, S.K. *The Complete Poems of Rabindranath Tagore's Gitanjali: Text and Critical Evaluation*. New Delhi: Sarup and Sons, 2006.

Potter, Jennifer. *Seven Flowers And How They Shaped Our World*. London: Atlantic Books, 2013.

Rajagopalachari, C. *Mahabharata*. New Delhi: Diamond Pocket Books. 1958.

Rajan, Chandra. ed and trans. *Kalidasa: The Loom of Time*. New York: Penguin Books, 1989.

Ramachandran. A. and Chameli. *Bhima and the Fragrant Flower*. New Delhi: Vadehra Art Gallery, 2012.

Riyaz, Tarannum. 'The Flowers of Love'. Translated by Jaipal Nangia. *Indian Literature* Vol XLVII No. 3 (2003): 82.

Row, Bangalore Suryanarain. *A History of Vijayanagar: The Never to be Forgotten Empire, Part 1*. New Delhi: Asian Educational Services, 1905.

Roxburgh, William. *Flora Indica, or, Descriptions of Indian Plants*, ed. William Carey. Serampore: Printed for W. Thacker. 1812.

Sadarangani, Neeti M. *Bhakti Poetry in Medieval India: Its Inception, Cultural Encounter and Impact*. New Delhi: Sarup and Sons, 2004.

Saguaro, Shelley. *Garden Plots: The Politics and Poetics of Gardens*. Burlington, VT: Ashgate, 2006.

Sahni, K. C. *The Book of Indian Trees*. Mumbai: Bombay Natural History Society and Oxford University Press, 1998.

Sharma, T.R.S., ed. *Ancient Indian Literature: An Anthology*. New Delhi: Sahitya Akademi, 2000.

Sivaramamurti, Calambur. *The Art of India*. New York: Harry N. Abrams, Incorporated, 1977.

Swami, A. C. Bhaktivedanta. *Srimad-Bhagavatam*. Fourth Canto: The Creation of the Fourth Order. Mumbai: The Bhaktivedanta Book Trust, 1974.

Tagore, Rabindranath. *The English Writings of Rabindranath Tagore, Volume 1: Poems*. Introduction by Mohit K. Ray. Atlantic Publishers & Distributors, 2007.

Topno, Sem. *Musical Culture of the Munda Tribe*. New Delhi: Concept Publishing Company, 2004.

Tulasidasa. *Shri Ramacharitamanasa*. New Delhi: Motilal Banarsidass Publishers, 1989.

Vaidya, Chintaman Vinayak. *Epic India, Or, India as Described in the Mahabharata and the Ramayana*. Asian Educational Services, 2001.

van Rheede, Hendrik, Ranga Bhat, Vinayaka Pandit, Appu Bhat, et al. *Hortus Malabaricus*. Amsterdam: 1678-1693.

Vanita, Ruth. *Gender, Sex, and the City: Urdu Rekhti Poetry in India 1780-1870*. Palgrave Macmillan, 2012.

Verma, Som Prakash. *The Illustrated Baburnama*. New Delhi: Routledge India, 2016.

Viswanathan, S. ed., *Abhijñana Shakuntala:* With English notes and translation by C. Sankara Rama Sastri. Mylapore, Madras: Sri Balamanorama Press, 1961.

Wescoat, Jr, James L. and Joachim Wolschke-Bulmahn, eds. *Mughal Gardens: Sources, Places, Representations, and Prospects*. Washington D.C.: Dumbarton Oaks, 1996.

Wilson, Kax. *History of Textiles*. Boulder, Colorado: Westview Press, 1979.

Zimmer, Heinrich. *Myths and Symbols in Indian Art and Civilization*. Edited by Joseph Campbell. Princeton: Princeton University Press, 1972.

Image Credits

Chapter 1: P.2. Floral offerings: Kyley Rumohr/Shutterstock **pp.4-5.** Pink Frangipani: Rachel Moon **pp.6-7.** Marigold flower decorations: pjhpix/Shutterstock **p.8.** Parijat flower: Rashbihari Dutta/Shutterstock **p.10.** 'Krishna uproots the parijata tree': LACMA, Public domain, Wikimedia Commons **p.11.** Buddha: Pinterest/Ramsesosirus; Lotus: pixabay **p.12.** 'Sita shies away from Hanuman': The David Collection, Copenhagen, via Wikimedia Commons **p.13.** 'Sita in the Ashoka Vatika': Public domain, Wikimedia Commons; 'Babur watching men altering the course of a stream': Public domain, Wikimedia Commons; Garland of rose, marigold and jasmine: 123RF **p.14.** 'Devi wearing a garland of red hibiscus': Pinterest/Radha Skhi; Jasmine flowers adorn the hair of a bride: IVASHstudio **p.15.** Marigold flowers for worship: Pixabay; Jasmine flower garlands: peter verreussel/ Shutterstock **p.16.** Frangipani flowers: Coloa Studio; jasmine with gerbera flowers: 123RF **p.17.** Title page of the *Hortus Malabaricus*: Wikimedia Commons; Detail from the *Hortus Malabaricus:* DrCyCoe, Wikimedia Commons **p.18.** Flower garlands: Tul R./ Shutterstock **p.20.** Lotus flower offerings at Ramabhar Stupa: Rachel Moon/Shutterstock; Flowers as auspicious items: irina d'elena/Shutterstock **p.21.** 'Goddess Lakshmi', Raja Ravi Varma: Public domain, Wikimedia Commons; Chloris: Public domain, Wikimedia Commons; Narcissus: Public domain, Wikimedia Commons; Immaculate Heart of Mary: Nancy Bauer/ Shutterstock **p.22.** Bodhisattva Padmapani, Ajanta: Abdulsayed/ CC BY-SA 3.0, Wikimedia Commons; Engraved stone wheel: Subhrajyoti07/CC BY-SA 4.0, Wikimedia Commons; Sanchi Stupa gate: Deepak Bishnoi/Shutterstock **p.23.** White floral wreath: **p.23.** vilax/Shutterstock/Shutterstock **pp.24-5.** Thukral & Tagra: courtesy Alka Pande

Chapter 2: P.26. Clemantis: FAG/Shutterstock **p.28.** Thangka: Public domain, Wikimedia Commons **p.29.** Priest-king: www.harappa.com; Khasarpana Lokesvara: Hyougushi/CC BY-SA 2.0, Wikimedia Commons **p.30.** Female figure: Photo Dharma/CC BY-SA 2.0, Wikimedia Commons; Bodhi tree: Biswarup Ganguly/ CC BY 3.0, Wikimedia Commons; Lion capital: Vincent Arthur Smith, Public domain, Wikimedia Commons **p.31.** Buddhist divinities: Kevin Standage/CC BY-SA 2.0, Wikimedia Commons; Heliodorus Pillar: Dilipkumarftii1977/CC BY-SA 4.0, Wikimedia Commons **p.32.** Goddess Hariti: Ism^^on/CC BY-SA 4.0, Wikimedia Commons; Dancing Shiva: Vasukrishnan57/ CC BY-SA 4.0, Wikimedia Commons; Mukteshwara temple: Thamizhpparithi Maari/CC BY-SA 3.0, Wikimedia Commons **p.33.** *Baburnama*: Public domain, Wikimedia Commons **p.34.** *Nimat Nama*: Public domain, Wikimedia Commons **p.35.** Torana: Thamizhpparithi Maari/CC BY-SA 3.0, Wikimedia Commons; Aurangzeb: Public domain, Wikimedia Commons **p.36.** Great Gate: Tosujoy/CC BY-SA 3.0, Wikimedia Commons **p.37.** Itimad-ud-Daulah's tomb: Royroydeb/CC BY-SA 4.0, Wikimedia Commons; Inlay work: Leochen66/Shutterstock and Savvapanf Photo/Shutterstock **pp.38-9.** Itimad-ud-Daulah's tomb: JeremyRichards/Shutterstock **p.40.** Maharaja Gulab Singh: Public domain, Wikimedia Commons; Tulip: Public domain, Wikimedia Commons **p.41.** *Najum-al-ulum*: Public domain, Wikimedia Commons **p.43.** Emperor Jahangir: Public domain, Wikimedia Commons; **p.44.** *Pabuji ka phad*: Nomu420/ CC BY-SA 3.0, Wikimedia Commons **p.45.** Kohbar of Mithila: Sumanjha1991/ CC BY-SA 4.0, Wikimedia Commons; Warli painting: Jean-Pierre Dalbéra/CC BY 2.0, Wikimedia Commons.

Chapter 3: P.46. Lotus petal: Tian Gan/Shutterstock **p.48.** Krishna and Radha: public domain/The Met NY **p.49.** Krishna revels with the gopis: public domain/The Met NY **p.50.** The gopis: public domain/The Met NY; Kanchana flower: Ho Van Ty/Shutterstock **p.51.** Draupadi: Ramanarayanadatta astri, Public domain, Wikimedia Commons; Tulsi: PosiNote/ Shutterstock **p.52.** Rama, Sita and Lakshmana: public domain/ The Met NY; Rama and Sita: Public domain, Wikimedia Commons **p.53.** Pandavas: Ramanarayanadatta astri/Public domain, Wikimedia Commons **p.54.** 'Shakuntala', Raja Ravi Varma: Public domain, Wikimedia Commons; *'Shakuntala patralekhan'*, Raja Ravi Varma: Public domain, Wikimedia Commons; Kadamba: Swapan Photography/Shutterstock **p.55.** 'Shakuntala and her friends', Raja Ravi Varma: Wellcome Collection/CC BY 4.0 **p.56.** Manku: Wellcome Collection/CC BY 4.0; Lotus petals: Boonchuay1970/ Shutterstock **p.57** 'The Monsoon Flowers', A. Ramachandran: courtesy Alka Pande; Ramesh Gorjala: courtesy Alka Pande **pp.58-9.** 'All is gray when the black is washed away', Aisha Khalid: courtesy Alka Pande **p.60.** Flowers in the rain: SantaLiza/Shutterstock **p.61.** Heap of Jasmine: sta/Shutterstock **p.62.** Jasmine garland: Patty Chan/ Shutterstock; champak flowers: Ohhlanla/ Shutterstock.

Chapter 4: P.64. Array of roses: masalskaya/Shutterstock **p.66.** Kamadeva: LACMA, public domain, Wikimedia Commons

p.67. Champa: Sommai/Shutterstock; Water lily: Beetle/Pixabay **p.68.** Attar bottles: Stanislav71/Shutterstock **p.69.** Sandalwood: wasanajai/Shutterstock; Rose water: denio109/ Shutterstock **p.70.** Empress Nur Jahan: LACMA, public domain, Wikimedia Commons; Hibiscus oil: StockImageFactory.com/Shutterstock; **p.71.** Lalla Rookh: tomato.to/photo/11053197314/; Calendula: rawf8/ Shutterstock **p.73.** Spa treatment: Shebeko/ Shutterstock.

Chapter 5: P.74. Hindu woman: reddees/Shutterstock **p.76.** Chidambaram temple: Destination8infinity/ CC BY-SA 4.0, Wikimedia Commons; Krishna mural: Tommy/CC BY-SA 2.0, Wikimedia Commons **p.77.** Malati flowers: sabza/Shutterstock; Bakul flowers: Noppharat Studio 969/Shutterstock; Akund flower: Butterfly Hunter/Shutterstock **p.78.** Palash flowers: sabza/Shutterstock; Yuthika flowers: Gagan/CC BY-SA 3.0, Wikimedia Commons; Punnag flowers: Yongkiet Jitwattanatam/ Shutterstock **p.79.** Shore temple: nkumarm/Shutterstock **p.80.** Floral offerings: Natalia Mikhalchuk/Shutterstock **p.81.** Hindu deity: Aisha Nair/Shutterstock **p.82.** Durva grass: StockImageFactory.com/Shutterstock **p.83.** 'Goddess Lakshmi', Raja Ravi Varma: Public domain, Wikimedia Commons **p.84.** Banyan tree: reddees/Shutterstock; Lord Shiva: Ramanarayanadatta astri, Public domain, Wikimedia Commons **p.85.** Savitri: The Calcutta Art Studio/public domain, Wikimedia Commons **p.86.** Pink water lily: Laura_Vn/Shutterstock; Bharatnatyam dancer: jai7678/Shutterstock; White water lily: Koldunov Alexey/Shutterstock **p.87.** Jasmine: Tukaram.Karve/ Shutterstock **p.88.** Hibiscus: wadeoak/Shutterstock; Rows of containers: rudram/Shutterstock **p.89.** Floral tributes: nevenm/ Shutterstock; HelpusGreen: Karan Rastogi.

Chapter 6: P.90. Bride's hands: Vinogradova Elena/Shutterstock **p.92.** Rama, Lakshmana and Sita: Public domain, Wikimedia Commons **p.93.** Shakuntala/Ravi Varma School: Public domain, Wikimedia Commons **p.94.** Sultan Shah/Shutterstock.com; Rudra Narayan Mitra/Shutterstock.com; kkssr/Shutterstock.com **p.95.** 'Lady with flower garland', Raja Ravi Varma: Public domain, Wikimedia Commons; 'Nair woman', Raja Ravi Varma: Public domain, Wikimedia Commons **p.96.** *Veni*: Manju Mandavya/Shutterstock; Jasmine garlands: Santhosh Varghese/ Shutterstock **p.97.** *Mundan*: clicksabhi/Shutterstock **p.98.** Jaimala: SyedAliAshraf; Yellow marigold flowers: PattayaPhotography/ Shutterstock **p.99.** *Mangalsutra*: StockImageFactory.com/ Shutterstock; rings: MGhosh_silentwings/Shutterstock **p.101.** Nuptial bed: Alla Sravani/Shutterstock **p.102.** *Phoolon ka Gehna*: AliAshraf/Shutterstock **p.103.** Flower *teeka*: Alee Photographer/ Shutterstock; Jasmine *gajras*: Surbhi S/Shutterstock **p.104.** Buttonhole: Sultan Shah/Shutterstock **p.105.** Jewellery: Sofia Zhuravetc; Arts Illustrated Studios; NehaS.

Chapter 7: P.106. Peony flower: Okrasyuk/Shutterstock **p.108.** Flower sellers: Denis.Vostrikov/Shutterstock **p.110.** Priest creating rangoli: Mazur Travel/Shutterstock; Lit oil-lamp: reddees/Shutterstock **p.111.** Floral rangoli: StockImageFactory.com/Shutterstock; Malayali woman: Santhosh Varghese/ Shutterstock; Kolam: creators studio/Shutterstock **p.112.** *Pookalam:* yogesh_more/Shutterstock **p.113.** *Ixora coccinea:* Thanakorn Hongphan/Shutterstock **p.114.** Screw pine: aomsopida/Shutterstock; Lotus pond: Pranay Chandra Singh/ Shutterstock; Neem flowers: pukao/Shutterstock **p.116.** Orchids: Studio Grand Web/Shutterstock; Narcissus: Ole Schoener/Shutterstock; Peonies: Nella/Shutterstock **p.117.** Protea: lcswart/Shutterstock **p.118.** Orchid tree: Layue/ Shutterstock; Marigold: de2marco/Shutterstock; Flame of the Forest: Phongphiphat/Shutterstock; Golden Shower: TigerStock's/Shutterstock; Madonna lily: Scisetti Alfio/ Shutterstock; Pride of India: sakhorn/Shutterstock; **p.119.** Lady's Slipper orchid: Natalia van D/Shutterstock; Red Vanda: Jadezmith/Shutterstock; Rhododendron: Nipitphon Vetayavong/Shutterstock; Ashoka flowers: addkm/Shutterstock; Cannonball tree's flower: drpnncpptak/Shutterstock; Glory lily: max777/Shutterstock; Brahma Kamal: StockImageFactory.com/Shutterstock **p.120-21.** Vijay Chowk: diy13/Shutterstock

Chapter 8: P.122. Bird of Paradise flower: evantravels/ Shutterstock **p.124.** Matancherry Palace mural: Mark Hills/ CC BY 2.0, Wikimedia Commons **p.125** 'Mohini', Raja Ravi Varma: Public domain, Wikimedia Commons; 'Arjuna Subhadra', Raja Ravi Varma: Public domain, Wikimedia Commons **p.126.** Flowers from Heaven/Manu Parekh: courtesy Alka Pandey **p.127.** Krishna and Radha: public domain/The Met NY **p.128.** *Samudra manthana*: Public domain, Wikimedia Commons **p.129.** Rishi, parijata tree and Kamadhenu: Public domain, Wikimedia Commons **p.130.** Besnagar sculpture: Public domain/Royal Netherlands Institute of Southeast Asian and Caribbean Studies and Leiden University Library/Wikimedia Commons **p.131.** Pawon Temple relief: lrosebrugh/Shutterstock **p.132.** *Baburnama*: Public domain, Wikimedia Commons **p.133.** Babur's Garden: IM3847/ Wikimedia Commons **p.134.** Fatehpur Sikri: Ruslan Kalnitsky/Shutterstock **p.135.** Humayun's Tomb: Iryna Rasko/ Shutterstock **p.136.** Trees of India: public domain/The Met NY; 'Sunlight Radiance'/Maite Delteil: courtesy Alka Pandey **pp. 138-39.** Nishat Bagh: saiko3p/Shutterstock **p.140.** Emperor Jahangir and Shah Abbas: Public domain, Wikimedia Commons **p.141.** Garden in Amer Fort: Aleksei Kornev/Shutterstock

Chapter 9: P.142. Tree of life embroidery: Joat/Shutterstock **p.144.** Tribal woman: CRS Photo/Shutterstock **p.145.** Paithani sari: LACMA/Public domain, Wikimedia Commons; Shawl fragment: Cooper Hewitt, Smithsonian Design Museum/Public domain, Wikimedia Commons **p.146.** Akbar: Chester Beatty Library, Dublin/ Public domain, Wikimedia Commons **p.147.** Mughal carpet images: public domain/The Met NY; **p.148.** Hand block printing: Kalcutta/Shutterstock; Palampore: Ji-Elle/CC BY-SA 3.0, Wikimedia Commons **p.149.** Chikankari: LACMA, Public domain, Wikimedia Commons; Kashmiri embroidered bag: courtesy Misha Oberoi **p.150.** Kantha: Honolulu Museum of Art/ Public domain, Wikimedia Commons **p.151.** Kalamkari: public domain/The Met NY **p.152.** Kalamkari: Kasubabu/CC BY-SA 3.0, Wikimedia Commons **p.153.** Palampore: public domain/The Met NY **p.154.** Banarsi saris: AjayTvm/Shutterstock; Indian woman: Niks Ads/Shutterstock **p.155.** Sambalpuri sari: Public domain, Wikimedia Commons

Chapter 10: P.156. Indian woman: Pikoso.kz/Shutterstock **p.158.** 'Sa'di in a Rose garden': newsdesk.si.edu/Public domain, Wikimedia Commons **p.159.** Rose plantation: Penka Todorova Vitkova/Shutterstock; Dried rose buds: Chursina Viktoriia **p.160.** *Rose gallica*: Tony Baggett/Shutterstock **p.161.** *Rosa Centifolia*: Steiner Wolfgang/Shutterstock; Alba rose: Mariya Dimova/Shutterstock; Chinese roses: Jun Mu/Shutterstock **p.162.** Binding of a Divan of Hafiz: Public domain, Wikimedia Commons **p.163.** An Allegory of Empress Josephine: The Met NY/Public domain, Wikimedia Commons **p.164.** Roses: astral232/Shutterstock **p.166.** Damask Rose water: wasanajai/ Shutterstock **p.167.** 'Babur reading in a garden': Public domain, Wikimedia Commons **pp.168-9.** Shalimar Bagh: SB Stock/ Shutterstock **p.170.** Gulbadan Begum: Helena Bx/CC BY-SA 4.0 CC, Wikimedia Commons; Empress Nur Jahan: Public domain, Wikimedia Commons; rose petals: joannawnuk/Shutterstock **p.171.** National Rose Garden: Dheeraj Arora/Shutterstock **p.172.** Turkish delight: Creative Family/Shutterstock **p.173.** 'Krishna and Radha conversing': public domain/The Met NY **p.174.** Freshly plucked roses: Paolo Guarneri/Shutterstock

Chapter 11: P.176. Pink lotus: Praew stock/Shutterstock **p.178.** Lotus Temple: Nadezda Murmakova/Shutterstock; Blue Lotus: foto76/Shutterstock; Lajja Gauri: public domain/The Met NY **p.179.** God Upulvan/Ramanarayanadatta astri, Public domain, Wikimedia Commons **p.180.** 'Universal Gateway': public domain/The Met NY; pink lotus flowers: Lotus Images/ Shutterstock **p.181.** Guanyin as the Nine-Lotus Bodhisattva: public domain/The Met NY **p.182.** 'Sita finds Rama among lotus blooms': Donald Alexander Mackenzie/Public domain, Wikimedia Commons **p.183.** 'Lotus-clad Radha and Krishna': Public domain, Wikimedia Commons; Lord Brahma: d_odin/ Shutterstock **p.184.** Sapta Chakra: Public domain, Wikimedia Commons **p.185.** Lotus motif: Photo Dharma/ CC BY 2.0, Wikimedia Commons **p.186.** 'Vishnu and Lakshmi on Shesha Naga': Public domain, Wikimedia Commons **p.187.** Trays of colourful flowers: Pete Burana/Shutterstock; Lotus: Noppharat Studio 969/Shutterstock **p.188.** Lotus offerings: kasarp studio/Shutterstock; Goddess Lakshmi: VedSutra/CC BY-SA 4.0, Wikimedia Commons **p.189.** Floral decorations in the Mahabodhi Temple Complex: Jesse33/Shutterstock **p.190.** Brahma, Vishnu and Shiva seated on lotuses': Public domain, Wikimedia Commons **p.191.** 'Lord Vishnu holding lotus flowers': Yann/Public domain, Wikimedia Commons **pp.192-3.** Lotus flowers at night: Konstantin Baidin/Shutterstock **p.194.** Mani stones at Kunzum Pass: Gerd Eichmann/CC BY-SA 4.0, Wikimedia Commons; Dharmachakra on Jokhang temple: Chris Redan/Shutterstock.com **p.195.** 'Buddhist Goddess Green Tara seated on a lotus': Public domain, Wikimedia Commons; Four-armed Tibetan form of Avalokitesvara: ShahJahan/CC BY-SA 3.0, Wikimedia Commons **p.196.** Sculpture of Lord Vishnu: Wmpearl/Public domain, Wikimedia Commons **p.197.** 'A representation of the Gayatri Mantra': Public domain, Wikimedia Commons; 'Indrakshi seated on a lotus flower adored by the gods': Public domain, Wikimedia Commons **p.198.** 'Goddess Saraswati': Public domain, Wikimedia Commons **p.199.** Golden Lotus in Meenakshi Amman Temple, Tamil Nadu: IM3847/ CC BY-SA 4.0, Wikimedia Commons **p.200.** Lotus motifs at Dilwara Temple: Surohit/CC BY-SA 3.0, Wikimedia Commons; Lotus-like floral motifs at Red Fort: Ashok Dilwali **p.201.** Lotus Mahal: jorisvo/Shutterstock **p.202.** Lotus fountain, Red Fort: Ashok Dilwali **p.203.** Lotus-like floral motifs in paintings in the Ajanta Caves, Maharashtra; 'Goddess Chinnamasta': Public domain, Wikimedia Commons **pp.204-5.** Lotus Gate: itch13/ Shutterstock **p.206.** Padmasana: joesayhello/Shutterstock **p.207.** Radha and Krishna: Public domain, Wikimedia Commons **p.208.** 'Legend of Vasudhara': LACMA, Public domain, Wikimedia Commons

Chapter 12: P.210. Marigold flowers: Roop_Dey/Shutterstock **p.212.** Marigold archway: Jen Watson/Shutterstock; Idol of Lord Vishnu: d_odin/Shutterstock **p.213.** Marigold flowers at the entrance: Dmitry Galaganov/Shutterstock **p.214.** Goddess Saraswati: Natalia Mikhalchuk/Shutterstock; Traditional *pooja thali*: Dipak Shelare/Shutterstock **p.215.** 'Cupid Disturbs Krishna's Penance': Public domain, Wikimedia Commons **pp.216-7.** Golden champa flowers: Channel M2 **p.218.**

Portraits of Jahangir and Shah Jahan: Public domain, Wikimedia Commons **p.220.** Marigolds and champa flower, stainless steel sculptures, Alex Davis studio: courtesy Alka Pande **p.221.** 'Portrait of Jahangir': public domain/The Met NY **p.222.** Foot spa: Rido/Shutterstock; Fragrances and oils: KMNPhoto/Shutterstock **p.223.** Red hibiscus and white champa: DAwee/Shutterstock **p.224.** Narada: Public domain, Wikimedia Commons; yellow champak flowers: Noppharat Studio 969/Shutterstock **p.225.** 'Companion persuading Radha': LACMA, Public domain, Wikimedia Commons **p.226.** Champa flowers: hayachanta/Shutterstock

Chapter 13: P.228. Saffron flowers: Fresnel/Shutterstock **p.230.** Phirni: dseengal/Shutterstock; Peacock flower: Laura_VN/Shutterstock **p.231.** Kachnar ki kali: Arunpal Khati/Shutterstock **p.232.** Bundles of fresh herbs: LiliGraphie/Shutterstock **p.233.** Sugar-candied rose petals: chicca N/Shutterstock **p.234.**Frozen rose ice cubes: Y Photo Studio/Shutterstock **p.235.** Gulkand: StockImageFactory.com/Shutterstock; Saffron strands: Kozlenko/Shutterstock; Rawa kesari: StockImageFactory.com/Shutterstock **p.236.** Red nasturtium flower salad: Bernd Schmidt/Shutterstock **p.237.** Candied pansy: Heike Rau/Shutterstock **p.239.** Herbal tea infusion

Endmatter

p.252. Flower-seller festooned with marigold garlands: Mazur Travel/Shutterstock.

Acknowledgements

To my parents, Jagdish and Kamla Pande, who always upheld Dharma, through which they taught me *Saundarya*.

To my publishers Bikash and Tultul Niyogi, who commissioned me to create this unusual book. In my exploration of the landscape of Indian art and aesthetics, the floral motif in particular has been an integral part and has featured in many art historical and visual culture publications, which I have authored. It was the vision of Bikash and Tultul Niyogi and the Editorial Director, Nirmal Kanti Bhattacharjee, who packed me off into the hitherto uncharted territory of the Indian flora and gardens where the Indian flowers have been valourised.

I could not have covered such an expanse and landscape of the Indian florae without my dedicated team of researchers, who I in turn sent to explore the little-known myths and cultural connotations of Indian blossoms. It began with Surbhi Bahl and continued with Dilpreet Bhullar, Neha Mathrani, Ekta Niranjan. Neha Mathrani in particular who sat with me sometimes till the wee hours of the morning learning about the innumerable myths of the lotus. Saurabh Rai, Ravinder Singh and Kushal Singh, for their patience and dogged perseverance, in making innumerable photocopies, spiral-bound booklets, and chasing books. My office staff, of Suprabha and Aditi who bore with my sudden disappearance from the office, when I needed to refresh a thought in the middle of the day.

To my husband Mukul Joshi, who bore with the bedroom light when I was up reading late. To my sister Tripti, whose sage advice helped me celebrate my work. To my daughter Mandakini, whose perceptive eye provided some startling images and also tweaked my way of thinking.

To my friends, the artists who have generously shared their art works; to Tulika Kedia of Must Art Gallery, who has graciously given images from her personal collection of some extraordinary tribal art.

To the editor Priyamvada and the designer Misha Oberoi, for their patience and painstaking re-doing of the many drafts, which has finally led to making this striking publication that brings together fragrances of the many flowers of India.

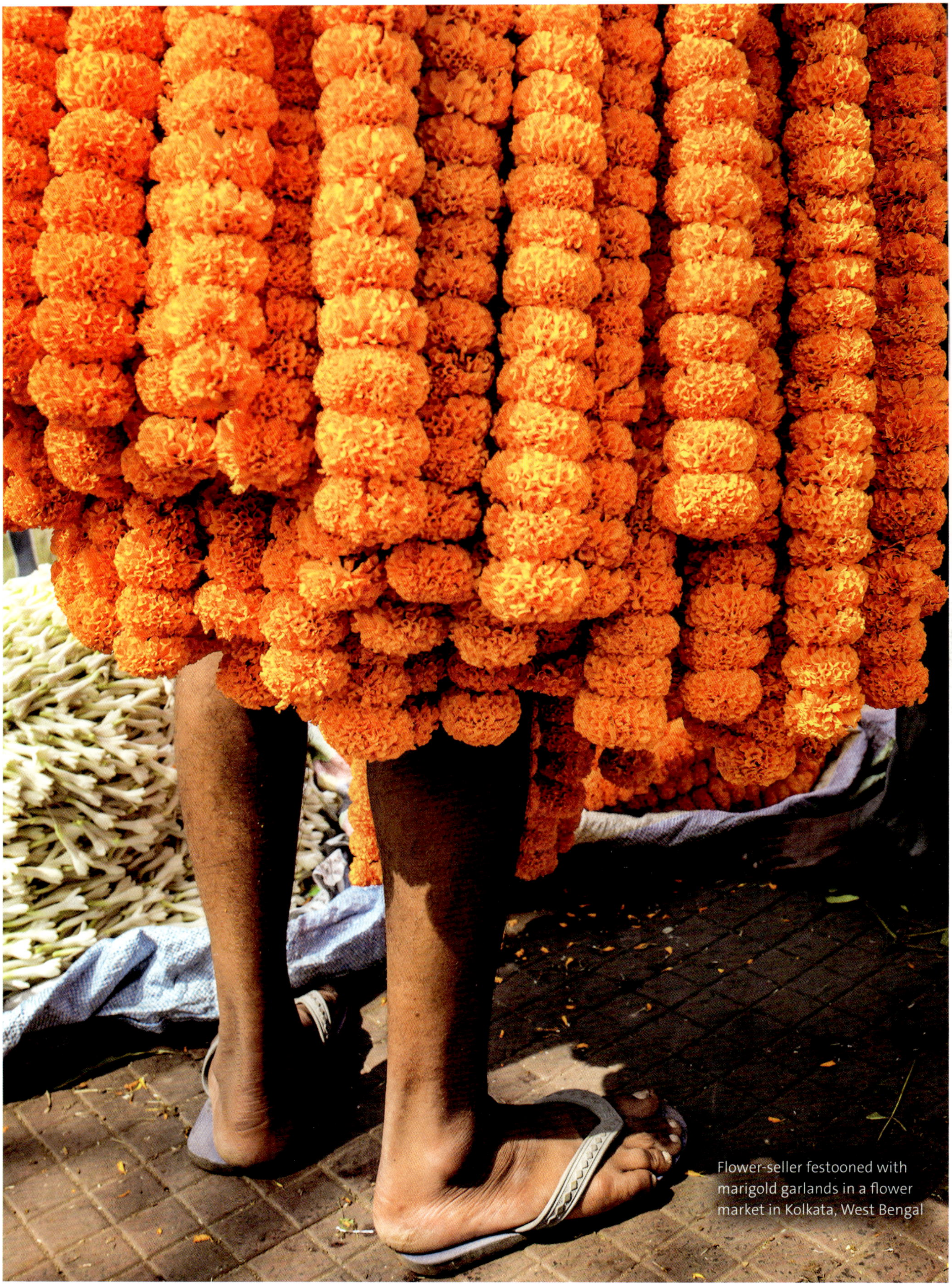

Flower-seller festooned with marigold garlands in a flower market in Kolkata, West Bengal

Index

Note: page numbers in italics denote images and captions

M

N